The Philadelphia Area
WEATHER BOOK

The Philadelphia Area WEATHER BOOK

Jon Nese
Glenn Schwartz

Foreword by Edward G. Rendell

TEMPLE UNIVERSITY PRESS
PHILADELPHIA

Temple University Press, Philadelphia 19122
Copyright © 2002 by Temple University
All rights reserved
Published 2002
Printed in the United States of America

♾ The paper in this publication meets the requirements of the American National
Standard for Information Sciences—Permanence of Paper for Printed Library
Materials, ANSI Z39.48-1984

Library of Congress Cataloging-in-Publication Data

Nese, Jon M.
    The Philadelphia area weather book / Jon Nese and Glenn Schwartz; foreword
by Edward G. Rendell.
        p.   cm.
    Includes bibliographical references and index.
    ISBN 1-56639-956-4 (cloth : alk. paper)
    1. Philadelphia (Pa.)—Climate.   2. Meteorology—Pennsylvania—Phila-
delphia—History.   I. Schwartz, Glenn, 1951– .   II. Title.

QC984.P4 P426   2002
551.69748'11–dc21   2001057991

For Gwen, John, Robert, and Katie—the sunshine of my life
_J.N._

For Wally, Francis, Herb, and Mr. Ehrlich, who inspired my dreams,
and to my parents and my wife, Wynette, who helped me achieve them
_G.S._

CONTENTS

FOREWORD

I hate snow. At least, that's the way I felt when I was mayor of Philadelphia from 1992 to 1999. We also had our share of ice storms, floods, hurricanes, and heat waves during my administration, and there was nothing good about them either.

I didn't always feel this way. When I was a child, I distinctly remember hoping for snow, listening to every broadcast on the radio from November until March and trying to find hints of snow. Our dining room on the eleventh floor of a Manhattan apartment building overlooking the Hudson River had the clearest view of the sky. When even the possibility of snow was mentioned in the forecast, I would go to the dining room three or four times a night, lean my head into the open air, and look up, hoping to see clouds and not stars. That was a constant part of my life during the winter. It's not as if I didn't like school, but I just loved the idea of having a free day.

Then, of course, I became mayor of Philadelphia, and learned the facts about snow. We were fighting for every penny, and storms could have disastrous budgetary consequences. A snowstorm could cost the city a million dollars a day in overtime. I began listening to weather reports constantly, even calling a weather forecast line from my car phone between speeches. I probably dialed that number ten or twenty times a week during the winter, but now I was rooting the other way. I became a weather watcher wishing for no snow, desperately hoping to see the stars at night and to hear that the temperature and air pressure were rising.

I remember that a couple days after John Street was sworn in as my successor, I got up and looked out the window. And for just a few seconds, I had that old familiar tingling, that dread of snow that I had as mayor. Then it hit me that it wasn't my problem anymore. Relief flooded my body. For the rest of my life, I'll remember the juxtaposition.

The first big weather problem during my administration was the heat wave of July 1993. We learned that excessive heat could devastate a city in which a lot of older people would stay indoors most of the time and were reluctant to even open their windows because of fear of crime. We kept air-conditioned senior citizen centers and city buildings open, and urged people to use them. I was surprised by the more than 100 deaths. Even though I knew Philadelphia pretty well, it never occurred to me that some people wouldn't open their windows or have access to a nine-dollar fan. I think we changed the way the nation reported heat-related deaths by the honesty of the way we reported them. And we were the national leader in developing a heat wave warning system that has saved hundreds of lives in the years since that deadly summer.

Just a few months after the heat wave, we had a horrible winter, with ice all the time. It seemed like one ice storm after another. From the city's perspective, the main story of that winter was the absolute inability to get enough salt. With ice storms, you need salt a lot more than for snowstorms. We were running out of salt, and every other city in the Northeast had the same problem. There just wasn't enough to go around. If you could find someone to sell it, it was at usurious prices. But we were willing to spend whatever it took.

My wife, Midge, and I went to Washington, D.C., for the Conference of Mayors that winter. She wanted to see the White House. While we were on the tour, we met the Chief of Intergovernmental Relations. We told her about our problems with salt, and she immediately started calling military bases all over the East Coast, looking for some. They didn't have any either. I had tried to use my clout with the Clinton administration, but that was the length we had to go. We ended up husbanding the salt that we had and persevered. It was "the winter without salt."

The worst episode I had with the press as mayor was after the January 1996 Blizzard. Generally, we got good press, but not this time. The first day wasn't the problem. It was snowing so fast and so hard, there wasn't much we could do. I thought we battled the storm as courageously as we could. Our biggest problem was where to dump the snow once the storm was over—there was simply so much and we had no place to put it all. We never got a firm go-ahead from the Environmental Protection Agency to dump the snow in the river, but we did get what we felt was a half-hearted nod of approval. After weighing the pros and cons of this situation, we felt it was in the best interests of the citizens and the City to continue dumping the snow in the river.

The second problem was how to deal with the narrow streets. At that time, we had no equipment that was small enough because it had never been a problem. In Philadelphia, the snow usually lasts for a day or two. People would shovel out their streets by themselves. But there had never been this much snow. We tried to organize shovel brigades. Some people complained about it, but overall, I think we did a great job. We did much better than the suburbs in getting the city back to life. I will never forget that the newspapers sent out reporters to ask people about the job the city did. One of those reporters was in tears in Chief of Staff David Cohen's office because she had interviewed 50 people, and 47 of them said the city did a good job. She told us that her editor made her go back and find more people who thought the city did a poor job, compromising her journalistic integrity. I remember being furious all through that time, because I thought the streets department did a great job and was being hammered unnecessarily for it. It was a bitter lesson for me in how the press conducts their business. It wouldn't be satisfied with anything but a negative story about the Blizzard.

Political careers are often shaped by events as much as they are by people, and weather has provided its share of political turmoil. A big and unexpected snowstorm in New York City in 1969 while John Lindsay was mayor led to a furor that is blamed for his loss in the Republican primary a few months later (though he did win re-election on the Liberal Party ticket). In Chicago, the city's poor handling of a huge snowstorm in 1979 was credited with helping elect Jane Byrne as the next mayor. We managed to get through the weather-turbulent 1990s in Philadelphia without such an upheaval. But it did lead John Street to make a major investment in snow-removing equipment to handle the smaller streets. In his

first two years in office, he's used it once. Basically, we're just not a big snow city. At least it doesn't seem so anymore—although the last time I said that publicly was a few months before the Blizzard of 1996, so I should be careful not to jinx us.

I am convinced that we are getting less snow because of global warming. The winters are certainly warmer than when I started as mayor. We never would have had sidewalk cafes operating from March to early December. When I was growing up, it snowed in December a lot, and sometimes even around Thanksgiving. It never seems to snow before New Year's Day any more. I wonder what it will be like ten, thirty, or even fifty years from now?

That's why I'm glad that Jon and Glenn have written this book. It's given me a chance to see for myself if my feelings about how our weather has changed are supported by the numbers. I want to know what changes I can expect at the shore, where I have a house two blocks from the beach. It was exciting to learn of Philadelphia's important role in weather advancements, from before Benjamin Franklin to the television age. It certainly was no surprise to learn that The Franklin Institute was one of the nation's leaders in the early days of weather science. Being a vital part of that organization gives Jon Nese great credibility. And I watch the weather on TV a lot, where I've gotten to know and trust people like Glenn. When I was growing up in the 1950s, TV was just developing, and I don't remember anybody who did the weather. Now I know many of them. As mayor, I tried never to miss a weather show. Even now, I don't miss many.

All of us are affected by the weather. We make small talk about it. We complain and speculate about it. More than ever, we depend on weather forecasts, and expect them to be accurate. Because we are such a highly weather-sensitive society, it pays to be, as Benjamin Franklin once put it, "weatherwise." And this book will satisfy anyone's curiosity about all the types of weather that affect the Philadelphia area—past, present, and future—and what meteorologists go through to forecast them.

Edward G. Rendell
January 2002

PREFACE

Beginning in grade school, children all over the country learn about the Philadelphia area's importance to American history. They become familiar with the names, places, and events associated with the nation's beginnings such as William Penn, Independence Hall, Valley Forge, and Washington crossing the Delaware. Less well known is the region's rich legacy of meteorology, the science of the atmosphere and its phenomena. Although "official" Philadelphia weather observations—the ones considered reliable enough to be recognized by the National Weather Service in their record keeping—did not begin until the 1870s, weather observations were made in this area as far back as the mid-seventeenth century. Benjamin Franklin was one of the first to combine observations with scientific theories in the eighteenth century, and for a time in the nineteenth century, Philadelphia was the national center of weather theory and observation.

Philadelphia's early engagement with weather science and observation is more than an additional source of local pride; it also provides a great deal of information that we can compare to present-day observations. How often have you heard people say that the winters were colder or the summers hotter when they were children? Just about everyone can recall a big snowstorm, a soaking hurricane, a blisteringly hot summer, and cool breezes at the shore. But memories are not precise. That huge blizzard of our youth may not have been as big as we remember; perhaps it occurred in a different year than we recall. But official records of these events as well as all the unremarkable, "normal" days that have faded from memory do exist. These records form the basis of this book as we put the important snowstorms, hurricanes, floods, heat waves, tornadoes, cold spells, droughts, and other important weather and climate events and extremes that have affected the Philadelphia area into the broader picture of typical seasonal weather for this area.

Jon's position at The Franklin Institute Science Museum and Glenn's at NBC-10 give us the opportunity to talk to many groups of students and adults about the principles of weather and weather forecasting. We are always gratified to see the consistently high level of interest in meteorology, so we have been sorry to disappoint people who ask us to recommend books they could read about the region's weather. Although there are recent books about weather-related topics of local interest, the last book specifically devoted to the weather and climate of the Philadelphia area was published in 1847, decades before a national weather service was created. We think it is time for an update and have developed this book to respond to the many and varied questions we are frequently asked about the weather and climate of this region. Though our focus will be on the immediate Philadelphia area, where reliable "official" weather observations are available back to 1872, this book also includes information about the weather and climate from the Poconos to the shore. This area of coverage includes the state of Delaware, central and southern New Jersey, the Lehigh Valley, and Chester and Berks counties in Pennsylvania. Although there are many similarities in the weather and climate across this area, there are also significant differences, and these are explained in this book.

We begin in Chapter 1 by describing the Philadelphia area's place in weather history, focusing on the key people, places, and methods of observation. Chapter 2 explains the basics of weather and weather forecasting. Then, in the next four chapters, we look at the characteristics of the seasons to provide an orderly march through the year. We present the highlights of winter, spring, summer, and autumn in the Philadelphia area, dedicating a chapter to each. Readers will immediately notice that meteorology divides the seasons a bit differently than astronomy. Both sciences partition the year into quarters, but the meteorological seasons are offset slightly from the solstices and equinoxes that define the astronomical seasons. For our purposes, winter includes December, January, and February; spring is March, April, and May; summer includes June, July, and August; and autumn is September, October, and November.

Many topics are truly seasonal and are easily placed into a particular chapter—for example, snowstorms are most common in winter, while heat waves clearly belong in the summer chapter. But as you certainly know, some weather events can happen in several seasons or throughout the year. Nor'easters, for example, form mainly from October to April. Flooding can occur in any season, from a variety of causes. Hurricanes occur in both summer and autumn. And many of us can remember big March snowstorms that violate the meteorological definition of winter. We had to decide where such topics fit best, and in the course of the book, we will explain why we take up a specific subject in a particular chapter. In general, though, weather events are discussed in relation to the season in which they most commonly occur.

As meteorologists, we are also regularly asked about what the climate will be like far in the future—next month, next season, even a decade from now. In Chapter 7 we discuss the techniques of monthly and seasonal forecasting, introducing El Niño and La Niña as important forces over those time scales. We also explore the forces that will help shape the climate of the Philadelphia area years and decades from now, focusing on the still-controversial issue of global warming. Is it real, and if so, what effect will it have on the Philadelphia area and the nearby shore? This final chapter offers an overview of how meteorologists and climate researchers look at such long-term trends.

Throughout the book we have included stories that illustrate what we do as meteorologists to prepare a tough forecast or report on a threatening storm at a television station. These "Stories from the Trenches" come from the best recollections of the people involved. They are not meant to represent the complete history of a particular storm or event, but to give the reader a feel for what it is like to try to give the public the best information possible in the midst of extreme events.

The appendixes include averages and records of temperature and precipitation for each day and each month for Philadelphia, along with similar climate information for Wilmington, Allentown, and Atlantic City (valid through December 2001). We also include a print bibliography for further reading as well as an extensive Web bibliography, since virtually all new, and a great deal of historical, information on weather is available on the Web.

We hope that *The Philadelphia Area Weather Book* will be useful to everyone who is curious about the atmosphere and the weather it produces—how snowstorms form, how weather forecasts are made, what happened during specific storms, or what may happen to the climate in the future. We know that there are many people like us who are passionate about the subject of weather. They have been an inspiration for this project, and this book is for them.

ACKNOWLEDGMENTS

No book on weather can be written without the contributions of the National Weather Service (NWS). For this project, meteorologists at the local office in Mount Holly, New Jersey have been particularly helpful, including Jim Eberwine, Joe Miketta, Walt Nickelsberg, Alan Cope, Bob Wanton, John Quagliariello, and Gary Szatkowski. We also thank Jason Franklin and Walter Drag at the NWS in Boston. The data, images, and forecasts provided by the NWS and NOAA (the National Oceanic and Atmospheric Administration, which oversees the NWS) are the foundation on which this book is built. The meteorologists and technicians of the NWS and NOAA get no publicity and work rotating shifts without high pay yet rarely receive an acknowledgment of their excellence and dedication.

Many other individuals have assisted us with research, entering data, drawing figures, providing images, or answering our many questions. These individuals include Christopher Roberts at the Delaware River Basin Commission; Fred Hauptman of the Philadelphia Department of Public Health; Jeffrey Gebert from the U.S. Army Corps of Engineers; Timothy Owen at the National Climatic Data Center; Dr. David Robinson, the New Jersey state climatologist; Dr. Dan Leathers, the Delaware state climatologist; Paul Knight, the Pennsylvania state climatologist; Kristeen Gaffney at the Environmental Protection Agency; Dr. Wendy Carey at the University of Delaware; Dr. Robert Livezey and Rich Tinker at the Climate Prediction Center; Dr. Bob Davis at the University of Virginia; and Roy Goodman at the American Philosophical Society. We are especially grateful to Irene Coffey and Virginia Ward, librarians at The Franklin Institute, for their tireless efforts on our behalf.

Special thanks to Herb Clarke, Wally Kinnan, Rita Ludlum, Dr. Francis Davis, Dr. Richard Weggel, Dr. Kenneth Spengler, and Dr. Jerry Mahlman for taking the time to be interviewed for the book.

We are also grateful to Jennifer Glovier, Bill McCabe, Kenneth Finkel, Pete DeCarolis, Joel Gratz, John Giampetro, Eric Kurth, Paul Torpey, Steven Strouss, Nicole Felici, Lis Cohen, Tomas Figueroa Jr., Elaine Crescenzi, Sarah Jennings, Lindsey Hench, Matthew Mehle, Justin Brolley, Kathleen Felter, Dorothy Weaver, Sarah Hopper, Dan Zarrow, Gannon Medwick, Ed Madigan, and John Krasting. Their assistance helped take this book from an idea to a reality.

The Philadelphia Area
WEATHER BOOK

CHAPTER 1

History of Weather Science and Observing in the Philadelphia Area

Some are weatherwise, some are otherwise.
—*Benjamin Franklin*

Imagine not knowing what this afternoon's weather will be like, let alone tomorrow's. Imagine planning a trip to New York City in March and unexpectedly running into a late-season blizzard. Or planting crops vital to your family's survival, unsure whether the current drought will break in time. Imagine being on a ship crossing the ocean and not knowing that a hurricane with 120-mph winds was stalled in your path. Or as you set out to sea, that a minor nor'easter is about to explode in strength to become the Colonial American equivalent of the "Perfect Storm."

The early European settlers of this part of the world sometimes found themselves in precisely these situations. They quickly observed that the weather in the new land was more varied and often more extreme than what they were used to in the Old World. These early settlers came from places where hurricanes did not occur and where temperatures below 0°F and wind chills of −20°F rarely, if ever, happened. Extended heat waves with high humidity and many days above 90°F? Also unheard of back home. The variety of weather in the New World both fascinated and threatened the colonists, and they soon began to take detailed notice of it.

This chapter traces the evolution of our understanding of weather and the process of weather observation in this country, emphasizing the Philadelphia area's role in the advancement of the science of meteorology. This story of progress combines individual passion with institutional commitment and scientific innovation. Technological advances have profoundly influenced weather observations, which are truly at the heart of weather forecasting. These observations may be as simple as looking at the sky, as common as measuring the temperature with a thermometer, or as sophisticated as electronically scanning the atmosphere with weather radar.

From before Benjamin Franklin's time to the dawn of television weathercasting in the mid-twentieth century, the Philadelphia area has produced many weather "firsts." Some are listed in Table 1.1. We begin our story a century before Franklin, in the mid-seventeenth century when some of the earliest weather observations made by settlers in the New World were recorded near Philadelphia. Franklin himself was at the center of eighteenth-century progress in weather science, while another local meteorologist, Dr. James Espy, and the organization he worked for, The Franklin Institute, kept Philadelphia at the forefront of meteorological advances in the nineteenth century. Starting in the middle of the twentieth century, technological innovations such as weather satellites and radar rapidly advanced the science of meteorology, and the new medium of television weathercasting emerged. Once again, Philadelphia produced several pioneers in this area, including television legends Dr. Francis Davis and Wally Kinnan, "The Weatherman." The "Story from the Trenches" in this chapter features recent

TABLE 1.1	Some Philadelphia-area Weather "Firsts"
1644–45	The first continuous weather observations in Colonial America are recorded by Reverend John Campanius near present-day Wilmington, Delaware.
1717–18	Cadwallader Colden of Philadelphia makes some of the first known weather observations in the Colonies with a thermometer and a barometer.
1731–32	Mr. De. S. takes thermometer measurements twice a day in Germantown—the earliest known surviving daily temperature observations.
1743	Benjamin Franklin was the first to suggest that big storms in the Northeast come from the southwest.
1836	Franklin Institute meteorologist James Espy publishes the first weather map in the United States that uses widespread weather observations to study a specific storm.
1837	One of the first government-funded weather observing networks is established in Pennsylvania as a joint venture of The Franklin Institute and the American Philosophical Society.
1882	Franklin Institute member W. N. Jennings takes the first photographs of lightning.
1946	The University of Pennsylvania unveils ENIAC, the world's first electronic, general-purpose, large-scale computer, which helps to pave the way for the era of numerical weather prediction.
1947	Dr. David Ludlum of Philadelphia, who later became America's foremost weather historian, is one of the earliest television meteorologists in the United States, and starts *Weatherwise* magazine while working at The Franklin Institute.
1959	Philadelphia television weathercasters Dr. Francis Davis and Wally Kinnan are two of the first three television meteorologists in the country to earn the Seal of Approval of the American Meteorological Society.

interviews with Davis, Kinnan, and Herb Clarke, another legendary Philadelphia weathercaster. They discuss the groundbreaking role that Philadelphia played in the 1950s and 1960s as a national leader in the early days of television weather.

We also chronicle the government's crucial role in the evolution of weather observation and weather forecasting in this country. From the U.S. Army Signal Corps observers in the 1870s to the present-day National Weather Service (NWS), the federal government has been the primary institution charged with taking and organizing weather observations and creating and disseminating forecasts and warnings. The federal government has also led the development and implementation of the satellite and radar technology that is vital to meteorologists today. Because information from these two weather observing systems is commonly shown on television and available to everyone via the World Wide Web, we describe satellite and radar imagery in detail.

OBSERVATIONS: IT ALL STARTS HERE

Weatherwise, Benjamin Franklin most certainly was. Although he is probably known best as an inventor, author, statesman, and postmaster, Franklin was also a shrewd weather observer. Throughout his life, he observed and studied the weather, offering theories as to the hows and whys of various atmospheric phenomena. Although his kite-flying encounter with lightning in 1752 gets the most attention, Franklin also pondered cold waves, changes in air pressure, tornadoes and waterspouts (see Illustration 1.1), and even climate change. Like modern weather forecasters, he was also interested in comparing weather observations from different parts of the country (although he could not do it while the weather was happening—the telegraph was invented half a century after he died). Franklin recognized the importance of observations for making deductions about how the weather works and how weather systems move. In fact, Franklin led one of the first attempts at organizing a network of weather observers in Colonial America, using the only countrywide organization in place in the eighteenth century: the post office. His wide-ranging interest in weather and skill as an observer made Franklin the leader in meteorology in this country in the eighteenth century, and his speculations became the starting points for weather theories of the next century.

In at least one fundamental way, weather science has not changed much since Franklin's time. As simple as it sounds, good observations are essential for good forecasts. We meteorologists often are asked how we predict the weather, and we sometimes answer with a grin, "First we look out the window." As simple as it sounds, good forecasts start with good observations. Experienced forecasters know that the first step in preparing a weather forecast is to look at observations; the next step is to look at more observations. By "observations," we mean not only what is happening outside your window, but also what is happening outside windows all over the country and beyond. We also mean measurements of the atmosphere above the earth's surface, cloud images obtained from satellites, and radar images of precipitation. There is so much data for modern meteorologists to consider that Franklin would be very envious. These observations are evaluated by forecasters as well as fed to weather computer models, huge software programs that turn weather prediction into a complicated math problem. Clearly, weather observations are still at the heart of understanding how the atmosphere works, though the modern-day process of using these observations occurs at a level of sophistication that even Franklin could not have imagined.

Though Franklin was one of the first in this country to attempt to scientifically study and explain the weather, the earliest daily records of weather observations by settlers of the New World preceded him by about a century. Some of those observations were taken in the Philadelphia area.

Early Colonial Weather Watchers

Instruments are not necessary for observing the weather. For example, the speed and direction of the wind; the sky cover; and sensations of warmth, chill, dryness, and humidity can be described qualitatively, without using numbers. Relative descriptions such as "Yesterday was cloudier than today" or "It turned warmer and more humid in the afternoon" are also useful weather observations. Some people can even sense changes in air pressure

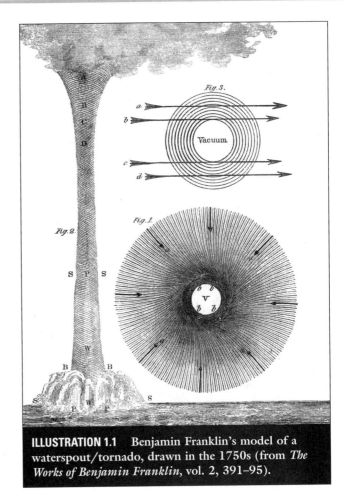

ILLUSTRATION 1.1   Benjamin Franklin's model of a waterspout/tornado, drawn in the 1750s (from *The Works of Benjamin Franklin*, vol. 2, 391–95).

in their "bones." And so it was with the earliest weather observers. Most of them had no instruments—they just recorded qualitative accounts of what they saw and felt.

Reverend John Campanius's daily weather observations during 1644–45 at Swedes Fort, a Swedish settlement near present-day Wilmington, Delaware, may have been the first continuous weather observations in Colonial America. Campanius had no instruments—he simply kept a descriptive journal of weather events. Although his original observations are lost, Campanius's grandson summarized them in an article published in 1702. An example translated from that summary shows that Campanius's observations could easily describe a recent Delaware Valley winter as well as one more than 350 years ago:

January 1644—The winter began about the 21st, with severe cold, and then much snow. Afterwards came rain and a thick fog, with occasional sunshine, until the end of the month. During this time the winds were NW. ESE. SE. S.

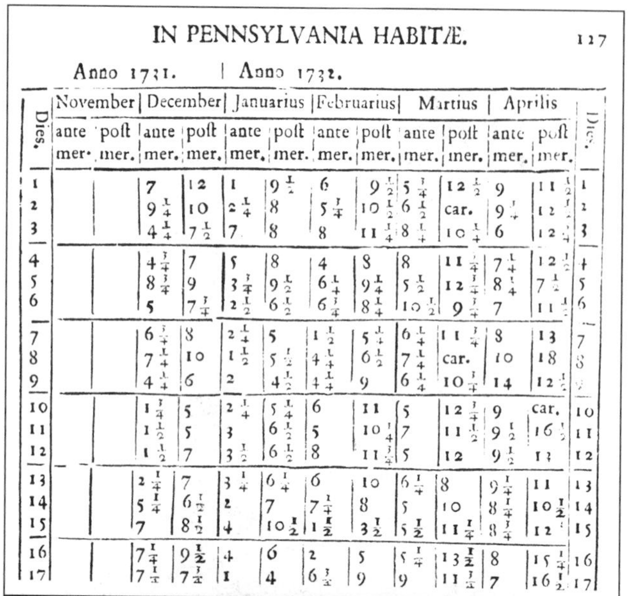

IN PENNSYLVANIA HABITÆ. 127

Anno 1731. | Anno 1732.

Dies.	November ante mer.	November post mer.	December ante mer.	December post mer.	Januarius ante mer.	Januarius post mer.	Februarius ante mer.	Februarius post mer.	Martius ante mer.	Martius post mer.	Aprilis ante mer.	Aprilis post mer.	Dies.
1			7	12	1	9½	6	9½	5¾	12½	9	11½	1
2			9¼	10	2¼	8	5¾	10½	6½	car.	9¼	12¼	2
3			4¼	7½	7	8	8	11¼	8¼	10¼	6	12¾	3
4			4¾	7	5	8	4	8	8	11¾	7¼	12½	4
5			8¾	9	3¾	9½	6¼	9¼	5½	12¾	8¼	7½	5
6			5	7¾	2½	6½	6¾	8¼	10½	9¾	7	11½	6
7			6¾	8	2¼	5	1½	5¼	6¼	11¾	8	13	7
8			7¼	10	1½	5½	4¼	6½	7¼	car.	10	18	8
9			4¼	6	2	4½	4¼	9	6¼	10¾	14	12½	9
10			1¾	5	2¼	5¼	6	11	5	12¾	9	car.	10
11			1½	5	3	6½	5	10¼	7	11½	9½	16½	11
12			1½	7	3½	6½	8	11¾	5	12	9½	13	12
13			2¼	7	3¼	6¼	6	10	6¼	8	9¼	11	13
14			5¼	6½	2	7	7¼	8	5	10	8¼	10½	14
15			7	8½	4	10½	1½	3½	5½	11¼	8¾	12	15
16			7¼	9½	4	6	2	5	5¼	13½	8	15¼	16
17			7¼	7¾	1	4	6¾	9	9	11¾	7	16½	17

ILLUSTRATION 1.2 The earliest known surviving continuous temperature observations, taken twice a day in the Germantown section of Philadelphia in 1731–32 by a man known only as "Mr. De S." He took readings to the nearest one-quarter degree Celsius (Columbia University Library).

With all due respect to Benjamin Franklin, Campanius is sometimes referred to as "America's First Weatherman." Campanius's contribution to weather observing in this country is considered so significant that the National Weather Service gives annual awards in his name to recognize outstanding volunteer weather observers.

Though purely descriptive weather observations have great value, weather instruments enable measurement and provide the quantitative standard that is necessary for comparing observations. Meteorology's emergence as a true scientific endeavor really began with the development of reliable instruments for measuring important atmospheric elements such as temperature and pressure. Though the thermometer and barometer (for measuring air pressure) were perfected in the seventeenth century in Europe, they were not really used much for taking weather observations for another century or so.

Some of the earliest instrumented weather observations in Colonial America date from the Delaware Valley in the early eighteenth century. Dr. Cadwallader Colden of Philadelphia took measurements of temperature and pressure during the winter of 1717–18. We know of his observations through a letter written to Colden by William Douglass of Boston. "I thank you for your account of the state of the Barometer and Thermometer at Philadelphia part 1717 and part 1718," Douglass wrote in 1721, acknowledging the measurements. Continuing the Philadelphia connection, the earliest known surviving temperature data taken in the Colonies came from the Germantown section of the city. The observer is known only as Mr. De S., and he took two thermometer readings each day from December 1731 to October 1732. His observations, taken in Celsius to the nearest quarter degree, were published in 1737 in a European journal (see Illustration 1.2).

The activities of Campanius, Colden, and Mr. De S. accurately portray the state of weather data collecting in early Colonial America. Weather observers were isolated individuals who were not part of any organized network. Many of these weather enthusiasts were physicians, ministers, and educators. They kept records in diaries and journals and not on standardized forms. The few who had instruments rarely had instructions to guide them. The most common instruments were a thermometer and a rain gauge, and they were not always reliable. And the observers did not always follow what are today considered standard rules of observation—for example, taking measurements on a daily basis and at the same time each day.

In the mid-eighteenth century, famous Philadelphian John Bartram was a diligent weather observer for a time. Bartram, who cofounded the American Philosophical Society (APS) with Benjamin Franklin in 1743, recorded temperature, wind, and weather conditions twice each day from August 1748 to January 1750. The APS, which is still headquartered in Philadelphia, was the premier learned society of the time, America's counterpart to the Royal Society of London. Members were enlightened men who believed that nature was knowable and behaved in consistent ways. In many ways, APS members were the earliest "natural philosophers."

Also around this time, Franklin's observations of wind and sky, combined with similar data from other colonial cities, helped establish that storms in the Northeast often approach from the southwest. Franklin's most celebrated observation involved a lunar eclipse that he planned to view from Philadelphia in the fall of 1743. In Franklin's words:

> There was an eclipse of the moon at nine o'clock in the evening, which I intended to observe; but before night a storm blew up . . . and continued violent all night . . . so that neither moon nor stars could be seen. . . . But what surprised me was, to find in the Boston newspapers an account of the observation of

that eclipse made there. . . . I wrote to my brother about it, and he informed me that the eclipse was over an hour before the storm began.

Piecing together the scanty information available, Franklin reasoned that the same storm that hit Philadelphia moved to the northeast and arrived in Boston several hours later. From this observation and others that he had made, Franklin proposed that many strong storms along the East Coast take this same track, approaching from the southwest. This was a bold leap at the time, because surface winds preceding such storms blew from the northeast, and the prevailing notion was that storms moved with the surface wind. More generally, Franklin's observations led to the realization that weather tends to move in a west-to-east fashion across the country.

Many other individuals were taking weather observations in the Philadelphia area during Franklin's time. These included William Bartram, son of John Bartram, and noted astronomer David Rittenhouse. Other observations survive from less well-known individuals such as Peter Legaux, a French lawyer and farmer living in Spring Mills, just northwest of Philadelphia; Reverend Henry Muhlenberg in Trappe, near Valley Forge; and prominent Philadelphians such as Phineas Pemberton, Thomas Coombe, and Dr. John Redman Coxe.

Thomas Jefferson, also a dedicated observer, recorded weather data from 1776 to 1818, usually several times a day. He even purchased a thermometer in Philadelphia on the morning of July 4, 1776, when he recorded temperatures of 68°F at 6 A.M., 76°F at 1 P.M., and 73.5°F at 9 P.M. in his weather journal. Like Franklin, Jefferson recognized that comparing weather observations from different locations was necessary for understanding how the weather worked, and Jefferson had access to Philadelphia weather data taken by Rittenhouse, Bartram, Legaux, and others. In recognition of Jefferson's weather-wise nature, the National Weather Service's highest award for outstanding volunteer weather observers is called the Jefferson Award.

By the late eighteenth and early nineteenth centuries, the potential value of organized networks of weather observers was becoming evident. Following an extraordinarily harsh winter in 1779–80 in the midst of the War of Independence, the American Philosophical Society publicly requested observations from "every part of the continent" (see Illustration 1.3). One item of interest was "what diseases prevailed most in the extreme cold." Indeed, physicians were one group that supported more

MARCH 22, 1780.   T H E   NUMB. 1332.

PENNSYLVANIA JOURNAL
AND
WEEKLY ADVERTISER

WEDNESDAY, MARCH 22, 1780.

Philadelphia, March 21, 1780.

The Committee appointed by the Philosophical Society, to make and collect observations on the effects of the severe & long continued cold of last winter, request the curious in every part of the continent, to communicate to them such remarks as they have already made, or may hereafter make on this subject, particularly such as may properly come under the following heads.

First, Meteorological observations, accurately made with good instruments.

Secondly, The effects of the cold on the earth and waters; such as the depth of earth frozen, the thickness of ice, &c. together with such remarkable circumstances, as may attend either freezing or thawing. Also its effects on spiritous, vinous and other liquors.

Thirdly, The effects of the cold, during and after the winter on animals, birds, reptiles, insects and their chrysalsies.

Fourthly, The same on vegetables, distinguishing the indigenous from foreign, the spontaneous from the cultivated.

Fifthly, What diseases prevailed most in the extreme cold weather, and after it. Accurate observations made on former winters, remarkable for cold will be accepted.

As the comparing together different climates and different seasons is not a matter of mere speculative curiosity, but real benefits may be derived to mankind, by improving this branch of natural knowledge, the Committee promise themselves the assistance of the ingenious, whether Members of the Society or otherwise, in their endeavours to unite in one common stock many valuable fragments of Philosophy, which must otherwise perish with individuals.

The Committee do not propose to make their report before the close of next summer. In the mean time such gentlemen as chuse to favour the Society with their observations, will please to direct their letters to Col. Lewis Nicola, of Philadelphia.

(The Printers of News-Papers on the continent, are requested to give the foregoing a place in their respective Papers.)

ILLUSTRATION 1.3 The American Philosophical Society circulated this request for weather observations following the brutal winter of 1779–80 for the purpose of "improving this branch of natural knowledge" (American Philosophical Society, in the *Pennsylvania Journal and Weekly Advertiser*, no. 1332, March 22, 1780).

systematic and regular observations, since the general health of patients and the occurrence of outbreaks of disease were thought to be related to the weather. In 1787, the prestigious College of Physicians in Philadelphia included a statement in its charter concerning the link between taking weather observations and investigating disease and epidemics. Dr. Benjamin Rush, a renowned professor at the University of Pennsylvania (and a founder of the College of Physicians), investigated the role of weather in the devastating yellow fever epidemic that was responsible for approximately 5,000 deaths in Philadelphia in 1793. In 1807, Dr. B. S. Barton of the University of Pennsylvania even suggested in an article in a Philadelphia medical journal that a national network of thermometers, barometers, and rain gauges be operated by one of the learned societies in Philadelphia.

During the War of 1812, Dr. James Tilton, Surgeon General of the Army (and a graduate of the College of Physicians), ordered his hospital surgeons to record weather data. This system of weather observers is generally considered the first in the country organized by the government. Starting with about twenty army posts with thermometers and rain gauges, the network grew to nearly 100 camps by the 1850s. In the Philadelphia area, observers at Fort Mifflin, the Frankford Arsenal in Bridesburg, and the U.S. General Hospital in Chester, among others, participated in this network at one time or another.

In the early nineteenth century, several extreme weather events helped to increase interest in creating new systems of weather observing. In September 1815, "The Great September Gale," the worst storm to hit New England in almost 200 years, devastated parts of Connecticut, Rhode Island, and Massachusetts. The next year, 1816, was the infamous "Year without a Summer" when snow fell in June in parts of the northeastern United States and killing frosts struck in July and August. (See Chapter 5 for more about that summer in the Philadelphia area.) These events demonstrated to the citizens and the government of the still relatively new nation just how critical the weather was to their livelihood and survival.

Several new weather observing networks sprang up in the decades to follow. In 1825, the New York State Legislature authorized funds to take temperature and rainfall observations at each state academy. Starting with about ten sites, the network grew to approximately forty-five stations by 1835, and functioned into the 1860s. A key figure in the history of American meteorology emerged from the New York network. Joseph Henry, a friend of many Philadelphia-area scientists, started his meteorological career compiling reports for the New York system. He later became the first secretary of the Smithsonian Institution and a driving force behind the creation of the first national "weather bureau." But before the Smithsonian would take its place in American weather history, the bright light of weather observation, research, and theory would first shine from Philadelphia.

James Espy and The Franklin Institute

What Benjamin Franklin was to Philadelphia's place in weather history in the eighteenth century, James Pollard Espy (Illustration 1.4) and The Franklin Institute (TFI) were in the nineteenth.

ILLUSTRATION 1.4   James Pollard Espy, first meteorologist at The Franklin Institute (in *Popular Science Monthly* 34[1889]: 834–40).

In view of Franklin's interest in weather, meteorology had a special status in TFI's early years. Chartered in 1824 to honor Franklin and advance the usefulness of his inventions, the institute hosted Espy's volunteer lectures on meteorology in its first year of operation. By the early 1830s, Espy had formally joined TFI and devoted himself entirely to studying the weather. Over the next decade, he was the driving force who would turn Philadelphia into the hub of meteorological activity in the United States. Espy's weather interests were nearly as diverse as Franklin's; he lectured and published on a variety of topics, including the importance of humidity in weather forecasting, the formation of clouds in mountainous regions, the mechanisms for tornado formation, and weather modification. Espy even enlisted the help of the Franklin Kite Club to verify his calculations of the heights of clouds.

But Espy's greatest meteorological passion was studying storms. For the rest of his life, he focused his research on understanding the causes and characteristics of all types of storms, from the routine rains that came every few days to hurricanes and tornadoes. This desire to understand storms led Espy to push for new and expanded weather observing systems, and also earned him the title "The Storm King."

Though Franklin had proposed almost a century earlier that storms in the Northeast approached from the southwest, there had been little (if any) formal study of how storms formed and moved, mainly due to the lack of organized weather data. Espy firmly believed that meteorology in general had been a neglected science:

> Why is it that this highly interesting and useful branch of human knowledge makes such slow advances? . . . Do philosophers think that . . . the weather never can be predicted?

With the study of storms in mind, Espy helped organize a committee of meteorology at TFI in 1831. One of its tasks was to keep a "meteorological register" of weather in Philadelphia. The data—twice-daily temperature, pressure, and wind measurements, plus daily precipitation observations and remarks on the day's weather—were

published in the *Journal of The Franklin Institute*. The first set of monthly data, collected in January 1831, is shown in Illustration 1.5. Espy knew, just as Franklin did, that observations like these were needed from many observers in distant locations in order to gain insight into the structure of storms and the paths that they took across the country. In Espy's words, such knowledge could be gained only "by simultaneous observations over a wide extent of territory."

Over the next few years, Espy urged The Franklin Institute to petition the federal government for funding to establish a national network of weather observers. Meanwhile, Espy organized his own group of volunteers to record weather conditions and send them to him for compilation.

Espy's volunteer network found a home in September 1834 when TFI teamed with the American Philosophical Society to establish a Joint Committee on Meteorology to compile weather observations from around the country. Espy chaired the committee, which also included Alexander D. Bache, great-grandson of Benjamin Franklin

Meteorological observations for January, 1831.

Moon.	Days.	Therm. Sun rise.	Therm. 2 P.M	Barometer. Sun rise.	Barometer. 2, P.M.	Wind. Direction.	Wind. Force.	Water fallen in rain and snow.	State of the weather, and Remarks.	
				Inches.	Inches.			Inches.		
	1	33°	36°	29.70	29.80	West.	Blustering.		Clear.	Cloudy.
	2	28	38	30.10	30.20	do.	Moderate.		Clear.	Clear.
	3	27	46	.20	.20	W. S.	do.		Clear.	Cloudy.
	4	51	59	29.70	29.70	S. S. E.	do.	.90	Cloudy.	Rain.
	5	57	42	.70	.70	S. W.	Lighter.	1.00	Rain.	Rain.
☾	6	26	34	30.10	30.12	West.	do.		Clear.	Clear.
	7	26	41	.10	.12	do.	do.		Clear.	Clear.
	8	32	40	.13	.9	N. E.	do.		Cl'dy--cl'dy--hail at night.	
	9	30	31	29.85	29.85	do.	do.	.65	Sleet.	Snow.
	10	24	27	.73	.75	do.	do.	.50	Snow.	Snow.
	11	23	32	.90	.85	do.	do.		Snow.	Drizzle.
	12	26	26	.85	30.00	W. N. W.	do.		Clear.	Clear.
⊗	13	8	16	30.03	.35	West.	do.		Clear--clear--very cold.	
	14	20	20	.20	·10	N. E.S.	do		Cl'dy-cl'dy--rain in night.	
	15	20	23	29.33	29.30	N. E.'	High.	.85	Snow--much drifted.	
	16	18	24	.40	.40	West. E.	Blustering.	.45	Snow cont'd--18 in. deep.	
	17	20	20	.40	.70	do.	Moderate.		Flying cl'ds--*nav. obstr'd.*	
	18	8	24	.70	.60	do.	do.		Cl'dy--cl'dy--slight snow.	
	19	21	32	.60	.53	do.	do.		Clear--good sleighing.	
	20	23	39	.50	.50	S. W.	do.		Clear.	Cloudy.
☽	21	14˙	20	.80	.80	S. E.	do.		Clear--cloudy--snow 2 in.	
	22	11	26	.40	.40	West.	Blustering.	.24	Cl'dy--clear--fine sleighing.	
	23	10	16	.60	.60	do.	do.		Clear day.	
	24	7	20	.75	.75	do.	Calm.		Cloudy--clear--*Delaware*	
	25	0	14	,70	70	do.	Moderate.		Clear day. [*frozen.*	
	26	6	24	.70	.70	do.	do.		Do.	
	27	14	26	.70	.73	do.	do.		Do.	
☉	28	16	30	30.00	30.00	do.	do.		Do.	Cloudy.
	29	24	34	29.90	29.80	do.	do.	.60	Cloudy--slight snow.	
	30	18	26	.87	.90	do.	do.		Clear.	Clear.
	31	12	32	.90	.83	W. South.	do.		Cloudy	Cloudy.
				mdn't	.23			.90	Snow, hail, and rain in night.	
	Mean	21	1 29 6	29.48	29.48			6.09		

Thermometer.
Maximum height during the month, . . 59. on the 4th,
Minimum do. . . . . 0. on the 25th, .
Mean do. . . . . . 25.3 . . .
Water fallen in snow and rain,   6.09 inches.

Barometer.
30.70 on the 2nd.
29.30 on the 15th.
29.48

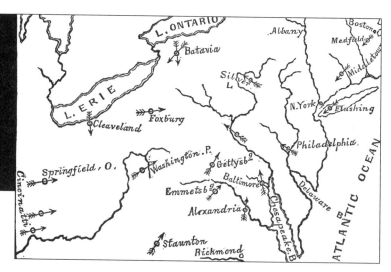

ILLUSTRATION 1.6   This chart depicting a June 1836 storm in Pennsylvania and surrounding states shows wind directions using arrows and may be the first weather map published in the United States that focuses on a specific storm. In compiling such maps, Espy noted that the winds converged toward a central point where the heaviest rain tended to fall (in this case, somewhere between Foxburg, Batavia, and Silver Lake, New York) (from the *Journal of The Franklin Institute*, n.s., 19 [1837]: 19).

and a prominent member of TFI. Bache would later become the first president of the prestigious National Academy of Sciences. Within a month, the Joint Committee circulated a request for observers:

> The prime object of this circular is to obtain a complete knowledge of all the phenomena accompanying one or more storms of rain or hail . . . For this purpose, you are requested, immediately on receiving this circular, to commence a journal of the weather, noting the direction of the wind at the surface of the earth . . . the appearance of the heavens as to clear or cloudy, and the character of the clouds, according to your own mode of description, be noted at least three times a day, as near the following hours as convenient—7 A.M., 2 A.M., and sunset. . . . Please to forward your observations, monthly, to the Joint Committee of the American Philosophical Society and Franklin Institute.

In response to this request, volunteers from points as close as Bucks County Academy and Gettysburg, Pennsylvania, and as distant as Baton Rouge, Louisiana, and Portland, Maine, sent observations. Led by Espy, the Joint Committee analyzed and published detailed reports about the most significant storms, the first time this had been done in the United States. A map of the storm of June 19–20, 1836, covering an area from the Mississippi River to New England, is shown in Illustration 1.6. This chart is probably the first weather map published in the United States showing widely collected data focusing on a single storm. For the first time, fundamental characteristics such as the tendency for winds to converge toward the center of a storm were confirmed using observations. In Espy's words, "By casting an eye on the wood-

cut, it will be seen at a glance that the wind blew on all sides toward the point of greatest rain."

In its reports, the Joint Committee also lobbied for government funding of a countrywide weather network. In 1837, The Franklin Institute took the bolder step of authorizing the committee to "memorialize Congress for the purpose of obtaining national aid in furtherance of this interesting object." The letter from the Joint Committee, which appears in the records of the House of Representatives on April 20, 1838, was the first official request to Congress for a government-funded national weather service. More than thirty years would pass, though, before the federal government would create one.

But progress was made at the state level. On April 1, 1837 the Pennsylvania Legislature appropriated $4,000 (in 1837 dollars) to TFI to organize and run a statewide meteorological research program. With these funds, TFI planned to equip one observer in each Pennsylvania county with a barometer, several thermometers, and a rain gauge. Instrument maker L. C. Francis made more than fifty sets of instruments for $16 per set in his workshop on Fetter Lane in Philadelphia. The standard barometer, which was used to calibrate the other sets, was designated "No.1" and presented to the city of Philadelphia. This original is on public display today at The Franklin Institute (Illustration 1.7).

Making the instruments turned out to be the simplest part of the project. Finding competent, dependable observers and delivering and maintaining the instruments in working order was the biggest challenge. By October 1839, all but four counties had received instruments, but twenty counties with instruments had never sent in any observations. In fact, only about thirty counties ever reported during any single month. Espy took charge of compiling and summarizing the data. Starting with January 1839, statewide monthly summaries were published in the *Journal of The Franklin Institute*. A portion of the "Meteorological Report for the State of Pennsylvania" for August 1840 is shown in Illustration 1.8. Other Philadelphia-area observers listed in the report included L. H. Parsons of Newtown, Bucks County; William Jefferies of West Chester, Chester County; and an unidentified observer at the Haverford School in Delaware County. William Hough, a schoolmaster in Hatboro, Montgomery County, was also an observer; the barometer he used, one of the few remaining from the original network, is today in possession of the Historical Society of Montgomery County.

In addition to the observers participating in TFI's statewide system, several other weather observing sites were active in the Philadelphia area around this time. At the Frankford Arsenal in Bridesburg, Captain Albert Mordecai of the U.S. Army was a faithful observer of temperature and precipitation from 1836 to 1843. His observations were also published in the *Journal of The Franklin Institute*. Alexander Bache established a pri-

vately funded weather observatory at Girard College in 1840, where meteorological observations were taken until 1845. Charles Peirce was a faithful observer of the weather in Morrisville, Pennsylvania, across the Delaware River from Trenton, from 1790 to 1846. He summarized each month of observations in *A Meteorological Account of the Weather in Philadelphia*, which he wrote in 1847. To our knowledge, Peirce's book is the last written specifically about the weather of the Philadelphia area, and as such it is a valuable resource for reconstructing that era of local weather.

By 1840, Espy's reputation as "The Storm King" had grown (by some accounts, so had his ego), and he was lecturing around the country to promote his theories about storms and build support for expanding weather observing systems. Espy also traveled abroad—in France, he was so highly regarded that he was compared to Isaac Newton when introduced to the prestigious French Academy of Sciences. When Espy published *The Philosophy of Storms* in 1841, a popular book describing his theories about storms, he reinforced his standing as the premier American meteorologist of his time.

It was this success, and his still-unfulfilled goal of establishing a truly "national" weather service, that soon led Espy to Washington, D.C.

The Smithsonian Institution and a National Weather Service

During the 1840s, meteorology's center of gravity in the United States began to shift to Washington, D.C. This transition occurred partly because Espy headed there to lobby for a federally financed, countrywide weather observing service. Once in Washington, he served as a professor for the Navy Department and also oversaw improvements in the Army Surgeon General's weather network. In essence, these roles made him America's first "national meteorologist" and allowed him to continue compiling weather observations from across the country to pursue his studies of storms. He also expanded observing to some navy posts, including the U.S. Navy Yard in Philadelphia.

Washington's growing prominence in weather science was also related to the opening of the Smithsonian Institution in 1846 and the appointment of Joseph Henry as its first secretary. Henry was a veteran of the New York weather-observing network and familiar with Espy's in Pennsylvania. For Henry, establishing "systems of extending meteorological observations for solving the problem of American storms" was a primary research objective for

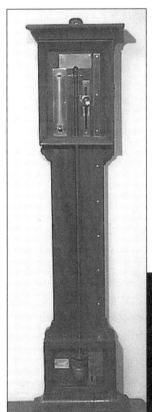

ILLUSTRATION 1.7 Barometer "No. 1," made in 1837 and used by James Espy and other observers at The Franklin Institute (courtesy of the Historical and Interpretive Collections of The Franklin Institute, Philadelphia, Pa.).

ILLUSTRATION 1.8   A portion of a monthly "Meteorological Report for Pennsylvania" compiled from data collected on the weather observing network established in 1837 by The Franklin Institute and the American Philosophical Society. Counties, towns and observers are listed, as well as temperature data for August 1840 (from the *Journal of The Franklin Institute*, 3d ser., 1 [1841]: 144).

METEOROLOGICAL REPORT

FOR THE STATE OF PENNSYLVANIA,

Collated from returns made to the Committee on Meteorology of the Franklin Institute of the State of Pennsylvania, for

AUGUST, 1840.

	County.	Town.	Observer.	7, A. M.	2, P. M.	9, P. M.	Maximum.	Minimum.	Mean.	Days omitted.
1	Philadelphia,	Philadelphia,	J. M. Hamilton,	74.47	78.85	70.96	86.00	66.00	74.76	3
2	Montgomery,									
3	Bucks,	Newtown,	L. H. Parsons,	71.63	80.67	66.30	88.00	47.00	72.87	.
4	Lehigh,									
5	Northampton,	Easton,	Charles Elliot,	65.74	70.00	67.57	83.00	57.00	67.77	1
6	Monroe	Stroudsburg,	A. M. Stokes,	66.65	81.05	70.20	88.00	50.00	72.63	11½
7	Pike,	Milford,	Ralph Bull,	60.80	78.90	62.50	87.00	51.00	67.40	.
8	Wayne,	Honesdale,	W. Richardson,							
9	Susquehanna,	Silver Lake,	E. Rose,							
10	Luzerne,	Wilkesbarre,	W. F. Dennis,							
11	Schuylkill,	Port Carbon,	John G. Hewes,	64.27	80.81	66.25	90.00	54.00	70.44	1
12	Berks,	Reading,	C. F. Egelmann,	69.10	78.06	68.35	84.00	60.00	71.84	.
13	Chester,	West Chester,	Wm. W. Jefferies,	72.46	80.38	76.92	87.25	65.00	76.59	.
14	Delaware,	Haverford,	Haverford School,	69.87	79.68	68.06	86.00	60.00	72.54	½
15	Lancaster,	Lancaster,	Conservatory of Arts,	69.15	81.57	70.61	89.00	59.50	73.78	½
16	York,									
17	Lebanon,									
18	Dauphin,	Harrisburg,	J. Heisely,	71.65	79.90	75.94	86.00	64.00	75.83	.
19	Northumberland,	Northumberland,	Andrew C. Huston,	68.68	77.44	70.77	85.00	60.00	72.30	.
20	Columbia,	Danville,	C. H. Frick.							
21	Bradford,									
22	Tioga,									
23	Lycoming,									
24	Union,									
25	Mifflin,									
26	Juniata,	Mifflintown,	J. A. Rinkead,	61.94	81.63	66.94	92.00	53.50	70.17	5½
27	Perry,									
28	Cumberland,	Carlisle,	Prof. W. H. Allen,							
29	Adams,	Gettysburg,	Prof. M. Jacobs,	67.39	80.58	70.35	86.50	59.00	72.77	.
30	Franklin,									
31	Huntingdon,	Huntingdon,	Jacob Miller,	67.87	83.03	69.45	91.00	55.00	73.45	.
32	Centre,	Bellefonte,	John Harris,	63.83	78.97	66.96	86.50	52.50	69.92	7
33	Potter,									
34	M'Kean,	Smithport,	Richard Chadwick,	61.23	75.55	56.87	86.00	48.00	64.55	.
35	Clearfield,									
36	Cambria,	Ebensburgh,	Richard Lewis,	63.13	76.72	62.87	86.00	49.00	67.57	2¾
37	Bedford,	Bedford,	Samuel Brown,	70.87	78.68	74.15	87.20	60.50	74.57	.
38	Somerset,	Somerset,	George Mowry,	63.65	73.52	63.64	82.00	52.00	66.94	.
39	Indiana,	Indiana,	Richard White,	72.94	77.74	66.00	85.00	58.00	72.23	.
40	Jefferson,	Rose Cottage,	C. C. Gaskell,							
41	Warren,	Warren,	C. S. Brown,	65.82	73.62	71.38	82.00	55.00	70.27	6¾
42	Venango,	Franklin,	Wm. Connely,	59.97	82.81	68.26	94.00	48.00	70.35	.
43	Armstrong,									
44	Westmoreland,									
45	Fayette,	Uniontown,	J. P. Weithers,	68.87	83.58	69.34	94.00	54.00	73.93	.
46	Green,									
47	Washington,	Cannonsburg,	A. H. Campbell,	66.29	78.78	70.13	86.00	52.00	71.73	.
48	Alleghany,	Pittsburgh,								
49	Beaver,	Beaver,	Wm. Allison,							
50	Butler,	Butler,	Jacob Mechling,	66.26	77.48	71.32	86.00	52.00	71.69	.
51	Mercer,	West Greenfield,	S. Campbell,							
52	Crawford,	Meadville,	J. Limber,	68.97	77.21	68.48	92.00	60.00	71.55	1¾
53	Erie,	Erie,	Park & Reid,							

the new institution. With a $1,000 appropriation in 1848, the Smithsonian Meteorological Project came to life. Henry consulted closely with Espy on the new project, using some aspects of the Pennsylvania weather observing system as a model. For example, the first Smithsonian observers used the same reporting forms as Espy used in Pennsylvania a decade earlier.

Henry, as secretary of the Smithsonian, and Espy, as national meteorologist, co-signed an announcement for volunteer observers. The request was sent to more than 400 people in thirty states. About 150 observers volunteered to keep weather journals and submit monthly reports by mail. They were to take observations at sunrise, 9 A.M., 3 P.M., and 9 P.M. local time. The first monthly reports arrived in March 1849, and Espy used the data to construct weather maps to aid in his storm studies. Professor James Coffin, a prominent meteorologist and mathematician at Lafayette College in Easton, Pennsylvania, summarized the data by computing daily, monthly, seasonal, and annual averages—essentially producing some of the earliest climate data for the United States.

Henry was also a hearty supporter of the use of the telegraph for relaying weather information. The telegraph would finally allow weather information to move faster than the weather itself, giving the possibility of advance warning of approaching storms. Telegraph lines had linked Washington with Baltimore in 1844 and expanded to Philadelphia by 1845. David Brooks, a telegraph operator in Philadelphia around that time, described the practical benefit of sharing weather information by telegraph:

> About the year 1849, I became manager in Philadelphia, and was in the habit of getting information about the condition of the lines westward every morning. If I learned from Cincinnati that the wires to St. Louis were interrupted by rain, I was tolerably sure a "northeast" storm was approaching (from the west or southwest). For cold waves we looked in Chicago.

Many observers from the Philadelphia area participated in the Smithsonian Meteorological Project at one time or another, including volunteers at West Chester, Valley Forge, Haverford, Norristown, Germantown, the Pennsylvania Hospital, Burlington, and Trenton. One dedicated Philadelphia observer was Professor J. A. Kirkpatrick, who taught civil engineering at both Philadelphia and Central High Schools. With a thermometer, barometer, rain gauge, and psychrometer (for measuring humidity), Kirkpatrick observed the weather on Vine Street for more than three decades. From 1857 to 1869, summaries of his observations were published in the *Journal of The Franklin Institute,* and for a few years he even wrote a monthly column titled "The Meteorology of Philadelphia."

While the Smithsonian network was growing in importance and stature, Espy retired from his posts in 1857. He died three years later without seeing his dream of a national weather service realized. But after the disruption of the Civil War, Henry and other prominent meteorologists of the time resumed the lobbying campaign Espy had started. Probably the most vocal supporter was Professor Increase Lapham of Milwaukee, a former observer for Espy. Lapham proposed a storm warning service for the Great Lakes, and he sent frequent clippings about casualties and property damage on the lakes to General Halbert Paine, Congressman for Milwaukee. Lapham pointed out that in 1868 alone, 209 lives were lost in storms on the lakes, with property damage valued at $4.1 million (in 1868 dollars). In one letter Lapham asked if it were not "the duty of the government to see whether anything can be done to prevent, at least, some portion of this sad loss in the future?" Congressman Paine recognized the importance of Lapham's cause, and on February 2, 1870 he introduced a Joint Congressional Resolution

> . . . to provide for taking meteorological observations at the military stations in the interior of the continent and at other points in the States and Territories of the United States . . . and for giving notice on the northern Lakes and on the sea coast by magnetic telegraph and marine signals, of the approach and force of storms.

The resolution was passed by Congress and signed into law on February 9, 1870 by President Ulysses S. Grant, with little notice from the news media. But an agency had been born that would affect the daily lives of most of the citizens of the United States through its forecasts and warnings. The new service was placed under the direction of the Secretary of War, because "military discipline would probably secure the greatest promptness, regularity, and accuracy in the required observations." Because the entire operation depended on a reliable communication system, it was assigned to the Army Signal Service Corps under General Albert J. Myer, who named the new agency the Division of Telegrams and Reports for the Benefit of Commerce.

On November 1, 1870, observer-sergeants at about twenty-five locations took the first meteorological reports of the new weather service. (Philadelphia was added two months later.) Each observer had a thermometer, barometer (for pressure), hygrometer (for humidity), anemometer (for wind speed), wind vane (for wind direction), and a rain gauge. Three times a day, around 8 A.M., 6 P.M., and midnight (Washington, D.C. time), each station telegraphed an observation to Washington. There, meteorologists interpreted the data, prepared forecasts, and sent them back to the observers, to railroad stations, and to interested news media.

Although the military took most of the weather observations for the new weather service, Myer also hired civilians. The first, Professor Lapham, took responsibility for the Great Lakes region. On his first day on the job, November 8, 1870, Lapham issued the first storm warning in the history of United States weather services, sending this message to observers on the Great Lakes:

> High wind all day yesterday at Cheyenne and Omaha; a very high wind this morning at Omaha; barometer falling with high winds at Chicago and Milwaukee today; barometer falling and thermometer rising at Chicago, Detroit, Toledo, Cleveland, Buffalo, and Rochester; high winds probable along the Lakes.

Some eighteen months later, on June 10, 1872, Congress extended the warning service beyond the Great Lakes and the coasts to all of the United States, making it truly a "national" weather system.

The Army Signal Corps and U.S. Weather Bureau in Philadelphia

The Army Signal Corps arrived in Philadelphia with little fanfare. With Sergeant A. W. Eastlake in charge, observations began on January 1, 1871 on the fourth floor of the Board of Trade Building at 505 Chestnut Street. A year later, after a move to the Chamber of Commerce Building on South Second Street, the Corps was still learning how to arrange a practical setup for the instruments, as illustrated by these comments of the observer on December 31, 1871:

> Shelter for hygrometer and thermometers rather too open. When it snows the room is filled with snow. When it rains the instruments become thoroughly wet, so that the drying of the bulbs must be done before each observation.

From 1871 to 1873, only three temperature readings were taken each day, not frequently enough to give reliable daily maximum and minimum temperatures. So from the standpoint of "official" record-keeping purposes, reliable high and low temperature records for Philadelphia did not start until January 1, 1874. Official precipitation measurements begin a little earlier, on April 1, 1872. "Precipitation" means rain plus melted snow and ice.

While the Army Signal Corps was settling into Philadelphia, The Franklin Institute continued to be a national leader in meteorology, with several institute members engaged in innovative weather observations. In the 1870s, Professor John Wise made more than 450 ascensions in a hot-air balloon and took measurements of atmospheric conditions above the ground. W. N. Jennings's photographs of lightning (see Illustration 1.9) in the early 1880s showed for the first time that a lightning bolt occurred as more of a wavy line than a zigzag, as was commonly depicted at that time. (His photographs were even published in *Scientific American* in 1885.) And The Franklin Institute organized the Pennsylvania State Weather Service in 1887 to assist the Army Signal Corps Weather Service in providing additional weather information at the local level, particularly for agricultural needs.

The growing recognition that weather conditions were critical to agriculture, and therefore to the nation's food supply, led to a reorganization of the government's weather system in 1891. All weather services were combined into a single agency, renamed the United States Weather Bureau (USWB), which was housed in the Department of Agriculture. With the USWB established,

ILLUSTRATION 1.9   First photograph of lightning, taken September 2, 1882 by W. N. Jennings (courtesy of the Historical and Interpretive Collections of The Franklin Institute, Philadelphia, Pa.).

The Franklin Institute gave up ownership of the Pennsylvania State Weather Service. For the next forty years or so, Philadelphia's presence on the national weather scene consisted primarily of the USWB office, which by that time had moved to the Post Office Building at Ninth and Chestnut Streets. There, in the heart of downtown Philadelphia, the official record warmest day (106°F, August 8, 1918) and coldest day (−11°F, February 9, 1934) in Philadelphia history were measured. The original records of the USWB during this period are on public display today at The Franklin Institute.

Nearly fifty years after the United States Weather Bureau was established, and nearly a century after Espy put Philadelphia in the meteorological spotlight, The Franklin Institute once again revitalized Philadelphia's national role in meteorology. In 1934, TFI moved to its present location on the Benjamin Franklin Parkway and opened as one of the first two public science museums in the country. (Chicago's Museum of Science and Industry was the other.) At first, there was no exhibit devoted entirely to meteorology, but public interest and the recognition of TFI's place in weather history spurred a plan for such an exhibit. With contributions of funds, equipment, and time from the USWB, private and corporate donors, and the American Meteorological Society (the nation's premier professional organization of meteorologists), the *Hall of Weather* opened in June 1942. The exhibit featured sections on the history of weather science, the preparation of weather maps, and the application of weather knowledge to industry and daily life, as well as the largest (at that time) collection of weather instruments in the country.

Two other events of the 1940s enhanced Philadelphia's role in U.S. weather science. In 1946, just after World War II had ended, the University of Pennsylvania formally dedicated the world's first electronic, general purpose, large-scale computer, dubbed ENIAC (for Electronic Numerical Integrator And Computer). Although ENIAC was an army project and used primarily for plotting ballistic trajectories, some primitive weather calculations were also run on the new computer. The military certainly recognized the potential benefits of better weather forecasting for their operations because critical plans during the war, such as the invasion of Normandy, depended on weather conditions. In 1950, the first successful computer weather forecast was made using ENIAC, ushering in the era of numerical weather prediction. (For more on ENIAC, see Chapter 2.)

While the work on weather modeling was underway, a man who would later become America's foremost weather historian, Dr. David Ludlum, joined The Franklin Institute in 1947 as its Director of Meteorology. Ludlum had been a battlefield weather forecaster in World War II and, as such, was responsible for predicting the weather for the 1944 assault on German forces at Cassino, Italy. This complicated land and air operation succeeded, in part, because of Ludlum's weather forecasts and came to be known as "Operation Ludlum," the only World War II action named for its weatherman. Like Espy a century before, Ludlum had a passion for weather and worked in an institution that supported that enthusiasm. At The Franklin Institute, Ludlum founded the Amateur Weathermen of America, a club for weather enthusiasts, and headquartered it at TFI. Soon, branches of the new organization appeared all over the country. He started *Weatherwise,* a popular nontechnical weather magazine that is still published today. And, in 1947, Ludlum worked briefly as Philadelphia's first (and one of the nation's first) television meteorologists.

Ludlum's short stint on local television just preceded the era when three personalities would make Philadelphia *the* place in early television weather: Dr. Francis Davis,

Wally Kinnan, and Herb Clarke. Francis Davis had started weather forecasting on the radio in Philadelphia in 1946 and added television to his duties in 1948. He remained on Philadelphia's Channel 6 until 1971, when he left to become a full-time dean at Drexel University. In the 1950s, another legend-to-be—Wally Kinnan, "The Weatherman"—joined Channel 3, where he would remain for a decade. Kinnan also opened a weather forecasting service at The Franklin Institute and, with Davis, helped to start a national accreditation program for meteorologists. The third television weather pioneer arrived in 1958, when Herb Clarke joined WCAU-TV. Clarke had the longest tenure of the three, remaining on-air for nearly forty years. Meteorologists and viewers alike recall the period from 1958 to 1965, when the three were on-air at the same time, as the heyday of Philadelphia television meteorology (see Illustration 1.10). During that period, three innovations in meteorology—computer models, satellites, and radar—were just coming into use, so it was a time when technology met personality to make Philadelphia a national leader in television weathercasting. (For more on these Philadelphia television legends, see the "Story from the Trenches" in this chapter.)

ILLUSTRATION 1.10 Philadelphia's three legendary television weather personalities in the early 1960s: *(l–r)* Wally Kinnan, Herb Clarke, and Dr. Francis Davis (courtesy of Herb Clarke).

Philadelphia's Early Television Weather Legends

Research studies throughout the country confirm that the number-one reason that people watch local television newscasts is for the weather report. Weather combines mass appeal (we all want to know what will happen tomorrow) with compelling pictures of local, national, and international interest.

Television weathercasts began in the United States around 1940. Some used fashion models or mascots to present the forecast. Weather, and other segments, were sponsored in ways not seen today: the weathercaster (or a cartoon character) would actually do the commercials. In New York City, for example, the weather on the station that is now WNBC was sponsored by Botany wrinkle-proof ties. An animated creature named Wooly Lamb started the weather segment by looking through a telescope, then singing: "It's hot, it's cold. It's rain, it's fair. It's all mixed up together. But I, as Botany's Wooly Lamb, predict tomorrow's weather." Then a slide would appear showing the latest forecast. Wooly Lamb presented the weather for seven years.

In Philadelphia, the commercial connection drove one of the great meteorologists right off the small screen. Dr. David Ludlum, after a very brief stay on local television in 1947, was fired because he refused to do the commercials required by management. "I couldn't croon the weather ditty they wanted to use to introduce the program, so they hired a crooner and made him the weatherman. I was just as glad," he later said.

About the time that Ludlum left the air, the first Philadelphia legend in television weather reporting arrived. As was the case with many other meteorologists of that era, Dr. Francis Davis had studied other sciences before becoming involved in weather. He was recruited to be a meteorologist during World War II. (See the story of his involvement in the D-Day forecast in Chapter 2.) With degrees in physics and math, he taught physics at Drexel University after the war and gave no thought to further work in weather.

Enter Walter Annenberg. The owner of newspapers and television and radio stations around the Philadelphia area, Annenberg saw the need to improve weather coverage and contacted the American Meteorological Society. He asked Ken Spengler, the Executive Director of the AMS, if he knew anyone in the Philadelphia area who could help. Spengler, who had been Davis's boss at the Pentagon during the war, recommended Davis for the job. This was the first step in Philadelphia's rise to national prominence in television weather over the next two decades.

Davis auditioned and was offered a job on radio in 1946. His salary was $10,000 a year to forecast at 7 A.M., 7 P.M., and 10 P.M. By 1948, he added television (Channel 6) to his weather duties, and his salary was doubled. Forecasting at that time was complicated by the limited amount of available weather information and a certain amount of guardedness about sharing it. Weather data at that time was distributed by teletype, essentially a typewriter that received and printed messages carried by telegraph or telephone. As Davis remembers, "I didn't have anything to work with. The U.S. Weather Bureau was jealous of private meteorologists taking over their terrain. They [the national office of the USWB] refused to allow us to get teletype data. They said it was dangerous to put teletype data in a television station because anyone could look at it." However, the local Philadelphia office of the USWB did allow Davis to visit their office each morning, look at the maps, and make some copies. And he did the same in the afternoon. He also got an evening phone update from the Weather Bureau.

In those days, the thirty-minute Channel 6 newscast was divided into three segments: fifteen minutes of news, ten minutes of sports, and finally five minutes of weather. Commercials took up about a minute and a half of that five minutes, leaving three and a half minutes for the weather. Davis was told that weather was last because it was "the most important thing people want, so we want to keep the viewers to the end." He did the weather by rear projection, using acetate overlays, so he could draw all the weather features individually. The forecast extended to three days in the future. Davis's philosophy of weather forecasting sounds similar to some used today: "I talked about all the different possibilities. I always found that even if the forecast wasn't right, there would be people the next day who would say: 'You got it right—you said it might rain.'" Davis's style was authoritative and somewhat professorial, and indeed he still taught physics full time at Drexel while he was working in television. Davis continued to forecast the weather on Channel 6 until 1971, when he left to become a dean at Drexel.

In 1957, another meteorologist, Wally Kinnan, joined Davis on the Philadelphia television weather scene. Kinnan's early training was in engineering, and, like Davis, he became involved in weather forecasting while in the military during World War II. After the war, Kinnan stayed in the Air Force and was stationed in Oklahoma City, where in 1950 he started doing local television weather. But within six months, he was sent overseas to serve in the Korean War. According to Kinnan, he was "in charge of forecasting the upper air for the entire Pacific Ocean." When his tour of duty ended, he returned to Oklahoma City and did weather at the NBC station there from 1953 to 1957. When his boss left to become manager of NBC stations in Philadelphia and New York City, Wally was asked to audition for weather jobs at both stations. "I didn't like the atmosphere in New York, but I did like Philly," he recalls (referring to his feelings about the stations, not the weather in either city).

At Channel 3 in Philadelphia (which at the time was WRCV, an NBC station), Wally replaced a fashion model who, in his words, "swished down the stairs à la Loretta Young, with a new dress each night." With Wally Kinnan, "The Weatherman," things changed immediately. "I believed in explaining what was going on, and why," says Kinnan. The team of Kinnan and Vince Leonard "in a span of two years became the number one news team in the city." Part of the reason for Wally's success was his personable, humble, self-deprecating style on air. He looked and talked like a regular guy.

While establishing his reputation on Philadelphia television, Kinnan also put his mark on local and national meteorology in several other ways: he started a weather forecasting center at The Franklin Institute; he was one of the first (if not the first) meteorologist to issue a five-day forecast; and he helped create the Seal of Approval program of the American Meteorological Society. The Franklin Institute Weather Service opened in May 1963. Kinnan designed the weather center and staffed it full time with six other meteorologists. Kinnan's assistant director was Colonel Robert C. Miller, who fifteen years earlier had issued the first successful tornado forecast while an Air Force meteorologist in Oklahoma. The weather center was an observation room behind windows (see Illustration 1.11); people could

ILLUSTRATION 1.11 Meteorologists Maximillian Kozak, Charles Umpenhour, and Wally Kinnan study maps in the newly opened Franklin Institute Weather Center in May 1963 (Urban Archives, Temple University, Philadelphia, Pennsylvania).

watch the meteorologists work and ask questions. Kinnan would prepare his television program at the institute and then go to the station for the early newscast.

A few years earlier, around 1960, Kinnan had introduced the five-day forecast on Channel 3. "I believe it was the first effort in that regard in the country," he recalls. To make the five-day, he used his experience from forecasting upper-air patterns during the Korean War. "It was my belief that you could take that upper-level flow and predict surface conditions out to five days with a certain degree of success," he says. Kinnan would compare the current upper-air weather map with charts from the past that were similar and use those past comparable situations to guide his forecasting, a technique known as the "analog method." (This method is still used today for seasonal forecasting—see Chapter 7.)

Many of his colleagues instantly condemned his innovation. At the time, the American Meteorological

Society's guidelines for meteorologists recommended against forecasting beyond three days. According to Kinnan, other meteorologists "fought me at every turn because no one wanted to try to do it, or even knew how to do it . . . the Weather Bureau opposed it, although never publicly." And so did Dr. Francis Davis, though he too never made his protest public. The conflict over the five-day strained the relationship between the Philadelphia weather pioneers. Kinnan recalls, "Francis and I were good friends. We did have that disagreement over the five-day forecast for awhile, and I don't know if he ever really forgave me." The irony is that some of the "analog" maps that Kinnan used for his five-day forecasts may have been the ones drawn up years before by Davis when he was part of one of the D-Day forecast teams.

The two did work together on the Seal of Approval program of the American Meteorological Society. In the mid-1950s, the AMS convened a committee to find ways to promote and accredit professional meteorologists on radio and television. Davis chaired the committee, and in an article in *TV Guide* in 1955, he summarized the AMS rationale for doing this:

> If TV weathermen are going to pose as experts, we feel they *should* be experts. We think the weather should be discussed with dignity. . . . We think many TV "weathermen" make a caricature of what is essentially a serious and scientific occupation, helping foster the notion that forecasters merely grab forecasts out of a fish bowl.

Davis's committee recommended a procedure and a set of standards for certifying television meteorologists. An applicant would send a film clip of a weather program that would be reviewed by members of the Board of Broadcast Meteorology. Davis, Kinnan, and three others served on the initial board. The men on the committee applied for the seal first, and to avoid a controversy about who would receive the first seal, AMS Director Ken Spengler decided they would be given in alphabetical order. That's how Francis Davis received AMS Seal Number One and Wally Kinnan AMS Seal Number Three. To date, the AMS seal has been awarded to more than 1,200 people, and it is required of television weathercasters by many stations throughout the country.

Kinnan inspired many other Philadelphians to become meteorologists, including Glenn Schwartz. Another young Kinnan disciple from Philadelphia, Joel Myers, went on to found AccuWeather, currently the world's largest commercial weather service. AccuWeather's forecasts were combined with the talents of the late Jim O'Brien in the 1970s on Channel 6, leading to more success for both in the Philadelphia television market.

In the mid-1960s when the ownership of the station changed, Wally relocated with NBC to Cleveland. He is still remembered as a key contributor to Philadelphia's reputation for television meteorology. Kinnan and Francis Davis were two of the three personalities that made the period from the late 1950s to the early 1960s the heyday of Philadelphia televi-

THE MODERN NATIONAL WEATHER SERVICE

It was true when Francis Davis and Wally Kinnan were doing the weather on Philadelphia television, and it is true today. Most of the raw weather information that meteorologists use—including surface and upper-air weather observations, computer model guidance, and satellite and radar images—comes from, in some way, an observation or prediction system maintained by the federal government. In 1970, the U.S. Weather Bureau was renamed the National Weather Service, and today the NWS is part of the Department of Commerce as an agency of the National Oceanic and Atmospheric Administration (NOAA). The mission of the modern National Weather Service is "to provide weather, hydrologic, and climate forecasts and warnings for the United States, its territories, adjacent waters and ocean areas, for the protection of life and property and the enhancement of the national economy."

Today, the National Weather Service maintains about 120 forecast offices across the United States. Every year, these offices issue more than 700,000 weather forecasts, 800,000 river and flood forecasts, and nearly 50,000 potentially life-saving severe weather warnings. NWS meteorologists issue forecasts twice each day, around 4 A.M. and 4 P.M. (local time) for individual counties or groups of counties, and the forecasts are updated at other times as needed. (The "local forecasts" shown on the Weather Channel are NWS forecasts.) NWS offices also issue a variety of other products, including aviation and marine forecasts and river and drought statements. The

sion weather. But the Philadelphia legend with the longest tenure is Herb Clarke, whose on-air weather career spanned some forty years.

Herb had been news director at a television station in Richmond, Virginia, but his on-air warmth and communication skills led to a career as one of more than forty "Atlantic weathermen" on television stations along the East Coast. The program was sponsored by the Atlantic Oil Company, which was headquartered in Philadelphia. In October 1958, Herb auditioned for the job in the company's home city. He was hired at $200 per week on WCAU, where he did a three-minute-and-twenty-second weathercast, a little longer than today's shows, using four sliding boards that showed various maps (see Illustration 1.12). His anchorman on the news was the legendary John Fascenda (known later as the voice of NFL films).

Since Herb was not a meteorologist, he used forecasts from the U.S. Weather Bureau, giving credit to the government forecasters and calling it the "official forecast." This acknowledgement earned him the enduring respect of his competitors and other meteorologists in the area. His style could be described as folksy and reassuring. Viewers would not be frightened of impending stormy weather after hearing a calm, collected Herb deliver the forecast. After Wally Kinnan's departure, Herb and Channel 10 relied on Franklin Institute meteorologists for forecasts, giving them full credit.

Herb's television career in Philadelphia spanned five decades; he started with drawing on maps and

ILLUSTRATION 1.12   Herb Clarke as an Atlantic weatherman in 1959 (courtesy of Herb Clarke).

closed using three-dimensional computer graphics. But according to Herb, the main change is: "Compared to what we're getting now, there was time for a bit more togetherness with the audience. The viewer gets closer to the person doing the weather than anyone else on the news . . . we both care for the same reason." As evidence of this connection with viewers, Herb received between forty and sixty letters a day. "I tried to answer them all," he says. Herb retired from Philadelphia television weather forecasting on New Year's Eve, 1997. ❖

local NWS office was located in Center City Philadelphia until 1993 when it moved to Mount Holly, New Jersey. Illustration 1.13 is a view inside that office, which is a typical NWS office. The Mount Holly office is responsible for ten counties in eastern Pennsylvania, sixteen counties in New Jersey, five counties on the upper eastern shore of Maryland, and all of Delaware (see Illustration 1.14). This area of responsibility includes the cities of Philadelphia, Allentown, Reading, Atlantic City, Trenton, Wilmington, and Dover.

The National Weather Service performs other essential activities as well. The NWS oversees several networks of weather observation—on land, at sea, and above the ground. They maintain a national system of radars for detecting precipitation and potentially severe weather. And the NWS runs the computer weather models used

by every forecaster—not just NWS meteorologists. All this information is made available to other government agencies, the private sector, the media, and the public. The cost for these services amounts to about $2.50 for each American citizen every year.

Today, the primary nationwide weather observing system consists of about 1,000 "first-order" observing locations, most at airports or military installations. The official Philadelphia first-order observing site has moved several times since the Army Signal Corps observers arrived in 1871. (See Table 1.2 for a complete list of these sites.) Since 1940, the official Philadelphia observation has been taken at the airport a few miles southwest of Center City. Other local first-order observing sites include the Northeast Philadelphia Airport, Dover Air Force Base, and the Mercer County Airport in Trenton,

ILLUSTRATION 1.13   Inside the National Weather Service office in Mount Holly, New Jersey (courtesy of National Weather Service).

New Jersey. Illustration 1.15 shows the locations of the first-order observing stations within the forecast area of the NWS Mount Holly office and some adjacent counties. Meteorologists identify each site with a unique three-character code—the same city codes used by the airlines to route luggage. Often, the code is easy to decipher—for example, PHL is the Philadelphia International Airport. Note the lack of observing stations in some areas, including Chester, Delaware, and Schuylkill counties in Pennsylvania and Cecil County in Maryland. These gaps in the data-observing network are problematic for local forecasters, because the weather in the Philadelphia area typically approaches from a westerly direction.

First-order observing sites routinely report weather conditions on the hour, but can report more frequently if weather conditions are changing rapidly. Typically, the following are observed: temperature, humidity, air pressure, wind speed and direction, cloud cover and cloud height, horizontal and vertical visibility, and weather (such as rain, snow, or fog). The National Weather Service, in partnership with the Federal Aviation Administration and the Department of Defense, has nearly completed automating the first-order observing network. Human observers have been replaced with a suite of electronic monitoring devices known as ASOS (for Automated Surface Observing System). Illustration 1.16 is a picture of an ASOS. The change to ASOS allows more frequent and more remote weather observations, and the new automated system has performed well for measuring basic weather elements such as temperature, humidity,

and wind. But ASOS cannot measure snowfall depth (among other things), so a human observer still augments the reports at some ASOS sites. (In fact, many meteorologists still long for the days when humans took all the observations.) The observation site at the Philadelphia International Airport was automated on December 1, 1995.

In addition to these first-order observing stations, the National Weather Service supports other countrywide observing systems, including a network of thousands of automated rain gauges. Another nationwide observing system, called the "cooperative observing network," is staffed by citizen volunteers at over 10,000 sites. Data at these locations is collected only once per day—typically, an observer might take readings at 8 A.M. local time, noting the high and low temperature and any precipitation over the previous twenty-four hours. Because of this observing schedule, data from this network is primarily used for climatological purposes and not for short-term forecasting. The Franklin Institute has been the cooperative observing site for Center City Philadelphia since

ILLUSTRATION 1.14   The thirty-four counties that are shaded and labeled are in the service area of the Mount Holly office of the National Weather Service (courtesy of National Weather Service).

TABLE 1.2 Location of the Official Philadelphia Weather Observing Site, 1871 to the Present

Location	Period
Philadelphia Board of Trade 505 Chestnut	Jan. 1, 1871 to Sept. 20, 1871
Chamber of Commerce Building 133 South Second	Sept. 21, 1871 to Jan. 31, 1882
Mutual Life Insurance Building Tenth and Chestnut	Feb. 1, 1882 to March 31, 1884
Post Office Building Ninth and Chestnut	April 1, 1884 to Dec. 17, 1934
U.S. Custom House Second and Chestnut	Dec. 18, 1934 to June 19, 1940*
Southwest (International**) Airport Administration Building	June 20, 1940 to Dec. 22, 1954***
International Airport New Terminal Building	Dec. 23, 1954 to Nov. 30, 1995
International Airport Automated Surface Observing System	Dec. 1, 1995 to present

*When the official observation site for Philadelphia moved to the airport in 1940, the U.S. Custom House (home of the U.S. Weather Bureau) continued its observations as the official reporting site for downtown Philadelphia. That downtown site moved to the Philadelphia Electric Building at Tenth and Chestnut on May 16, 1959 and then to the Federal Office Building at 600 Arch Street on Dec. 3, 1973, where it resided until 1993 when the National Weather Service moved to Mount Holly, N.J.

**Name changed to International Airport on April 1, 1948.

***From June 23 to November 30, 1945 the airport office was closed, and official observations were taken at the U.S. Custom House.

1993. Other local cooperative observing stations include Valley Forge and Washington Crossing in Pennsylvania, Moorestown and Glassboro in New Jersey, and Newark and Lewes in Delaware.

There are other weather observing networks that are not administered by the National Weather Service. For example, the Department of Environmental Protection in Pennsylvania maintains a statewide system of several dozen weather and pollution monitoring stations. The New Jersey Department of Transportation operates a similar network of over thirty roadside weather information systems. And some television stations install weather equipment in local schools and businesses, and then show the live observations during their weathercasts. These weather observing systems often provide very useful data, especially in areas where there are few first-order stations, but the observations usually are not quality controlled, so the data varies in reliability.

Weather observations at the earth's surface are a small part of the observational puzzle that meteorologists must build to understand the current weather pattern and predict future ones. Weather data from above the ground comes primarily from instrument packages called **radiosondes** that are carried aloft by weather balloons. During their ascent, they transmit back observations of temperature, pressure, and humidity to a ground-based receiving station until the balloon bursts at around fifteen to twenty miles altitude. (The radiosonde then parachutes back to earth—some are returned to the NWS and recycled.) Tracking the position of the radiosonde as it ascends also enables meteorologists to infer the wind speed and direction at various levels. Presently, radiosondes are launched twice a day from about seventy-five sites in North America and hundreds more worldwide.

First used by the U.S. Weather Bureau in 1936, radiosondes (many made by VIZ Corporation, based in

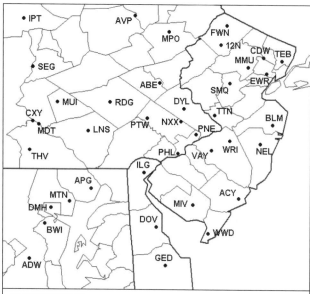

PENNSYLVANIA

ABE	Allentown (Lehigh Valley Intl. Airport)
AVP	Wilkes–Barre/Scranton Intl. Airport
CXY	Harrisburg (Capital City Airport)
DYL	Doylestown Airport
IPT	Williamsport (Lycoming County Airport)
LNS	Lancaster Airport
MDT	Harrisburg Intl. Airport
MPO	Mount Pocono (Pocono Mountains Municipal Airport)
MUI	Muir Army Air Field/Indiantown
NXX	Willow Grove Naval Air Station
PHL	Philadelphia Intl. Airport
PNE	Northeast Philadelphia Airport
PTW	Pottstown-Limerick Airport
RDG	Reading Regional Airport
SEG	Selinsgrove (Penn Valley Airport)
THV	York Airport

MARYLAND

ADW	Camp Springs/Andrews Air Force Base
APG	Phillips Army Air Field/Aberdeen
BWI	Baltimore–Washington Intl. Airport
DMH	Baltimore, Inner Harbor
MTN	Baltimore, Martin

DELAWARE

DOV	Dover Air Force Base
GED	Georgetown (Sussex County Airport)
ILG	Wilmington (New Castle County Airport)

NEW JERSEY

12N	Andover (Aeroflex–Andover Airport)
ACY	Atlantic City Intl. Airport
BLM	Belmar–Farmdale
CDW	Caldwell (Essex County Airport)
EWR	Newark Intl. Airport
FWN	Sussex Airport
MIV	Millville Municipal Airport
MMU	Morristown Municipal Airport
NEL	Lakehurst Naval Air Station
SMQ	Somerville (Somerset Airport)
TEB	Teterboro Airport
TTN	Trenton (Mercer County Airport)
VAY	Mount Holly (South Jersey Regional Airport)
WRI	Mcguire Air Force Base
WWD	Wildwood

ILLUSTRATION 1.15 First-order weather observing sites, with official three-character identifiers.

Philadelphia) are relatively low-tech devices. Probably the most technologically advanced weather observations come from an armada of orbiting satellites and ground-based radars (two more NWS/NOAA services) that are continuously monitoring the skies for clouds, precipitation, and much more. Because they revolutionized the science of meteorology and the way weather information is presented to the public, weather satellites and weather radar deserve a closer look.

Weather Satellites

Satellites, man-made objects put into orbit around the earth, are used for communications, military monitoring, scientific research, and weather observing. U.S. weather satellites are maintained and operated by NOAA. Along with radar and computer models, weather satellites are one of the three great technological advances in meteorology of the latter half of the twentieth century.

The first weather satellite was TIROS-1, launched on April 1, 1960. (TIROS stands for Television InfraRed Observation Satellite). Early satellite images were black and white and rather grainy (see Illustration 1.17), but at the time such images were a monumental advance. Previously, cloud patterns could be visualized only in theory; with satellite imagery, meteorologists could finally see the sharp line of thunderstorms along a cold front, the spiral bands around a hurricane, and the comma shape of a mature low-pressure system. TIROS-1 and other early weather satellites orbited too low to see much of the earth at once. To get a wider view, meteorologists had to piece together the individual images taken as the satellite circled the earth. The result was a collage of pictures taken at different times, with some of the images hours old.

To solve the problem, scientists put a satellite higher up, where it had a wider view. And if the satellite was placed over the equator at an altitude of about 22,500 miles, it could orbit at the same rate as the earth rotated. In this way, the satellite would essentially remain fixed above the same point on earth and thus always have a view of the same area. Then, images of that area could be played one after another to create loops, or animations, to show the movement of weather systems. The first such "geostationary" weather satellite, GOES-1, was launched in October 1975. (GOES stands for Geostationary Operational Environmental Satellite.) Today, several generations later, GOES-8 monitors the eastern United States and the Atlantic Ocean, while GOES-10 tracks the weather in the eastern Pacific and western U.S. GOES-11

ILLUSTRATION 1.16 An ASOS (Automated Surface Observing System). This array includes instruments to measure temperature, humidity, wind direction, wind speed, visibility, cloud height, rain amount, and type of precipitation. Human observers no longer take most official hourly surface weather observations (courtesy of National Weather Service).

taken with a camera. People and objects reflect varying amounts of light—the light comes from the sun or from a flash bulb. Of course, there is no flash bulb in the sky to substitute for the sun's absence, so visible satellite images are not useful to meteorologists at night.

Fortunately, meteorologists have another way to "see" clouds, even in the absence of sunlight. Everything—including clouds—gives off invisible energy, which meteorologists refer to as infrared radiation. The hotter something is, the more infrared radiation it emits. Since clouds are usually high in the sky, they tend to be colder, so they emit less radiation than the warmer ground. Special sensors on board weather satellites can measure this invisible radiation and, with the help of computers, create infrared images based on the differences in the amount of radiation emitted.

Infrared satellite images can not only differentiate between clouds and the ground but also between clouds at different heights in the atmosphere. High clouds, such as cirrus or the tops of tall cumulonimbus, are colder, so they emit a relatively low amount of radiation. Low clouds, such as puffy popcorn-like **cumulus** or layers of

was launched in May 2000, while GOES-12 went into orbit in July 2001. They are ready to take over if GOES-8 or GOES-10 fails. Other geostationary satellites owned by Japan, India, and the European Space Agency help extend the view around the globe.

The three main types of weather satellite images—visible, infrared, and water vapor—each has strengths and weaknesses. Used together they give meteorologists an unprecedented ability to probe the atmosphere for clouds and much more.

Visible images depend on sunlight that reflects off clouds and the earth. Brighter objects reflect more sunlight than darker objects, so clouds (and snow) appear white while land and oceans appear darker. The thicker that clouds are, the more sunlight they reflect, and the brighter white they appear. So visible satellite images can distinguish between thin and thick clouds. In Illustration 1.18a, the line of tall thunderstorm clouds (called **cumulonimbus**) moving off the southeast and mid-Atlantic coasts appears bright white. In contrast, the streamers of wispy high-altitude **cirrus** clouds being blown eastward out into the Atlantic off the tops of these thunderstorms are a duller white, indicating that these clouds are relatively thin. The clouds over the Great Lakes, which appear an intermediate brightness, are an intermediate thickness. In a way, visible satellite images are similar to pictures

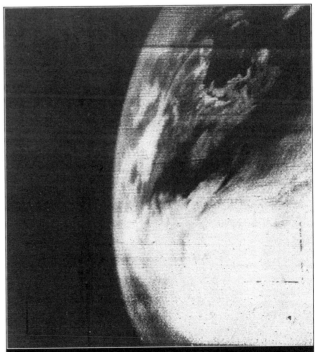

ILLUSTRATION 1.17 The first "television picture" from space, taken by the Tiros I satellite on April 1, 1960. Note how fuzzy the clouds are, and how difficult it is to tell where you are looking (courtesy of NOAA).

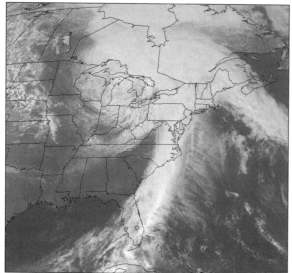

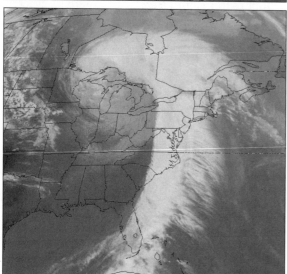

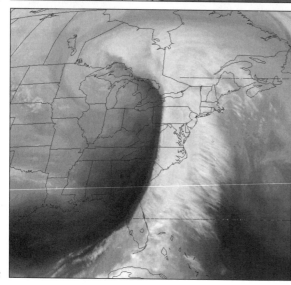

dreary **stratus**, are warmer and therefore emit greater amounts. On infrared satellite images, the warmest low clouds are typically "colored" a shade of gray, cooler mid-level clouds dull white, and the coldest high or tall clouds bright white.

Illustration 1.18b is an infrared image taken at the same time as the top panel. The grayish shading of the clouds that stretch from West Virginia and western Pennsylvania back into Michigan indicates these clouds are relatively low, with tops only a few thousand feet above ground. In contrast, the bright white representing the thunderstorms along the eastern seaboard and the cirrus clouds streaming east off their tops indicates that these clouds are very high in the sky.

Infrared imagery does have a few shortcomings. When clouds are close to the ground, they may be about the same temperature as the earth's surface. When that happens, infrared imagery may not be able to show the difference. Or sometimes on winter nights, snow-covered ground can be just as cold as a patch of nearby clouds. On infrared imagery, they both will look about the same. In these cases, forecasters turn to observations taken at the ground to decide what is a cloud and what is not.

Water vapor imagery is a third type of satellite imagery that is rarely shown on television but is immensely useful to meteorologists. **Water vapor** is the gaseous form of water and thus is invisible. But it is possible to "see" water vapor with a special kind of sensor on board the satellite. One limitation is that the sensor can detect water vapor only high up in the atmosphere, above 15,000 feet or so, so water vapor imagery cannot measure humidity near the ground. Still, being able to see high-altitude water vapor has great value. Illustration 1.18c is the water vapor counterpart to the visible and infrared images in the top two panels. It is traditional in water vapor images to depict relatively dry air as dark, while air with lots of high-altitude water and water vapor is shown as bright white. The dark area stretching from western Pennsylvania into the Carolinas marks dry air at high altitudes that is surging northward behind the line of bright-white thunderstorms.

ILLUSTRATION 1.18 *(a)* **Visible satellite image: the thicker the clouds, the whiter they appear. Ground and ocean appear dark.** *(b)* **Infrared satellite image taken at the same time as the visible image in part** *a.* **The coldest cloud tops are white.** *(c)* **Water vapor satellite image made at the same time as the images in parts** *a* **and** *b.* **Moist high-altitude air is shown in white (courtesy of NOAA).**

But probably the greatest advantage of water vapor imagery comes when the images are animated. Then, the water vapor acts as a tracer to show upper-level winds even where there are no clouds. This allows meteorologists to see important high-altitude weather features such as the jet stream, the fast river of air that encircles the globe high over the midlatitudes (see Chapter 2). This capability has been a boon to forecasters, especially those tracking tropical storms and hurricanes over the oceans where there are few upper-level wind observations.

Though many satellite images are simply shown in black and white, meteorologists often colorize the images to emphasize certain features and to make the images more visually appealing for television. For example, the tallest and thickest clouds, which typically produce the heaviest rain or snow, are commonly colored orange or red to make them stand out.

Finally, weather satellite imagery is used for more than just detecting clouds. In fact, when clouds do not block the way, infrared images can be used to sense differences in the temperature of the surface. So ocean water temperatures can be measured from space, a particularly useful capability for beachgoers and tropical forecasters trying to predict where hurricanes will form and intensify. On land, infrared imagery can detect the relative warmth of cities over surrounding countryside, the so-called "urban heat island" (see Chapter 2). And satellites are increasingly being used to estimate the temperature, humidity, and wind above the ground. This technology will go a long way in filling in the gaps between radiosonde observations and thus will improve weather forecasts.

Weather Radar

In addition to the vigilant satellites looking down on earth from space, meteorologists also depend on ground-based radars that probe the atmosphere for precipitation and wind. Radar, which is an acronym for Radio Detection And Ranging, was used successfully during World War II to detect ships and airplanes rather than rain and snow. But after the war, some weather enthusiasts aimed radars at the sky, starting an era of radar meteorology that has led to the sophisticated, high-tech colorized radar displays that you see today.

All radars send out pulses of energy, or radiation. Weather radar actually uses microwave radiation, not radio waves as implied by the name. This radiation travels through the atmosphere and reflects off targets such as raindrops and snowflakes. The more energy returned to the radar, the more intense the precipitation. By mea-

suring the time it takes for the energy to return to the radar, meteorologists can determine how far away those radar "echoes" are. And with a full 360-degree rotation of the radar, a complete picture of the precipitation pattern around the radar can be constructed.

In the late 1950s, the first national network of weather radars became operational. The black and white images produced by these WSR-57s (Weather Surveillance Radar, 1957) were static. Different intensities of rain were displayed in various shades of gray, and meteorologists outlined storm locations with a grease pencil on a plastic overlay. With multiple images, the direction and speed of movement of the precipitation could be estimated. Even though the coverage was not uniformly good for the whole country, and the radars did not do a very good job of detecting light precipitation (especially snow), these WSR-57 radars provided meteorologists with an unprecedented tool for short-term forecasting. The National Weather Service updated many of the radars in the 1970s, when the demand for animations and color radar images increased.

But the big leap in radar technology came in the form of **Doppler radar**, which offered more than just an improvement in pinpointing precipitation and its intensity. Doppler radar could actually measure winds, and thus had the potential to identify which thunderstorms might produce a tornado, something that the older radars did not do well. After nearly two decades of research, the National Weather Service (in partnership with the Department of Defense and the Federal Aviation Administration) unveiled a nationwide network of Dopplers in the late 1980s and the 1990s to take the place of the old, outdated 1950s- and 1970s-style radars. These WSR-88D radars (1988 technology, *D* for Doppler) were a huge technological advance, and an expensive one. At a cost of almost $2.5 million per radar, and with about 140 of them needed in the lower forty-eight states to provide nearly complete coverage, the WSR-88D radar program was threatened by budget cuts in the 1980s. Its importance to the military saved the project. That connection also explains why so many radars are located at or near military bases. Two of these Doppler radars, located at Fort Dix, New Jersey and Dover Air Force Base, Delaware, provide continuous radar coverage of the skies over the Philadelphia area. Illustration 1.19 shows the Fort Dix radar. The actual radar unit sits inside the forty-foot diameter dome, which protects the equipment from the elements.

For severe weather warnings, Doppler radar's primary advantage is its ability to detect the motions of air inside

ILLUSTRATION 1.19 The National Weather Service Doppler radar at Fort Dix, New Jersey, one of two NWS WSR-88D radars that cover the Philadelphia area (courtesy of National Weather Service).

thunderstorms. The principle that underlies this capability was first described in the mid-nineteenth century by Christian Johann Doppler, an Austrian mathematician. Doppler explained why a train whistle sounds different depending on whether the train is moving toward or away from you. Sound travels in waves, and when a source of sound is moving toward an observer, more waves pass per second. This higher frequency yields a higher-pitched sound. The opposite is true if the source of the sound is moving away: fewer waves of sound pass per second, and this lower frequency yields a lower pitch.

The same principle applies to radiation and, thus, to radar. The frequency of the radiation reflected from a target is changed if the target is moving toward or away from the radar. The frequency of the returned energy is then easily calibrated into a speed. Police officers use this principle when they point radar guns at fast cars. Meteorologists do the same thing with Doppler radar, deriving "pictures" of wind motions inside storms. This capability of Doppler is known as the "velocity" mode of the radar.

The WSR-88D radars are largely automated. In about six minutes, a WSR-88D can make detailed scans of all elevations in all directions around the radar. The data is fed into computer programs that calculate the storm movement, intensity, and the potential for severe weather. Collectively, the radars and the software and hardware that interpret the radar data are known as NEXRAD (for NEXt generation RADar). The NEXRAD system can estimate what time a storm will hit specific towns, whether hail is likely, how much rain has fallen, and how strong wind gusts might be. The velocity mode of the radar can detect whether the raindrops or snowflakes in the storm are moving toward or away from the radar site. Winds blowing toward the radar in one part of a thunderstorm,

right next to winds blowing away from the radar, are a telltale sign of rotation and a possible tornado. If the NEXRAD system detects this pattern, it plots a circle on the radar image and can even trigger an alarm.

The NEXRAD system is a tremendous advance for meteorologists charged with issuing severe thunderstorm, tornado, and flood warnings (see Illustration 1.20). Instead of sometimes having to wait for severe weather to be reported before issuing a warning for strong winds, large hail, a tornado, or flash flooding, the National Weather Service can now (in most cases) issue a warning *before* the severe weather strikes. With specific towns mentioned in the warnings, people are also more likely to take appropriate protective action. Nationwide, the average warning time for tornadoes has increased from six minutes in 1993 to about twelve minutes in 2001 (though it is probably less than that for weak tornadoes, like most of those in the Philadelphia area, because weaker twisters are harder to detect).

Doppler radar has other improved capabilities over earlier radars. It is more sensitive and thus better at detecting light precipitation, such as drizzle. The radar images have a higher resolution, allowing finer details to be seen not only in thunderstorms but also within bands of rain or snow. The Doppler radars also have a larger range, at times able to detect precipitation more than 250 miles away. (However, the useful range for detecting severe weather and flash flooding is much smaller, generally under 100 miles.) And the old problem of "false echoes," where the radar picked up objects such as buildings or mountains near the radar site, is largely resolved too. The NEXRAD system can eliminate echoes that never move, thus removing such "ground clutter" from the images.

There is at least one problem that even NEXRAD has not solved: when the air between the clouds and the ground is very dry, rain or snow often evaporates before reaching the surface. The Doppler detects the precipitation leaving the cloud and depicts it on the radar image, leaving confused weather watchers at the ground wondering what happened to the rain or snow. In such instances, surface observations are indispensable for identifying where precipitation is reaching the ground and where it is not.

Television stations can access the WSR-88D network and show the images on the air. A few companies take the radar data and do their own storm tracking on colorful, extremely detailed maps, which they then sell. In either case, because of the six-minute automated

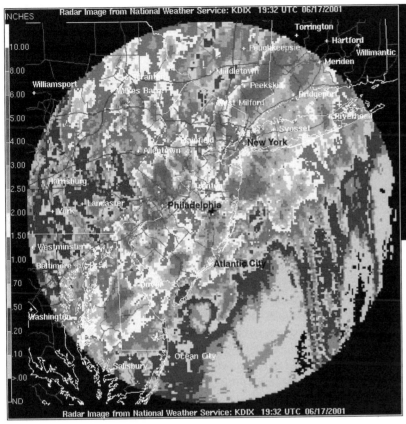

Radar Image from National Weather Service: KDIX 19:32 UTC 06/17/2001

Radar Image from National Weather Service: KDIX 19:32 UTC 06/17/2001

ILLUSTRATION 1.20 (*See plate.*) An example of rainfall estimated by Doppler radar from the remnants of Tropical Storm Allison, which brought flash flooding to parts of the Philadelphia area on June 16–17, 2001. Philadelphia is approximately in the center of the image. The color legend on the left indicates radar-estimated rainfall amounts. Parts of Montgomery and Bucks counties received more than six inches of rain based on this estimate. Doppler-radar derived precipitation estimates allow meteorologists to more effectively issue flash flood warnings (courtesy of National Weather Service).

sequence of the radar, and the relaying and processing time from radar to television screen, images from the WSR-88D network shown on television are not actually live. There's a delay of at least six minutes and sometimes more. Most of the time, that delay has little practical impact, but in cases of severe weather it can make a big difference. A fast moving line of thunderstorms can race ten miles in ten minutes, quickly outpacing the ability of the on-air radar to keep up.

To get around the delay, some stations buy their own Doppler radar so they can say that their radar is truly "live," and also to have control over the radar itself. More than 100 television stations currently have their own Doppler radars, and the number is increasing. Some utility companies and other weather-sensitive businesses are also purchasing them. These privately owned Doppler radars augment the National Weather Service's WSR-88D system. In January 2000, NBC-10 was the first Philadelphia-area television station to purchase its own Doppler radar.

A television innovation, not a meteorological one, allows modern weathercasters to seemingly without effort incorporate satellite and radar images into their presentations. The process is called chroma-key, also known in Hollywood as "blue screen." In blue-screen, a person is photographed in front of a solid blue background, usually a large wall. Then, using computers, all the blue in the picture is "keyed out" (removed) and replaced with another image. The result: it looks like the person is standing in front of the new image. For a television weathercaster, that image might be a weather map, a satellite loop, or a radar shot, and it is a simple matter to switch from one to another. With the proper lighting, viewers cannot tell that the weathercaster is not standing in front of the image. The weathercaster knows where to point by looking at the finished product on a monitor that sits just off camera, or by looking into the camera itself, which can serve double duty as a monitor. A weathercaster can go from current conditions to live shots to radar to satellites to computer graphics and even to Internet graphics without moving. And it all appears seamless.

Chroma-key works best with the color blue, and early chroma-key walls were nearly all blue. However, a blue wall means that the weathercaster cannot wear blue clothing because the color is too close to that of the wall. If an article of clothing "keyed out," the viewer could see the maps through parts of the weathercaster's body. Green was the next best color, and since few television people wear green clothing, that is the color most used today.

Observational capabilities have come a long way since the time when a weather observing station consisted of a thermometer and a rain gauge. Modern weather observing systems, including satellites and radar, produce so much data that in most cases computers are needed to organize and analyze it to meet the deadlines of forecasters. To trained meteorologists, the observations reveal a picture of the atmosphere and patterns that can be used to predict the future. In Chapter 2, we investigate the patterns that meteorologists look for in weather data, and the process of weather forecasting itself.

CHAPTER 2

Basics of Weather and Weather Forecasting

For Age and Want save while you may;
No Morning Sun lasts a whole day.
—*Benjamin Franklin*

Meteorologists are continually asked why we cannot make better forecasts. Edward Lorenz, a prominent mathematician-meteorologist at MIT, gave one response: "Why should we be able to make any forecasts at all?" Lorenz was not merely avoiding or deflecting the question but instead pointing to the amazing complexity of the atmosphere. Only a combination of observation, experience, computer guidance, and a sound understanding of the dynamics of the atmosphere has brought forecast meteorology to its current state.

Before weather instruments, satellites, radar, and computer models, people attempted to forecast the weather. They did not understand the physical reasons why a certain day turned out cloudy or sunny, or rainy or snowy, in much detail until the nineteenth century. But an innate curiosity about the world and a need to know what comes next led humans to look for links between observations and the weather that followed. For example, a halo around the moon at night often foretold precipitation the next day. People did not know why the two events were connected, but they noted the sequence occurred over and over.

Such **weather folklores**—casual observations passed across generations—are the starting point for this chapter. Eventually science replaced folklore as formal observations, leading to reasoned explanations of why and how certain phenomena occur, explained most of the mysteries. This chapter presents many of those basic principles that help explain how the weather works. We will cover pressure and wind, clouds and humidity, and jet streams and fronts, essentially all the weather jargon you would

hear on any evening weathercast. We will also show how these weather elements fit into the overall, or general, circulation of the earth's atmosphere, and explain why the weather in the midlatitudes, where we live, behaves the way it does.

After this short overview, we will focus on the general weather and climate characteristics of the Delaware Valley, leaving the specifics of each season for Chapters 3 through 6. **Weather** refers to day-to-day changes in atmospheric conditions, the state of the atmosphere now and over the next few days. We can see and feel weather: the day-long rain, the cold slap of Arctic air, or the sudden snow squall. **Climate**, in contrast, is the long-term average of atmospheric elements over months, seasons, years, or even longer periods, along with information about weather extremes. Thus, climate data includes the mean temperature in April, the average winter snowfall, the hottest summers, and the longest stretches with no rain. When someone asks: "What's the weather like in Philadelphia in June?" they are really asking for a review of the climate of Philadelphia during that month.

But when people want to know what will happen tomorrow, or even next weekend, they are asking for a weather forecast. The process of weather forecasting is also described in this chapter, including a look at how forecasters use guidance from computer models to shape their predictions. Even with modern advancements in forecasting, and with evidence of overall improvement in weather forecasts in recent decades, there are still examples of big forecast "busts." Such a case is detailed in a "Story from the Trenches" in this chapter, which describes the "surprise" snowstorm of January 25, 2000. What led the forecasters astray, and how do meteorologists react when the weather is changing dramatically from what was predicted? It is a story from inside a television station when a storm is not behaving the way it was forecast. A second "Story from the Trenches" recalls the circumstances surrounding one of the most important weather forecasts in history—the prediction for D-Day, through the eyes of Philadelphia television pioneer Dr. Francis Davis, who participated in the forecast.

FROM FOLKLORE TO FUNDAMENTALS

The weather fascinates us with its variety, its splendor, and its sometimes spectacular and dangerous side. But on a day-to-day basis, much of our fascination with the weather stems from our need to predict it and limit our vulnerability to harsh weather conditions. Today, the growth in population and the size of populated areas, the increasing development along coastlines, and our dependence on high-tech communication and transportation systems, create situations in which severe weather can seriously disrupt and even endanger people's lives, and cause millions (even billions) of dollars in damage.

Our many protections that reduce our vulnerability include sturdy buildings to withstand the weather's pounding, the ability to efficiently cool and heat those structures, and the technology to warn us of potentially dangerous weather. For most of human history, none of these buffers existed—livelihood and even survival depended on the variations in weather. That extreme vulnerability forced people to find predictive information anywhere they could—for example, in their observations of the natural world. People looked to nature—signs from the sky, patterns in the behavior of animals and plants—to provide clues to future weather. Some of these observations became the basis for traditional weather folklores.

The early settlers of the Delaware Valley and surrounding regions, the Pennsylvania Germans and Dutch and the Quakers and others, had hundreds of weather sayings. Some have been borne out by sound science and have survived the test of time, while many are more amusing and colorful than useful. Popular among the Pennsylvania Germans were sayings such as "When the Indian can hang his powder horn on the moon, the weather will be clear," "Green Christmas, White Easter," and "The cry of the whippoorwill presages rain." Though these sayings were undoubtedly based on observations or recollections, they do not hold true often enough to provide any real forecasting usefulness.

On the other hand, some of these sayings are grounded in good science. The rhyme "A disk around the sun or moon, means that rain or snow is coming soon" describes a fairly common situation in which thin, icy, high-altitude **cirrostratus** clouds—which often precede the arrival of thicker clouds and precipitation by twelve to twenty-four hours—produce a ring around the sun or moon. The sensitivity of some people to decreasing air pressure is at the root of "Pain in a scar or the bones indicates rain," which links lowering air pressure to the approach of a storm. And "Rainbow in the morning, sailors take warning; rainbow at night, sailor's delight" describes the observation that rainbows form opposite the sun in the sky. For example, an evening rainbow ("night" in the folklore) means that it is raining in the eastern sky while the setting sun is poking through in the west (see Illustration 2.1b). Because weather systems typically move from west to east, this suggests that clearing is on the way—a sailor's delight. The opposite situation in the morning, with the sun in the eastern sky and the rain in the west, is shown in Illustration 2.1a.

The dependence on folklore began to lessen in the early 1800s when groups of weather observers started to organize. As we discussed in Chapter 1, several decades passed before these observers could share up-to-date weather information by telegraph; only then could larger-scale weather patterns be diagnosed on a routine basis. In 1870, the collection and analysis of weather data and the issuing of weather advisories came under the auspices of the federal government with the creation of the first U.S. weather service. The first half of the twentieth century brought more science into meteorology: the development of the theory of fronts and the regular launching of weather balloons to observe conditions above the ground. The application of radar to weather blossomed after World War II, while 1960 brought the launch of the first weather satellite. In the decades since, advances in computer modeling, new technologies in weather observation, and new theoretical insights into how weather

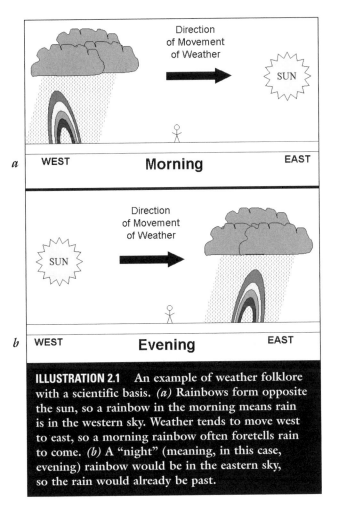

ILLUSTRATION 2.1   An example of weather folklore with a scientific basis. (a) Rainbows form opposite the sun, so a rainbow in the morning means rain is in the western sky. Weather tends to move west to east, so a morning rainbow often foretells rain to come. **(b)** A "night" (meaning, in this case, evening) rainbow would be in the eastern sky, so the rain would already be past.

and currents in the oceans work constantly to try to smooth out this imbalance in temperature. The differences in the surface on our planet—water versus land, mountains versus plains, desert versus rainforest—just complicate matters, creating other temperature and moisture differences in the atmosphere.

The Delaware Valley, located about halfway between the equator and the North Pole, is midway between the warmer tropics and colder polar region. As the winds and currents move to achieve a temperature balance, this "midlatitude" location not only means large seasonal variations, but also the potential for very changeable weather. Indeed, variability on all time scales—days, weeks, months, even years—is the most notable characteristic of midlatitude weather.

The most significant variations in our weather result from the change of seasons. The earth has seasons because its axis is tilted about 23.5 degrees with respect to the plane of its orbit around the sun (see Illustration 2.2). As a result, the intensity of sunlight received at any point on earth changes over the course of the year as the earth revolves around the sun. During our summer, the sun's direct rays are focused on the Northern Hemisphere (point A in Illustration 2.2). The sun gets higher in our sky, yielding more intense sunshine and more hours of daylight, and consequently higher temperatures. In winter, it is just the opposite. The sun's direct rays are focused on the Southern Hemisphere (point B in Illustration 2.2). For us, this means that the sun remains lower to the horizon, resulting in less intense sunlight and fewer hours of daylight.

The difference between summer and winter in the Delaware Valley, as measured by the sun's position in our sky and the number of hours of daylight, is dramatic. On the first day of winter (usually December 21), the sun is up for only nine-and-a-half hours in Philadelphia. It rises in the southeast sky and sets in the southwest sky, and in between it gets no higher than 27 degrees above the horizon. Illustration 2.3 shows the Philadelphia skyline looking west from Camden, with the location of the sun just before sunset at various times of the year superimposed. The leftmost circle represents the sun's position in the southwest sky just before sunset on the first day of winter. In contrast, the sunset location six months later on the first day of summer is shown by the rightmost circle in Illustration 2.3. On that day, the sun rises in the northeast sky and peaks more than 73 degrees above the horizon, before setting about fifteen hours later in the northwest.

Whereas seasonal variations lay the background for the Philadelphia area's overall climate, weather variations on

systems behave have yielded a gradual but steady improvement in the quality of weather forecasts.

We turn now to a discussion of the basics of weather and the patterns that meteorologists look for in weather observations. We will see how the Delaware Valley's weather fits into the larger-scale circulation of the atmosphere, and how the three-dimensionality of the atmosphere ties into the weather experienced at the earth's surface. This requires descriptions of fundamental concepts such as air pressure, wind, and humidity, which meteorologists use as diagnostic and prognostic tools.

BASIC BUILDING BLOCKS OF WEATHER AND CLIMATE

Ultimately, energy from the sun drives the weather on earth. No matter the season, the tropics receive much more sunlight than the poles, creating a large north-to-south temperature difference. Winds in the atmosphere

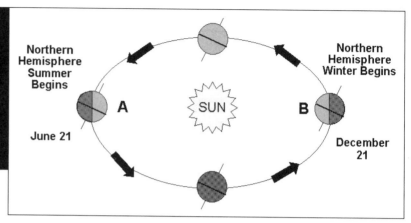

ILLUSTRATION 2.2 We have seasons because the earth is tilted on its axis with respect to the plane of its orbit around the sun. Around June 21 *(point A)*, the earth is at a point in its orbit such that the Northern Hemisphere is tilted toward the sun at the maximum angle, so the sun gets high in the sky and the days are long. The opposite is true on the first day of winter *(point B)*.

a day-to-day basis can usually be attributed to traveling areas of high and low pressure. These are the familiar "H" and "L" of the weather map. Forecasting the weather in the midlatitudes is largely a matter of tracking these highs and lows. In general, each has characteristic weather: low (or lowering pressure) typically means clouds and precipitation; high (or increasing) pressure favors fair weather. The highs and lows move in a general west-to-east fashion, in concert with upper-level westerly winds. These high-altitude winds and the low-level highs and lows are intertwined in a give-and-take manner. Although most of us, most of the time, care only about the weather at the earth's surface, meteorologists must always think in terms of three dimensions.

Clouds and precipitation are usually the headliners in any weather forecast. In most cases, clouds form where air rises, while sinking air inhibits cloud formation. Up and down movements of air are called **vertical motions** by meteorologists. At the most basic level, a forecaster's search for areas of clear and cloudy skies (and thus areas of rain or snow) comes down to determining the patterns of vertical motions. Unfortunately, these upward and

downward movements of air are typically very slow, subtle, and hard to measure directly (unlike horizontally moving air—the wind—which is relatively easy to measure). So meteorologists get at vertical motions another way—by means of air pressure. Variations in air pressure from place to place and over time are related to rising and sinking air and thus changes in the weather. Air pressure definitely deserves a closer look.

Pressure and Wind

It is easy to see how air exerts pressure if you think about inflating your car tires to the recommended level, perhaps 35 psi (pounds per square inch). Such a pressure measures the force with which energetic, constantly moving air molecules are colliding with the inside of the tire—in this case, 35 pounds on every square inch.

Average pressure in the free atmosphere is not as high as inside a tire. But air molecules are constantly in motion, exerting some force on each other and on surfaces they strike. To imagine the size of the force, consider that air is "stuff," so it has weight. Individually, air molecules

weigh very little. But there are enough of them (about 440,000,000,000,000,000,000 per cubic inch near the ground) that collectively they are a force to be reckoned with. The air above each square inch of earth's surface weighs about 14.7 pounds on average. So a typical air pressure is 14.7 psi, less than half of the air pressure inside a car tire.

In 1643, Evangelista Torricelli, a student of Galileo, invented the **mercury barometer** to measure air pressure. He put the open end of a long glass tube that had been filled with mercury in a container of the heavy liquid. The mercury did not run out of the tube but, rather, a column about 30 inches tall remained, supported by the force of air molecules striking the mercury in the container. This is the origin of the commonly used "inches" as a unit of air pressure. Average air pressure at sea level is about 29.92 inches of mercury (or just 29.92 inches, for short).

Most modern barometers do not use mercury, but instead consist of a tiny sealed coil of tubing with a spring inside. The tubing compresses and expands by small amounts as air pressure changes. These tiny variations are magnified and calibrated for reading from a pointer on a scale. Though these **aneroid barometers** have no mercury, "inches" is still typically used on their pressure scales. Many barometers also show another unit of pressure that is commonly used by meteorologists—the millibar. Just for reference: 1,013 millibars = 29.92 inches.

If you own a barometer and look at it from time to time, you will notice that the pressure changes from day to day, and even from hour to hour. Most aneroid barometers actually have a second pointer that can be set by hand to align with the current pressure, allowing changes to be easily noted later. At any given time, your barometer will not read the same as someone else's in another state, or even the next town over, for that matter. How do these pressure variations relate to changes in the weather?

Pressure differences from place to place exist for a few reasons. One is elevation: simply, the higher you are, the lower the pressure. But differences in pressure caused by elevation are nearly constant and not relevant to changes in the weather, so meteorologists first remove the elevation influence from pressure data. Once that is done, the remaining pressure differences are caused by weather-related factors such as differences in temperature, humidity, and the arrangement of winds at high altitudes. This modified pressure data is analyzed on a weather map by drawing lines of equal pressure called **isobars** (the solid lines in Illustration 2.4). With isobars in place, the highs (H) and lows (L) can then be located.

An inflated automobile tire can be used to demonstrate another reason why pressure differences are important. When the valve is opened, air rushes out, moving from the higher pressure inside toward the lower pressure outside. The difference in pressure sets the air in motion. The same is true in the free atmosphere—air tends to move away from highs and toward lows. This characteristic of the wind gives high- and low-pressure areas their characteristic weather. Winds in the vicinity of lows move inward, or converge, toward the center of lowest pressure. As air crowds together, the only escape is up, so air rises in the vicinity of lows. Rising air cools, making it more likely that clouds and precipitation will form. This is why lows are generally associated with inclement weather.

In reality, winds do not blow straight toward the center of a low. The earth's rotation deflects moving air to the right of this direct path in what is called the **Coriolis effect**. (See the list of web sites for more information on this effect.) As a result of this deflection, the convergence of air toward lows has a counterclockwise look to it. (See the arrows in Illustration 2.4.)

Around highs, winds behave in the opposite fashion. They circulate in a general clockwise fashion, with air

ILLUSTRATION 2.3 The Philadelphia skyline as viewed from across the Delaware River (so that north is to the right). Each circle represents the position of the sun just before sunset at a different time of the year. The northernmost circle is the position on the first day of summer, usually June 21. Each successive circle to the south represents the location of the sun two weeks later. The southernmost *(leftmost)* circle is the position on the first day of winter, December 21 or 22 (courtesy of Charles Penniman).

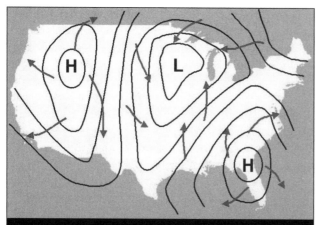

ILLUSTRATION 2.4   Weather Map, Part 1. Lines of equal pressure, called isobars *(solid lines)*, help locate the highs and lows. Troughs are elongated areas of low pressure, while ridges are elongated zones of high pressure. Winds *(represented by arrows)* blow approximately along the isobars but angle toward lower pressure. This gives the characteristic counterclockwise and converging pattern of winds around lows and the clockwise and diverging pattern around highs.

moving outward, or diverging, from the center of highest pressure. (See the arrows in the illustration.) With air spreading out horizontally near the surface in the vicinity of high pressure, there is a tendency for air above to gently subside to fill the space left by the diverging air. Sinking air warms and dries out, just the opposite of what is needed to make clouds. This is why high pressure favors fair weather.

The circulation of air around highs and lows also helps determine temperature patterns. Though high pressure is often associated with cold air masses in winter, highs and lows are not exclusively either warm or cold. Rather, the temperature characteristics of pressure systems depend on where you are with respect to the center of the high or low.

East of a center of high pressure (and west of a low), winds tend to come from a northerly direction, as seen in Illustration 2.4. And north winds typically bring colder air southward. In contrast, west of the center of a high (and east of a low), southerly breezes dominate, and warmer air tends to return northward. For this reason, you will sometimes hear meteorologists refer to a high as "two-faced." Such a high likely originated in central Canada and, thus, initially brought cold air into the United States via northerly winds on its eastern side. But as soon as the center of the high moved east of a partic-

ular location, winds shifted around to a more southerly direction, and milder air returned on the "back side" of the high.

A final important point about high- and low-pressure systems: experience shows that it is not enough to simply focus on the centers of high and low pressure—that is, the *H*s and *L*s. Meteorologists also need to find other territory being claimed by high or low pressure in a more subtle way—in the form of troughs and ridges.

Sometimes, areas of low pressure elongate away from the center of the low. These extended zones of low pressure are called **troughs**. In Illustration 2.4, a trough extends southward from the low in the northern U.S. into Texas, while another trough extends east from the low through the Great Lakes to the East Coast. Think of troughs as zones of lower pressure separating areas of higher pressure. Air converges at troughs just as it does at lows, leading to rising air and often clouds and precipitation. Similarly, highs often stretch into elongated zones called **ridges**. The high in the West in Illustration 2.4 has a pronounced ridge that extends south, while a ridge extends northward from the high in Florida. Like highs, ridges are associated with diverging air, downward vertical motion, and generally fair weather.

Clearly, there are plenty of good reasons for meteorologists to analyze air pressure and to monitor its variations over time. Low, or consistently lowering air pressure, usually means that cloudy, wet weather will continue or will soon arrive. High, or persistently rising pressure, typically signals fair weather, or its imminent arrival. And the wind circulation around highs and lows moves warm and cold air around in predictable ways. Meteorologists exploit all of these characteristics when preparing weather forecasts.

The General Circulation and Fronts

The arrangement of highs and lows on a midlatitude weather map might appear haphazard. But there is actually an underlying "big picture" that is tied to the larger global circulation of the atmosphere.

A view of the major components of this so-called **general circulation** of the atmosphere is shown in Illustration 2.5. A good place to start is in earth's boiler box, the tropics, where warm moist air is continually rising to create towering thunderstorms. The air does not ascend indefinitely. At high altitudes some of it heads northward and some of it southward. Because what goes up must come down, the air tends to sink in the subtropics, the zone adjacent to the tropics. This traffic jam of subsiding air in subtropical latitudes means that air masses there tend to have higher pressure (the "subtropical highs" shown in Illustration 2.5). Meanwhile, thousands of miles away in polar latitudes, cold air is in charge, especially during winter. Cold air tends to be heavy and dense, leading to sinking air and generally higher air pressure there as well (the "polar highs" shown in Illustration 2.5).

These zones of consistently higher pressure in polar and subtropical latitudes are natural breeding grounds for many of the areas of high pressure that you see on the evening weather map. High-pressure air masses bred in the subtropics are relatively warm and humid, while high-pressure air masses born in polar latitudes tend to be cold and dry.

Now step back from meteorology for a moment and consider a common-sense rule from geography: between any two mountains there has to be an area of lower elevation. This idea can be applied to the general patterns of air pressure. Between warm high pressure born in the subtropics and cold high pressure originating in polar latitudes there should be a zone with a penchant for lower pressure. And on a map of the globe, that in-between zone—the battleground of the air masses, so to speak—is the midlatitudes, where we live (see Illustration 2.5). Indeed, lows and troughs of low pressure are frequent sights on any midlatitude weather map.

And here is where fronts, those snaking blue and red lines on the weather map, fit in. A **front**, by definition, is a boundary between two huge masses of air of contrasting temperature and moisture. So, for example, a front might separate a warmer subtropical high-pressure air mass from a colder, polar-bred high-pressure air mass. In other words, fronts will be found in the zone of lower pressure that separates air masses of higher pressure. So on weather maps, you will find fronts in troughs, connected to centers of low pressure.

The vision of fronts as battlegrounds of air masses was first formalized by a group of Norwegian meteorologists around the time of World War I. They took their cue from the wartime front as the zone of conflict between opposing armies. In the atmosphere, the conflict of air masses leads to weather action—clouds and precipitation. This follows from the relationship between troughs and fronts. Air converges at troughs, so air rises at fronts. The warmer, lighter air is forced upward by the heavier, ground-hugging colder air. And the result can be a day-long rain, a quick burst of snow, or an hour long, gutter-gushing thunderstorm.

Generally speaking, you will see three types of fronts on weather maps: cold, warm, and stationary (see Illus-

tration 2.6). At **cold fronts**, the heavier cold air is advancing, claiming ground previously occupied by the warm air. Blue is the standard color for cold fronts. At **warm fronts**, the cold air is retreating, giving up territory to the warm air. Warm fronts are shown in red. Neither air mass has an edge at **stationary fronts**, which move either very slowly or not at all. Stationary fronts are drawn in alternating blue and red. (There is also a fourth variety of large-scale front—the **occluded** front—that is traditionally drawn in purple and has some characteristics of both cold and warm fronts.)

Illustration 2.7 is the same weather map as Illustration 2.4, but now the fronts have been drawn as well as the typical cloud shield associated with a midlatitude low-pressure system. The warmest air around the low is found between the cold front and the warm front in the **warm sector**. There, warm air moves northward and eventually ascends up and over the retreating cold air. This process, called **overrunning**, produces an extensive cover of

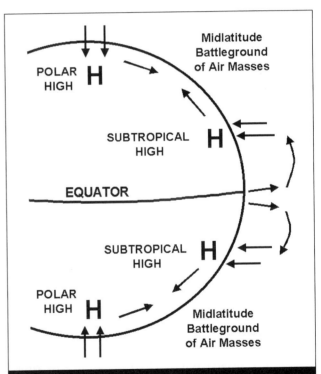

ILLUSTRATION 2.5 The general circulation of the atmosphere. Air tends to rise in tropical regions and sink in the subtropics and polar regions. Warm and cold air masses originating from relatively high pressure in the subtropics and polar regions commonly meet in the midlatitudes, resulting in changeable weather.

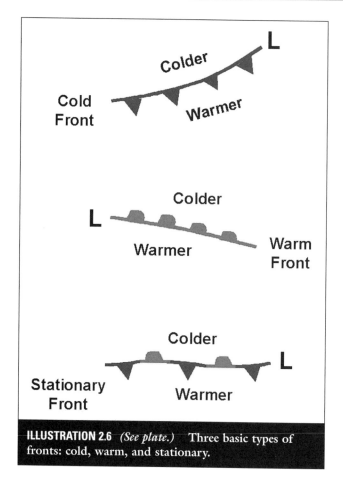

ILLUSTRATION 2.6 *(See plate.)* **Three basic types of fronts: cold, warm, and stationary.**

lower air pressure, which yields to higher air pressure, and so on, in a never-ending cycle. Generally, the average time between storms is on the order of three to five days (making it possible to get stuck in weather patterns in which it seems to rain every weekend). A graph of average daily air pressure at Philadelphia during the stormy month of January 1996 clearly shows how pressure rises and falls as the traveling highs and lows move by (see Illustration 2.8). Days when measurable rain or snow occurred are indicated with *R* or *S*—clearly, low (or decreasing) air pressure is correlated with precipitation, while high (or increasing) pressure and dry weather typically go together.

Upper Atmosphere Features

Because we live at the bottom of a deep ocean of air, it is easy to underappreciate that the atmosphere has three dimensions. But meteorologists have to be aware that weather conditions above the ground are intimately linked to what goes on at the surface.

At high altitudes over the midlatitudes, winds generally blow from west to east. This is true to such a degree that this dominant high-altitude flow is called the **westerlies**. Just as there will be currents of faster-flowing water in a river, a narrow channel of faster air typically runs through the westerlies. This speedier current of air is the midlatitude **jet stream**, or jet for short. The jet stream is typically found between 20,000 and 30,000 feet above the ground. It is usually a few hundred miles wide and

clouds and precipitation to the north of the warm front, particularly during the winter when the retreating cold air is heaviest and least willing to budge. Overrunning clouds can extend hundreds of miles north of the warm front, infringing on the high-pressure territory of the retreating cold air.

Meanwhile, behind the cold front, colder air is advancing south and east. With colder, heavier air barreling into the warmer air, the lifting at the cold front tends to be more forceful and more concentrated. As a result, precipitation along a cold front is usually more intense than at a warm front, and the rain or snow falls in a narrower band. Thunderstorms often form along cold fronts, especially in the spring and summer. Collectively, the cloud pattern associated with an intense low-pressure system often resembles a comma, with the clouds around the center of low pressure forming the head and those along the cold front forming the tail (see Illustration 2.7).

Midlatitude lows and their attendant fronts generally move west to east, maintained and steered by upper-level winds. At any given location, higher air pressure yields to

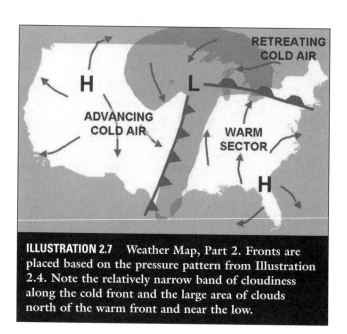

ILLUSTRATION 2.7 **Weather Map, Part 2. Fronts are placed based on the pressure pattern from Illustration 2.4. Note the relatively narrow band of cloudiness along the cold front and the large area of clouds north of the warm front and near the low.**

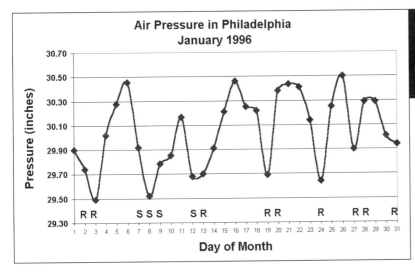

Air Pressure in Philadelphia January 1996

wraps nearly continuously around the globe. Its speed varies from place to place and from day to day, and at times the jet even splits into distinct northern and southern branches. There is no threshold of wind speed above which you are "in" the jet stream. Rather, the jet is simply the current of fastest-moving air—in winter, winds in the jet may reach 200 mph, while the lazier summer jet may slow to less than 50 mph.

In general, the jet stream is found in the zone of largest north-south temperature difference between colder air toward the poles and warmer air toward the tropics. That zone is always over the midlatitudes, but it shifts toward the pole in summer and toward the equator in winter. In a typical July, the jet stream spends much of its time over the northern border states and southern Canada, while in mid-winter the jet is commonly found over our latitude or to the south of the Delaware Valley.

Although high-altitude winds are westerly on average, the flow frequently bends and buckles (see Illustration 2.9). So the jet stream often has dips and bends in it too. The equator-ward dips are called troughs, while the pole-ward bulges are called ridges. An upper-air flow dominated by lots of troughs and ridges is referred to as **meridional**, while a relatively straight west-to-east flow is known as **zonal**. These upper-air troughs and ridges are related to the troughs and ridges on the surface weather map. Here is how.

Both surface highs and lows tend to be more intense in meridional flows because the curves in the upper-air pattern help, indirectly, to add spin and thus intensity to the surface weather systems. This intensification effect is most pronounced to the east of the corresponding upper-air feature. That is, the favorable area for surface lows and

troughs to strengthen is to the east of the upper-air trough, while surface highs and ridges are most likely to intensify to the east of the upper-air ridge (see Illustration 2.9). And because the fastest winds are found in the jet stream, the most intense surface highs and lows tend to be found beneath the jet.

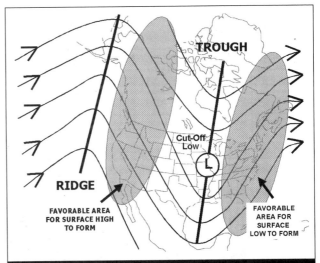

Anytime you see a highly meridional jet stream, look for strong high- and low-pressure systems and extremes in the weather—stormy in some places, perhaps record chill or heat in others. Sometimes, "cut-off lows" develop in highly meridional flows (see Illustration 2.9). These closed circulation centers at high altitudes are essentially separated from the main jet-stream flow, so they move very slowly. Get underneath or just to the east of a cut-off low and expect several days of cloudy, damp weather.

In summary, think of the westerlies, and in particular the jet stream, as the conductor of the lower atmosphere orchestra of highs and lows. These surface weather systems develop and move in concert with the upper-air flow. Just about any weather change—any heat spell or rainy period, any short-term or long-term weather pattern—can be tied to the strength and position of the jet stream. But the surface highs and lows are not just passive players. Their winds move colder air toward the equator and warmer air toward the poles, helping to shape temperature patterns in the lower atmosphere. And that helps control the location and intensity of the jet stream.

Moisture and Clouds

Changes in the jet stream and day-to-day variations in air pressure would not make much difference in changing the weather if the air had no moisture in it, because water is necessary to make clouds and precipitation. So knowing the air's moisture content, or "humidity," is essential to determine where clouds, rain, and snow will form. The humidity also determines how comfortable or uncomfortable the air "feels" to people. **Relative humidity** is the most popular measure of the air's moisture content. It works well for pinpointing when and where clouds and precipitation may form, but—contrary to popular belief—relative humidity cannot reliably quantify the air's "feel."

Two processes control the air's humidity. In **evaporation**, water changes into invisible water vapor. Nearly all of the water vapor in the air evaporated from oceans and lakes. **Condensation** is the reverse, water vapor becoming liquid. These two processes are always occurring, and in most of the atmosphere most of the time, evaporation exceeds condensation. When and where this is not true, clouds can form and the weather gets interesting.

When condensation exceeds evaporation near the ground, the result is dew. But water vapor can also condense in the air onto invisible suspended particles called **condensation nuclei**. If enough water vapor condenses onto enough of these microscopic specks of dust, salt, and smoke, the particles will grow large enough to be seen

and, voilà, a cloud is born! At the same time, however, water from these fledgling clouds evaporates back into the air, slowing the process of cloud making. The secret to ensuring that condensation keeps up with evaporation so that clouds can form is to cool the air. Cooling slows air molecules down, increasing the odds that water vapor molecules will stick to each other or to a condensation nuclei when they collide. In the atmosphere, the most common way for air to cool is by rising, and that is how most clouds form.

Where does relative humidity fit in? Relative humidity measures the relative rates of evaporation and condensation. When condensation equals or exceeds evaporation, the relative humidity is 100 percent or greater—clouds form and precipitation is likely. Observations show that a relative humidity of 100 percent is not even necessary for clouds to develop—they often start forming when a thick enough layer of air reaches a relative humidity of about 70 percent. Thus, relative humidity's great usefulness is that it points meteorologists to areas where clouds and precipitation are most likely.

But relative humidity cannot describe how humid the air feels because relative humidity does not depend solely on how much moisture is in the air—it also depends on the air's temperature. When temperature increases, relative humidity decreases, and vice versa, even if the actual amount of water vapor does not change. As a result, the relative humidity typically peaks in the morning when the temperature is lowest and then decreases during the day as the temperature rises.

To see how this can be deceptive, imagine a warm muggy July morning—70°F with 90 percent relative humidity. It seems reasonable to conclude that the relative humidity is high because the air is humid. Assume that later that day the temperature hits 95°F, and it still feels like a steam bath outside. But now the relative humidity will likely read only 50 percent, simply because the temperature went up (see Illustration 2.10). The decrease to this comparatively low relative humidity would undoubtedly be confusing to sweaty citizens.

Since relative humidity can be misleading as a gauge of how humid the air feels, meteorologists turn to another measure—the dew point. To understand dew point, think of a clear, calm morning with drops of dew on the grass. The dew formed overnight as air in contact with the chilling ground cooled, allowing water vapor to condense out of the air. The temperature at which the dew started forming is the dew point. So the **dew point** is a temperature—the temperature to which air must be cooled in order for condensation to equal evaporation.

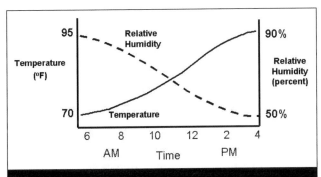

ILLUSTRATION 2.10   Relative humidity depends not only on the amount of moisture in the air but also on temperature. So when temperature increases from morning to afternoon, relative humidity tends to decrease.

Defined this way, the dew point is directly related to the way the air feels in terms of humidity: the higher the dew point, the more water vapor in the air, and the more humid the air feels. For most people, the air starts to feel a bit humid once the dew point reaches 60°F or 65°F. Dew points of 70°F are tropical, while dew points of 75°F or above are highly oppressive and rarely observed in the Delaware Valley.

Phrased in terms of dew point, the key to cloud formation is the cooling of a thick enough layer of air to its dew point. Because rising air cools, any mechanism that forces air to go up can potentially produce clouds and precipitation. We have already mentioned several of these mechanisms. Air tends to rise in the vicinity of low-pressure systems because air converges there. The same is true at fronts where warmer, lighter air rises over cold, relatively heavy air. When clouds and precipitation form in the Delaware Valley, lows and fronts are usually responsible, especially outside of summer. But there are other causes for rising air. Hot-air balloons are buoyant because the air inside them is warmer than the surrounding air. This mechanism works in the free atmosphere too, as heated air can rise and produce a field of puffy topped, flat-based cumulus clouds or bubble up into a full-fledged thunderstorm. Sometimes, the mechanism for making the air rise comes from high above, in the configuration of winds in the jet stream.

Air can also be forced to rise by the terrain. If the land slopes up, air blowing across the land has to go up too. In fact, many of the cloudiest, rainiest, and snowiest places are located on the windward sides of mountains— that is, on the side that the wind blows into the majority of the time. On the other hand, deserts or desert-like dryness are often found on the opposite, or leeward, side of a mountain where air tends to sink. An area of relatively low precipitation on the leeward side of a mountain is called a **rain shadow**. There is not much variation in elevation across the Philadelphia region, so topography is a relatively minor factor in helping to make clouds and precipitation here.

With an understanding of the basic ingredients for making weather in hand, we can now focus on how these forces work together to create the weather and climate of the Delaware Valley.

GENERAL CLIMATE FEATURES OF THE PHILADELPHIA AREA

The Delaware Valley lies about halfway between the equator and the North Pole. If you go due west you will pass south of St. Louis, close to Denver, and about 100 miles north of San Francisco. Head east across the Atlantic and you will nearly bisect Spain and pass an hour's drive south of Rome. This midlatitude location puts the Philadelphia area about equidistant from the reservoirs of cold air to the north and warm air to the south, a primary reason that the region has so many different types of weather. In addition, an unlimited source of moisture for storms lies just to the east, while 3,000 miles of land to the west supplies relatively dry air. This easy accessibility of both moist and dry air increases the weather's variability.

One way to summarize something that varies a lot is to determine its average, which then serves as a middle-of-the-road basis for comparison. Averages of weather elements such as temperature and precipitation help define a region's climate. Typically, to compute such averages, thirty years of data from the three most recently completed decades are used. (The averages being used now were derived from 1971–2000 data—see Appendixes A and B.) Thirty-year averages of measurements such as daily high temperature and monthly precipitation are often referred to as "normals," but that terminology is misleading if taken literally. In reality, it would be very unusual for a particular day's high temperature or a particular month's precipitation to exactly match the so-called "normal." Instead, it is typical for such quantities to fall within a range centered on the thirty-year average; but sometimes that range can be quite large.

As a typical example in winter, consider New Year's Day in Philadelphia, when the average ("normal") high temperature is 40°F. Historically, the high temperature on January 1 has been within 5°F of this average only 45 percent of the time, and within 10°F of the average

71 percent of the time. So, although 40°F is a reasonable first guess for the high temperature on New Year's Day, history suggests that in three out of every ten years you should expect the temperature on January 1 to stay below 30°F or rise above 50°F (that is, fall more than 10°F from the average). Similar variability holds for other days of the year, although the range is much less in summer.

The variability of weather on a given day in different years can be attributed to the traveling midlatitude highs and lows that bring day-to-day changes in cloud cover and wind direction. The basic state of climate that is left after averaging out that variability depends primarily on three factors: latitude, elevation, and proximity to large bodies of water. By considering these three influences, we can explain the patterns of average temperature and precipitation in the Delaware Valley.

Temperature

By far, latitude has the greatest effect on average temperature across the Delaware Valley. In general, the closer a place is to the equator, the warmer it is. Dover and Cape May are, on average, several degrees warmer than Trenton and Allentown simply because of the difference in latitude.

The overall effect of elevation on temperature is also straightforward: temperature typically decreases with increasing elevation, at the average rate of 3–4°F per 1,000 feet. At first, this may not seem logical, since increasing elevation brings you closer to earth's ultimate heating source, the sun. But a fundamental principle of meteorology is that the sun heats the ground, and the ground heats the air. As a result, the greater the distance from the ground, the less heating of the atmosphere.

The decrease of temperature with elevation is just an average, however. At times, the air near the ground can be colder than the air just above. This occurs most frequently in winter and at night, especially when winds are weak. In those situations, the natural tendency for cold air to be heavier than warm air takes over, and the coldest air tends to gather in the lowest elevations (such as valleys).

Philadelphia and adjacent areas to the south and east lie in a region known as the **Coastal Plain**. The land is

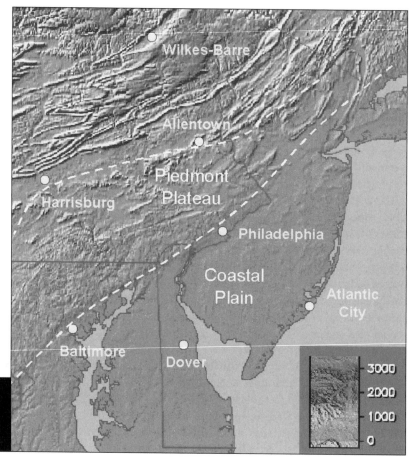

ILLUSTRATION 2.11 *(See plate.)* Topographic map of Philadelphia and surrounding areas. The color-coded elevation legend is in feet (courtesy of Ray Sterner).

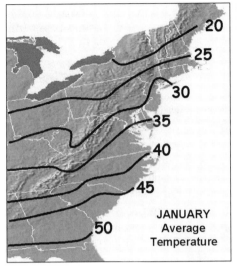

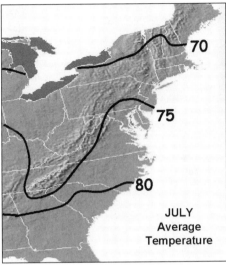

ILLUSTRATION 2.12 *(a)* Average January temperatures (in °F); *(b)* Average July temperatures (in °F). Latitude is the primary controller of average temperature.

a

b

low and flat, with elevations generally 100 feet or less (see Illustration 2.11). With so little variation in elevation on the Coastal Plain, very little temperature variation can be attributed to changes in altitude. North and west of the Coastal Plain is the **Piedmont Plateau**, which stretches to about a Harrisburg-Reading-Allentown line. This region features rolling hills, fertile valleys, and elevations ranging from 100 to 500 feet. Because of the larger variation in elevation across the Piedmont Plateau, the effects of elevation differences on temperature there are a lot more noticeable than on the Coastal Plain.

But topography's most significant effect on temperature in the Philadelphia area is more indirect. The higher elevations of the Appalachians to the west of the Piedmont Plateau and the Poconos to the north can slow or even block the progress of low-level chilly air. This reduces the sting of cold air masses southeast of the mountains (especially in winter). Without these mountains, cold northwest winds would have an unobstructed run into the Delaware Valley.

The influence of a large body of water on temperature depends on the time of year. Water is slow to warm but also slow to cool. So during winter, big bodies of water are generally warmer than adjacent land. Thus, locations near water tend to be a little milder during the winter than locations farther inland. The effect is reversed in summer, when water tends to be cooler than land—locations near water bodies tend to stay cooler than more inland sites. Along the East Coast, for example, an east wind always carries moisture inland, but it is a relatively mild wind in winter and a relatively cool wind in sum-

mer. In essence, proximity to water is a moderating influence on temperature. This moderating effect is much more pronounced on the west coasts of continents, where the typical west-to-east movement of weather systems means that the air has crossed the ocean before moving inland. Thus, the Pacific Ocean has a much greater moderating influence on temperatures along the West Coast than the Atlantic has on the East Coast.

Illustration 2.12 puts it all together, showing average January and July temperatures in the eastern United States. Latitude clearly has the dominant effect. The north-south temperature difference is much larger in winter because in summer warm air is in firm control of the U.S., and the main battleground with colder air is usually to the north in Canada. The effect of elevation on temperature is evident by the slightly lower temperatures in the Appalachians. The Atlantic Ocean's cooling effect in summer and warming effect in winter on coastal areas is not even noticeable on this broad scale.

Another important influence on temperature shows up only on localized scales. Cities tend to be warmer than surrounding suburbs, which in turn are warmer than nearby rural areas, a phenomenon called the **urban heat island**. This effect can be traced to the abundance of brick, stone, and asphalt in cities. These substances tend to be better absorbers of sunlight than the fields and forests they replace. As a result, cities heat up more during the day and release more energy to the air at night. In addition, the "concrete jungle" nature of cities and their efficient drainage systems help urban areas to dry out quickly. This means that less sunshine goes toward

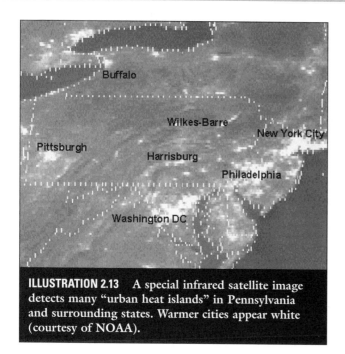

ILLUSTRATION 2.13 A special infrared satellite image detects many "urban heat islands" in Pennsylvania and surrounding states. Warmer cities appear white (courtesy of NOAA).

evaporating and heating water, leaving more energy to heat the ground. This too adds to the warmth of cities. (The urban heat island is not the same as "global warming"—see Chapter 7.)

Illustration 2.13 is a special infrared satellite image of Pennsylvania and surrounding states on a cloud-free night in July 1999—the brighter the shading, the warmer the ground. With this special infrared sensor, the large metropolitan areas, and even some smaller cities, "glow" when viewed from space. In the Philadelphia area, Center City is typically 3–6°F warmer than the surrounding suburbs, and a 10–20°F difference with nearby rural areas is not unusual on clear, calm winter nights. Illustration 2.14 shows low temperatures during one such night in January 2001 in Philadelphia and surrounding counties. Even if state and county boundaries were not drawn, the Philadelphia metropolitan area would be easy to locate as the zone of higher temperatures extending from southwest to northeast.

The average annual temperature in Philadelphia is about 55°F, with mean monthly temperatures varying from 32°F in January to 78°F in July. On the warm side, temperatures reach 90°F or above on twenty to twenty-five days a year, on average, but readings of 100°F are rare. Moist tropical air frequently barges in during the summer from the south and southeast, adding to the discomfort. July and August dew points average 62–64°F and sometimes rise above 70°F. On the cold side, tem-

peratures drop to 32°F or below fewer than 100 days a year in Philadelphia, on average, and readings below 0°F are just as rare as those above 100°F. The freeze-free season typically lasts more than 200 days.

These averages are valuable because they give a feel for Delaware Valley temperatures. But averaging hides the most fundamental feature: variability. Even when temperature is averaged over a full year, big swings occur from year to year (as you can see in Illustration 2.15). In fact, the temperature difference between Philadelphia's hottest and coldest years is more than 8°F. That may not sound like much unless the difference is translated into daily terms. During the warmest year (1998, which is tied with 1931), the temperature fell below 32°F on only forty-one days, while the temperature reached 90°F or higher on thirty-one days. In contrast, in the coldest year (1875), the temperature fell below 32°F on 105 days, and the temperature reached 90°F on just eight days. If you lived through those two years back to back, you would think that you were in two very different climates!

Deducing any long-term trends in temperature from Illustration 2.15 is tricky, in part because the observation

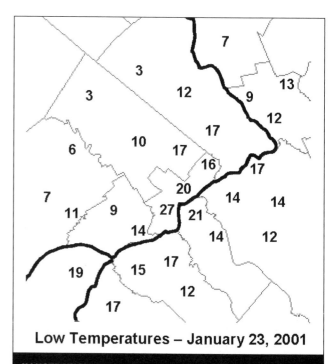

Low Temperatures – January 23, 2001

ILLUSTRATION 2.14 The urban heat island of Philadelphia is evident on a winter morning in January 2001. Temperatures in Center City were 10–20°F higher than in more rural areas of surrounding counties.

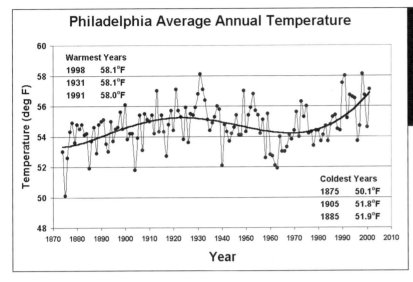

Philadelphia Average Annual Temperature

Warmest Years
1998   58.1°F
1931   58.1°F
1991   58.0°F

Coldest Years
1875   50.1°F
1905   51.8°F
1885   51.9°F

ILLUSTRATION 2.15   Philadelphia's average annual temperatures, 1874 to 2000. The thick line is mathematically "smoothed" making it easier to see long-term trends. The general cooling from 1940 to 1970 and warming from 1980 to 2001 is similar to overall global trends.

site has moved over the years. But the last two decades have certainly trended toward higher temperatures, and the warming would probably be greater in the latter half of the twentieth century had the official observation site not moved out of Center City to the less urbanized airport in 1940. We leave for Chapter 7 a discussion of whether this recent warming trend is just natural variability or a sign of global warming.

Precipitation

The amount of precipitation that falls in an area depends a great deal on its closeness to a large body of water (to provide the moisture) and the presence of mountains (to provide additional lift to the air). Average precipitation in the eastern United States generally follows these rules, decreasing as you move north from the Gulf of Mexico and west from the Atlantic Ocean. A stripe of higher precipitation coincides with the Appalachian Mountains, while localized higher amounts to the east and south of the Great Lakes are caused by lake-effect snows (see Chapter 3).

In the Philadelphia area, the tendency for higher amounts of precipitation caused by the proximity of the Atlantic Ocean is opposed by the prevailing westerly winds, which downslope out of the higher elevations to the west. This tendency toward sinking air is not conducive to precipitation, and it is common for areas of rain or snow to weaken as they move out of the Appalachian Mountains into southeastern Pennsylvania if winds in the lower levels of the atmosphere are westerly. Across the

Coastal Plain and the Piedmont Plateau, there is not much variation in average annual precipitation. The range is from 40 to 46 inches per year of rain plus melted snow and ice. It is hard to pin down any geographical patterns to the variation. However, the local National Weather Service has speculated that the slight rise in elevation at the **fall line** (the transition zone from the Coastal Plain to the Piedmont Plateau) may focus thunderstorm development in some situations, particularly when a tropical system approaches from the south. Supporting this theory are several localized heavy rain events in recent years in lower Bucks and lower Montgomery counties, such as the flash flooding associated with the remnants of Tropical Storm Allison in June 2001.

Precipitation is fairly evenly distributed throughout the year, though summer is the wettest time, on average (see Illustration 2.16). July is the rainiest month, averaging more than 4 inches, while February and October are the driest, both averaging about 2.75 inches. Measurable precipitation (that is, at least 0.01 inches of liquid—just enough to wet the ground) occurs on about 120 days each year, on average. Extremely wet days of one inch or more are rare, with typically about ten per year. Most rain in summer comes from thunderstorms, with an average of twenty to thirty days each year when thunder is heard. A slow-moving thunderstorm can dump a month's rain in a few hours, while a severe coastal storm or the soggy remnants of a tropical system can do the same in a day or less.

During the winter, precipitation runs the gamut from rain to snow to ice. An average season's snowfall ranges

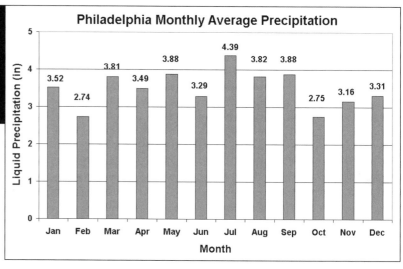

ILLUSTRATION 2.16 Average monthly liquid precipitation (rain plus melted snow and ice) for Philadelphia for the period 1971–2000. On average, February and October are the driest months, and July the wettest.

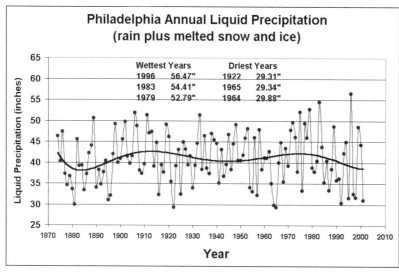

ILLUSTRATION 2.17 Philadelphia's annual liquid precipitation (rain plus melted snow and ice), 1874 to 2001. Note the extreme year-to-year variability. The mathematically smoothed thick line shows little evidence of any long-term trends.

from around 15 inches in Atlantic City and Dover to about 20 inches in Philadelphia and more than 30 inches north and west of the city. The biggest snows, occasionally a foot or more, come with intense lows that tap Atlantic moisture as they move northeastward up the coast. Winds off the ocean ahead of these "nor'easters" pull milder ocean air inland. So more snow often falls in the northern and western suburbs of Philadelphia than in Center City and locations to the south and east, where snow is more likely to mix with or change to ice or rain.

Relatively speaking, there is even more year-to-year variability in precipitation than in temperature (see Illustration 2.17). The average annual liquid precipitation in Philadelphia is about 42 inches of rain plus melted snow and ice, but the range is from less than 30 inches in the driest year (1922) to more than 56 inches in the wettest (1996). In the twentieth century, the 1920s and 1960s stand out as relatively dry decades, as do the 1990s (despite the wet 1996). At the other extreme, the 1900s and 1970s had many wet years. Overall, no long-term trend is evident in Philadelphia precipitation, though an increase of 15 to 20 percent over the last century has been detected in many other parts of the northeastern United States.

The Delaware Valley's midlatitude location not only means great variability in temperature and precipitation, but also a variety of severe weather threats that span the seasons. Heavy snow, extreme cold, and ice storms are the main winter worries. Thunderstorms, which can bring damaging winds, flash flooding, hail, and even tornadoes, are the primary hazard in late spring and summer. When tropical systems visit (mainly in late summer and early fall), their main threat is flooding and not wind, as evidenced by Floyd in September 1999 and Allison in June 2001 (see Illustration 1.20). At the other extreme, droughts occasionally develop and persist for months (and in the worst cases, years). In summer, excessive heat and humidity is particularly a hazard in highly urbanized areas. The single worst weather disaster of the last century in the Philadelphia area, as measured by loss of life, was a heat wave during July 1993.

Some extremes of Philadelphia weather and climate are given in Table 2.1. Ultimately, geography is responsible for the big difference between hottest and coldest, and wettest and driest. Being in the midlatitudes gives equal access to warm and cold air masses, while being near the east coast of a large continent means that both moist and dry air are never far away. This diversity in air masses also makes for some of the most challenging forecasting anywhere in the country.

Familiarity with a region's climate is a necessary first step to forecasting its weather, but it is far from sufficient. Predicting day-to-day weather variations requires a blend of observations, experience, climatology, and computer guidance. The challenging part is determining how much sway to allow each, because the weighting varies from forecast to forecast! We finish this chapter by taking a closer look at the process of weather forecasting.

WEATHER FORECASTING

"I think the Eagles will win this Sunday." That is a forecast. It might be based on a life-long knowledge of the game or perhaps on a careful analysis of statistics such as won-loss records and past performance against the upcoming opponent. Or the rationale behind the prediction might be very simple—the wishful thinking of a fan of the team.

"The stock market is going up, I tell you." That is a forecast too. It could be based on expertise and analysis, or it could be just a hunch.

Anyone can make a forecast. But chances are that those with the most thorough knowledge of the subject, the most years of experience, and the most detailed analyses have a better chance of being right than those who are only casual observers or those influenced by emotions. The same is true of weather forecasting.

Anyone can try to forecast the weather, and nearly everyone has at one time or another. Looking up at the sky and saying "It looks like rain" is a forecast. Seeing wet roads as the sun sets on a cold winter day might suggest that the water will freeze and make driving treacherous overnight—that is a forecast. Meteorologists get calls and E-mail regularly from the public, young and old alike, telling us what to expect from a coming storm. Many meteorologists, including the authors, started out

TABLE 2.1   Some Extremes of Philadelphia Weather and Climate

	Highest	Lowest
Average Yearly Temperature	58.1°F (1998, 1931)	50.1°F (1875)
Average Summer Temperature	78.5°F (1995)	71.0°F (1903)
Average Winter Temperature	43.3°F (1931–32)	28.0°F (1976–77)
Average Monthly Temperature	82.1°F (July 1994)	20.0°F (Jan. 1977)
Daily Temperature	106°F (Aug. 7, 1918)	−11°F (Feb 9, 1934)
Air Pressure	31.10 in. (Feb. 13, 1981)	28.43 in. (March 13, 1993)
Sustained Wind	94 mph (Oct. 15, 1954)	—
	Most	**Least**
Liquid Precipitation (day)	6.63 in. (Sept. 16, 1999)	—
Liquid Precipitation (month)	13.07 in. (Sept. 1999)	0.09 in. (Oct. 1924, Oct. 1963)
Liquid Precipitation (year)	56.57 in. (1996)	29.31 in. (1922)
Snow in a Day	27.6 in. (Jan. 7, 1996)	—
Snow in a Month	33.8 in. (Jan. 1996)	—
Snow in a Season (Oct. to April)	65.5 in. (1995–96)	Trace (1972–73)

as children thinking we could forecast as well as the "experts." Only later did we find out it was not as simple as it looked.

Good forecasts do start out simply enough. As Yogi Berra, the famous Yankees player and coach, once said, "You can observe a lot just by watchin'." Step One: look out the window. Indeed, all forecasts should begin with observations, and the more the better. At minimum, meteorologists analyze temperature, pressure, wind, and humidity data from all over the country, not only at the ground but up through the lowest few miles of the atmosphere. These observations paint a picture of the current state of the atmosphere and allow meteorologists to locate the highs, lows, fronts, and upper-air features that control the weather. Meteorologists then compare these analyses with satellite and radar images to make connections between the observations and the science. For example, are thunderstorms occurring with that strong cold front? If not, why not? How hard is it snowing to the north of that low, and where is the rain-snow line? And is that trough in the jet stream producing any clouds? The goal is to explain the weather's behavior in the present and the recent past, because without that understanding, there is little hope of predicting the future.

It takes a thorough familiarity with how the atmosphere works to even get this far. Some of the rules are fairly simple; for example, high pressure usually brings clear skies, or severe thunderstorms often develop along a strong cold front in spring. Then it takes a more specialized knowledge of a particular weather situation and the experience of seeing that situation before. A meteorologist who has forecasted for a certain area for many years has an advantage over equally knowledgeable forecasters who are less familiar with how the local topography or nearby bodies of water affect the local weather.

But observations, experience, and knowledge about how the atmosphere works can only get a forecaster so far. Fortunately, the physical laws that govern the atmosphere can be written as mathematical equations that, although very complex, can be solved (at least approximately) using the most powerful calculators available today—supercomputers. The process of weather forecasting by computer is called **numerical weather prediction**, and the lengthy software codes that produce the forecasts are called **computer models** (or just models, for short). These models take observations of the weather as input and then produce numerical weather predictions that meteorologists use to shape their forecasts. Generally speaking, the farther out in the future

the forecast, the more meteorologists rely on computer guidance.

As we mentioned in Chapter 1, numerical weather prediction has its historical roots in Philadelphia and the Moore School of Electrical Engineering at the University of Pennsylvania. In the early 1940s, John Mauchley, a physicist with a strong interest in both computing machines and meteorology, teamed up with a bright graduate student named J. Presper Eckert Jr. They led a team that in early 1946 unveiled the world's first electronic, large scale, general-purpose computer. ENIAC contained almost 18,000 vacuum tubes and 6,000 manual switches (see Illustration 2.18). It weighed thirty tons and was nearly 100 feet long and ten feet high. ENIAC could perform 5,000 additions and 300 multiplications per second, unbearably slow compared to today's fastest supercomputers (which can exceed a few trillion calculations per second). ENIAC was funded by the U.S. Army and used primarily to calculate trajectories for ballistic missiles. But Mauchley also used ENIAC for some primitive weather calculations.

Meanwhile, at nearby Princeton University's Institute for Advanced Study, mathematician John von Neumann and meteorologist Jule Charney were working on their own numerical weather prediction project. By 1949, their computer model was ready to be tested, but their computer was not yet finished. So the chief of the U.S. Weather Bureau helped them to obtain permission to run their weather model on ENIAC, which by then had been moved to Aberdeen Proving Grounds in Maryland. In April 1950, the first one-day numerical weather prediction was made on ENIAC. Largely because the computer kept breaking down, the forecast required more than twenty-four hours to execute (so the weather actually moved faster than the weather forecast) and required round-the-clock services of at least five people. But the computer forecast was reasonably accurate and encouraged the researchers. The first forecasts on Princeton's computer were made in 1952, and within a few years, many research groups worldwide were experimenting with "weather by the numbers." In May 1955, the first numerical weather predictions intended for use by operational weather forecasters were made on an IBM computer in a project jointly funded by the Air Force, Navy, and the U.S. Weather Bureau. Weather forecasting by computer had become a reality and it began in Philadelphia and nearby Princeton, New Jersey.

Today, there are more than a dozen different operational computer weather models, and meteorologists are

ILLUSTRATION 2.18 *(See plate.)* ENIAC occupied a thirty-by-fifty foot room. The controls are at the left, and a small part of the output device is seen at the right. The two men in uniform were being trained to maintain the machine after it was turned over to the Army (which sponsored ENIAC's development). The two women were assistants on the staff of the Moore School who helped program the ENIAC (courtesy of John W. Mauchly Papers, Rare Book and Manuscript Library, University of Pennsylvania).

continually fine-tuning them. All computer models use basically the same set of mathematical equations that describe the processes that can change the temperature, pressure, wind, and humidity. Computer models also include a virtual description of the underlying surface, be it ocean, ice, desert, plains, or mountains. A numerical forecast begins by "initializing" the model—that is, feeding current weather observations to the computer. The model equations then transform this initial data into a prediction of future conditions in the atmosphere, which can be displayed in text or map form. Basically, numerical weather prediction is a giant math problem, so computer models run on very fast supercomputers—in fact, meteorologists are some of the supercomputing industry's best customers.

Some computer models are short range and very detailed, producing forecasts only for the next few hours or days. Some longer-range models produce forecasts out to fifteen days. There are even climate models that produce probability-based outlooks for months or even seasons into the future. Some computer models forecast for just the eastern states, others for all of North America and nearby oceans, while some models are truly global. Forecasts from different models can be compared to see which ones are doing the best job in each situation. When all (or nearly all) models agree, meteorologists have greater confidence in the forecasts. When their solutions differ a lot, there is more uncertainty. Sometimes, a blend of several model solutions results in the best forecast.

The U.S. National Weather Service runs several models daily, but there are quality computer models available from sources outside the NWS, including the U.S. Navy, some universities and commercial weather companies, and weather services in other countries (such as Canada and the United Kingdom). The amount of computer model guidance available today is staggering. A meteorologist on a deadline might sit by two computer monitors, using one hand to manipulate the maps from one model and the other hand to look at the same forecast from a different model on the other monitor. And there is still usually not enough time to look at all the computer data before issuing a forecast. Illustration 2.19a shows examples of computer model forecasts in text form, while Illustration 2.19b shows some of the forecast maps produced by one of the models known as the "Eta." Almost all the model information that meteorologists use is available on the Web (see Web bibliography).

Since all of this information is potentially available to all forecasters, why do you sometimes hear different forecasts from different sources? While meteorology is a science, it is not an exact science. You could say there is still some "art" to weather forecasting. As in many situations involving the analysis of data, reasonable people can make different but reasonable conclusions from the same information. Meteorologists are the same way—one might focus on the intensification of a trough in the jet stream as the key to tomorrow's forecast, while another might conclude that a fast-moving cold front will be more important. Then

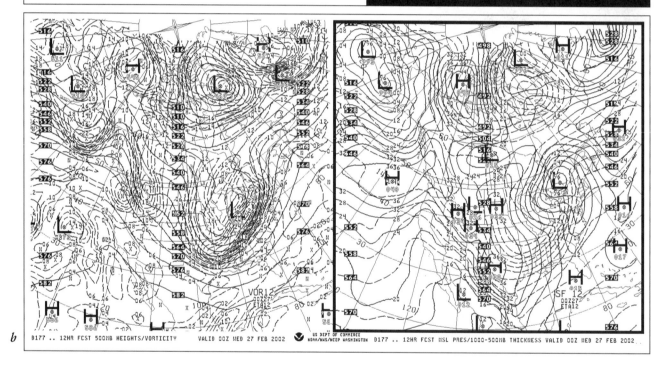

```
KPHL    AVN MOS GUIDANCE    8/17/2001   1200 UTC
DT /AUG  17/AUG  18                /AUG  19                    /AUG  20
HR   18  21  00  03  06  09  12  15  18  21  00  03  06  09  12  15  18  21  00  06  12
N/X                      69              88              70              84      70
TMP  88  89  84  77  73  71  73  80  85  86  81  76  73  72  73  78  81  77  72  73
DPT  68  66  65  66  66  65  66  65  65  65  67  69  68  68  69  70  70  70  70  70  70
CLD  OV  BK  SC  SC  CL  CL  SC  SC  BK  BK  BK  OV  OV  OV  OV  OV  OV  OV  OV  OV
WDR  23  23  22  24  26  28  31  24  25  24  23  20  23  24  03  16  24  21  18  22  21
WSP  16  15  10  07  07  06  04  02  05  06  06  03  03  03  03  05  06  06  04  03
P06          12       5       1       5      19      31      30      28      39  26  18
P12              8           20          44          56      40
Q06           0       0       0       0       0       0       0       0       1   0   0
Q12               0           0           1           2   1
T06       9/ 6  7/ 1  0/14  2/ 1 12/ 8 12/ 1 13/15 12/ 1 16/ 6  6/14
T12        12/ 6       5/14      19/ 8      19/15      38/ 6
CIG   7   7   7   7   7   7   7   7   7   6   6   6   5   5   5   5   5   5   5   5
VIS   7   7   7   7   7   7   7   7   7   7   7   7   7   6   7   7   7   7   5
OBV   N   N   N   N   N   N   N   N   N   N   N   N   N   N  BR   N   N   N   N  BR  BR


     PHL//847148 00299 172116 74232113
     06000616726 -1198 142412 74282013
     12000594022 -2198 142515 74282112
     18000542229 -1104 152811 73251911
     24000582544 00405 172809 71221811
     30000422959 00802 162410 72282011
     36000524548 -2900 142211 72282012
     42000877061 01799 152413 71231912
     48003947970 02998 141407 72221912
     54028817369 00497 131710 73251812
     60024927868 02298 121816 73231813
```

a

b   D177 .. 12HR FCST 500MB HEIGHTS/VORTICITY   VALID 00Z WED 27 FEB 2002   US DEPT OF COMMERCE NOAA/NWS/NCEP WASHINGTON   D177 .. 12HR FCST MSL PRES/1000-500MB THICKNESS VALID 00Z WED 27 FEB 2002

ILLUSTRATION 2.19   Examples of computer "numerical guidance." *(a)* The top block of forecast information, known as "MOS" (for Model Output Statistics), shows high and low temperatures and three-hour forecasts of temperature, dew point, cloud cover, wind, probability of precipitation, and other variables. The bottom block of numbers, known as "FOUS" (for Forecast Output Statistics), represents coded forecasts for each six-hour period out to forty-eight hours (for detailed explanations on how to decode MOS and FOUS, see the Web site bibliography). *(b)* Sample forecast maps from the "Eta" model. The map on the left is an upper-air chart, showing forecast conditions at approximately 18,000 feet above the ground. The map on the right is a surface chart, showing average temperature and highs and lows of pressure.

there is the human factor. A forecaster who loves stormy weather might tend to predict more rain and snow. Or a forecaster who underpredicted the last snowstorm might (consciously or unconsciously) try to avoid getting burned twice in a row by adjusting forecasted snow amounts upward a bit on the next storm. The best forecasters are aware of their biases and try to adjust for them.

Finally, there is the variety of computer models. Though their solutions are usually similar in a general sense, the details often differ greatly, and no particular model is right all the time. When computer models diverge in their solutions, forecasters often differ as to which computer model solution to accept. In a typical winter storm in our area, the differences between the models can lead to big differences in the forecast—for example, rain instead of snow, or no snow versus six inches. So far, human forecasters have been able to improve on the accuracy of computer forecasts, and we have a vested interest in keeping that edge. A forecaster who relies completely on computer guidance will never compensate for the times when the models perform poorly. Sometimes the best strategy is to completely disregard the computer model guidance and go with a forecast that seems to make the most sense given the current observations and the forecaster's experience, as we shall see.

Forecast Accuracy

When the Eagles are big favorites in a game, it is easy to predict a victory. Predicting upsets requires true skill. Or when the stock market is going up in all sectors (a "bull market"), it is easy to choose profitable investments. Picking winners in a "bear market" is how the pros earn their reputation.

The same is true in weather forecasting. Predicting a sunny day when there are no clouds within hundreds of miles is easy—true skill shows when the difficulty of the forecast is greater. Overall, there is no type of forecasting that is more accurate than weather forecasting. The accuracy of weather forecasts has gradually but steadily improved over the last few decades. However, with all the radars, satellites, and fancy graphics that meteorologists now use, the public's expectation level has risen as well. So in the end, the public perception of forecast improvement may not agree with the evidence. Forecasts have improved in two basic ways: greater accuracy of specific temperature and precipitation forecasts, and more detailed forecasts for individual counties.

Temperature forecasts are easily verified. We just compare the predicted highs and lows with what is actually observed. Illustration 2.20 shows the average errors in the National Weather Service's Day-1 ("today") and Day-2 ("tomorrow") high temperature forecasts over the last thirty years. It is clear that the errors have decreased over time. For example, the average error in a Day-2 temperature forecast in 1999–2000 was 3.2°F, which is less than the average error in a Day-1 temperature forecast in 1979–80 (which was 3.3°F). In other words, the NWS temperature forecast for "tomorrow" is now more accurate than the temperature forecast for "today" was twenty years ago. In addition, the percentage of Day-2 forecasts with a temperature error of 10°F or more (basically, very poor forecasts) has shrunk from nearly 9 percent in 1969–70 to only 2 percent in 1999–00. Longer-range temperature forecasts have improved as well—the average error in a five-day forecast in 1999–00 was 5.2°F, which is less than the average error in a three-day forecast twenty years ago.

Precipitation forecasts are more complicated to evaluate because they often include qualifying terminology such as "scattered" showers and "probabilities of precipitation." Even so, it is possible to document improvement in both the probability forecasts and precipitation amounts, though the improvement has not been as great as with temperature.

The other area of improvement has been in the forecast details. However, this improvement is difficult, if not impossible, to quantify. How do you compare a forecast from the 1970s that might have read "Tomorrow: Mixed

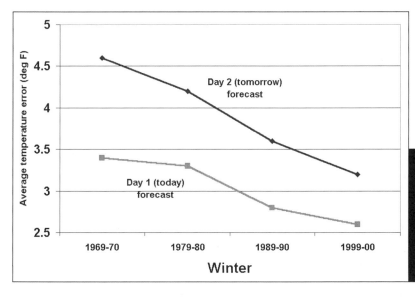

ILLUSTRATION 2.20 Average error in National Weather Service temperature forecasts during the winter for "Today" and "Tomorrow." The average error in a "Tomorrow" forecast issued in 1999–2000 is around 3°F, less than the average error in a "Today" forecast in 1979–80 (courtesy of National Weather Service).

Precipitation" with a recent forecast such as "Tomorrow: Snow beginning around noon, accumulating 1 to 3 inches, before changing to rain by evening"? Both forecasts might be right, but which one is more accurate? Obviously, the latter is better in that it is more specific, but there is a hitch: the more detail given, the greater the chance for imperfection.

Improvements in forecast details have also resulted from reducing the size of the area covered by a particular forecast. The National Weather Service has been making forecasts for groups of counties for many years. The trend has been to decrease the number of counties in each group, especially when severe or extreme weather threatens. It is not hard to imagine a day in the future when you will see specific forecasts for your own zip code.

Why have forecasts improved over the years? Three main reasons can be identified:

- *More weather observations of higher quality.* There are more observations today from more sources than ever before. A network of sophisticated radars scans the skies for precipitation. Satellites do more than detect clouds—they can measure air and ocean temperatures and estimate winds. Still, our picture of the atmosphere is, and always will be, imperfect. There is no way to observe all the essential weather

STORY FROM THE TRENCHES

The Snowstorm of January 25, 2000

John Schilling, a long-retired forecasting friend of ours, was fond of saying, "I forecast with great trepidation, and for that I have no regrets, because when I'm right no one remembers, and when I'm wrong no one forgets." For East Coast meteorologists, some of the most difficult and unforgettable forecasts involve the winter storms known as "nor'easters." One particularly challenging nor'easter surprised East Coast forecasters in late January 2000.

Prior to that storm, the 1999–2000 winter had been a lot like the preceding three: mild overall with below-average snowfall in Philadelphia. Only a few flakes had been sighted through mid-January. The calls and E-mail messages from the public had reached a fever pitch: "Is it ever going to snow this winter?" December had been relatively mild (about 4°F above average), and the first twelve days of January averaged an astonishing 13°F warmer than average. Cold air finally arrived on January 13, setting the stage for some real winter weather. The first measurable snow came on Thursday, January 20, with 3 inches in Philadelphia and Wilmington and 4 inches at Atlantic City.

With the cold air entrenched and a stormy weather pattern taking shape across the country, talk began about a potential big snowstorm for the following week. Most five-day forecasts on Friday, January 21, called for possible snow the following Wednesday. In fact, several computer models were suggesting a big storm. The main low-pressure system would track southeastward out of central Canada and redevelop along the East Coast, leading to a moist, cold, and intense storm moving up the eastern seaboard— a nor'easter.

But over the weekend, the models mostly backed off the big storm idea. On Sunday, the possibility of snow on Wednesday was still mentioned in most forecasts, although it was looking less and less likely. By Monday morning, some forecasters had even stopped mentioning snow as a possibility for Wednesday.

It turned out that the Canadian storm would not be the problem. Early Monday, a compact but potent trough in the jet stream was producing rain in northern Florida, with snow breaking out in Georgia. And the shield of precipitation was edging northward. According to virtually all of the computer models, this should not have been happening—this area of precipitation was projected to move east and out to sea.

Now, if the computer models from Sunday night were already wrong for their Monday morning forecast, their projections for Tuesday could not be trusted either. It was hard to believe that a storm predicted to move harmlessly out to sea would instead come up the coast into the cold air and dump tremendous amounts of snow. But if there was ever a time to throw

variables at every point at all times, and even where observations are made, there is always some tiny instrument or measurement error. When observations are imperfect or incomplete, the forecasts based on those observations will suffer. Computer models depend on these observations, and "garbage in, garbage out" is an accurate description of how computers work. Small errors in the weather observations fed to these models will grow larger as the computer forecast reaches farther out into the future.

- *A better understanding of how the atmosphere works.* Compared to other sciences such as astronomy and chemistry, meteorology has a short history. Some basic principles have been known for centuries, but much of modern meteorology was born out of necessity during World War II. The existence of the jet stream, for example, was confirmed during the war, while the concept of fronts was firmed up only around the time of World War I. With today's observational systems able to provide more detailed data, much current research focuses on better forecasting of smaller-scale phenomena such as thunderstorm complexes, lake-effect snows, and hurricanes.

- *Improvements in computer modeling.* As computers increase in capacity and speed, computer models can represent smaller and smaller details of the weather.

out the model predictions, this was it: the computer forecasts suggested partly sunny skies for Tuesday in Philadelphia with virtually no chance of snow. Yet a simple projection of the current radar pattern using the upper-air winds would bring snow as far north and west as Philadelphia. Unfortunately, most forecasters from the Carolinas to New England went "down the tube" with the poor computer guidance.

The new computer forecasts that came in Monday around noon still did not properly capture the northward drift of the storm. But now there was an indication that at least a little snow would fall in and around Philadelphia. At 3:30 P.M., the National Weather Service in Mount Holly issued a Winter Storm Watch for southern Delaware and southern New Jersey for the potential for 3 to 6 inches of snow, but the immediate Philadelphia area was not included in the watch. None of the local forecasters on the 6 P.M. news were predicting more than a few inches for the city. Many people went to sleep Monday night expecting only a light snow the next day.

Meanwhile, the pictures coming in from Raleigh, North Carolina told the story of one of the biggest forecast "busts" of all time in that city: at first, a forecast of no snow, then later a call for light snow. Instead, the city saw the biggest snowstorm in its history, an incredible 20 inches!

At Glenn's station, reports of that heavy snow convinced John Bolaris and Harry Holmes (the weekday evening meteorologists) to disregard the still-bad computer guidance. Their forecast on the Monday 11 P.M. news was for 6 to 12 inches in the Philadelphia area, with even 5 to 10 inches well west of the city. Other local stations adjusted their forecasts as well,

and the National Weather Service issued a Winter Storm Warning at 10:49 P.M. for the Philadelphia area, calling for 4 to 8 inches of snow. These updated forecasts turned out to be quite accurate; the problem was that they were issued only a few hours before the snow started and after many people had gone to bed. Anyone who did not see the late news on Monday was shocked the next morning as heavy snow made a huge mess out of rush hour. Officially, the storm totaled 8.5 inches in Philadelphia, 10 inches in Wilmington, and up to 14 inches in some nearby suburbs.

Area television stations broadcast continuous coverage of the storm on Tuesday, but that coverage did little to soothe many surprised viewers who were angry about the lack of warning for the big snow. For weeks afterward, local forecasters fielded E-mail messages and phone calls full of negative comments.

The January 25 "surprise" snowstorm illustrates two powerful and somewhat contradictory issues regarding weather forecasting. First, the public's expectation of accuracy is high, particularly for important storms. At least some of this expectation results from several highly publicized successes in forecasting the big East Coast storms of March 1993 and January 1996. Second, this forecasting success is, in large part, the result of guidance provided by computer weather models, which are still imperfect. One of the biggest challenges for weather forecasters today is to find the right balance of computer predictions, observations, and human experience to make the best forecast. In the case of the January 2000 "surprise" snowstorm, that balance was achieved too late to be of value to most of the people affected by the storm. ❖

Researchers are developing new methods to incorporate the increasing numbers of observations into the models. And more accurate representations of oceans, ice-covered surfaces, and even vegetation are being included in computer models to better reproduce the exchanges of heat and water between the surface and the air. In short, computer models become more sophisticated each year, and the general trend has been toward greater accuracy of computer forecasts.

Even with these improvements, there are limits to how far in the future the weather can be predicted. Think about dropping a stick into a stream and trying to predict where it will be in five seconds, five minutes, and five hours. The five-second forecast may be easy, but once you get to five minutes and then even more so at five hours, it becomes an impossible task, given the complexities of the motions of the water and the interaction with the stream bottom and banks. The farther in the future you try to predict, the less accurate you are likely to be.

In many ways, the atmosphere is like the stream (but without banks). The farther away in time a weather prediction is, the poorer its reliability. Currently, some computer models routinely produce forecasts of pressure, temperature, and precipitation for fifteen days in the future. But these models can rarely pin down the location and intensity of a specific storm or high-pressure system a week in advance, let alone fifteen days. In fact, if you closely monitor your favorite media outlet, you will find that the details and timing of the five- and seven-day weather forecasts are frequently off by days four and five, and sometimes sooner.

But some types of long-term forecasts are possible. For example, the overall weather pattern—such as the approximate location of the jet stream and the general placement of warm and cold air masses—is more predictable than the details of specific storms. So sometimes forecasters can say with reasonable certainty that a major change in the weather or a significant weather event is possible (or even likely) a week to ten days in the future. For example, a forecaster on a Sunday during summer might be able to say that a long heat wave will end the following weekend. Will the cooler air arrive next Saturday or next Sunday? That is too tough to call. Or on a Thursday in January a meteorologist might say that a powerful winter storm will probably develop in the East during the middle of next week. Where will it snow, and

how much? Will there be any ice? And will it be next Wednesday or Thursday? Again, these questions cannot be answered with certainty a week in advance.

To make forecasts in the three-to-ten-day range meteorologists rely almost exclusively on computer models. As in shorter-range forecasting, there is increased confidence in the longer-range forecasts if the various computer models agree with one another. In recent years, meteorologists have developed a new way to do model comparison, this time using just a single model. This major innovation is called **ensemble forecasting**. Ensemble forecasting is based on the idea that we never have a perfect set of initial observations to feed the computer models. And, as mentioned earlier, this is a primary reason why computer forecasts can quickly go astray. So meteorologists have devised a way to test how sensitive a model's forecast is to the observations that it is given. They tweak, or slightly alter, the observations and run the model again using these minutely different starting conditions. Then they start over and tweak the observations again, in a slightly different way, and run the model a third time. This process can be repeated many times to create an "ensemble" of runs of the same computer model—with one long-range model run by the National Weather Service it is done twenty-three times. With another long-range model developed by the Canadian Meteorological Centre, there are sixteen different runs. Meteorologists then compare the various forecasts produced by all the model runs. If most of the model runs come up with basically the same weather pattern for a certain day, meteorologists have a relatively high degree of confidence for that day's forecast. Illustration 2.21a is an example of a relatively consistent forecast—it shows the three-day prediction of a particular isobar for each of the sixteen runs of the Canadian model. The lines are bunched together in most areas, indicating general agreement between runs. If, however, the model has a variety of different solutions for a particular day, then that day's forecast is much less certain. Illustration 2.21b shows the forecast of the same isobar as in the top part of the figure, but now seven days in the future. Here the lines are much more widely separated, which would suggest much lower confidence in this longer-range forecast.

Basically, ensemble forecasting is a clever way of getting many different forecasts from the same model. Still, the ensemble approach does not help to pin down high and low temperatures for specific days or to determine precisely when and where it will rain. It does suggest to

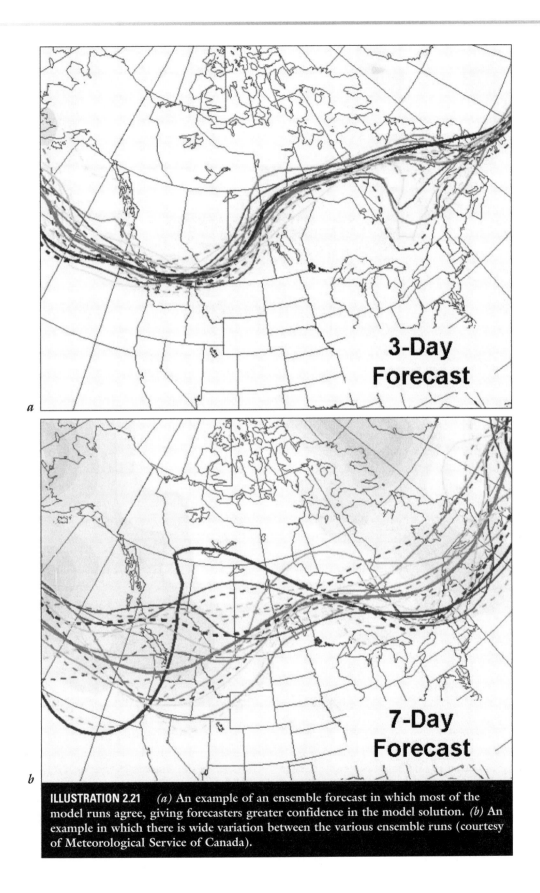

ILLUSTRATION 2.21   *(a)* An example of an ensemble forecast in which most of the model runs agree, giving forecasters greater confidence in the model solution. *(b)* An example in which there is wide variation between the various ensemble runs (courtesy of Meteorological Service of Canada).

meteorologists regions and time periods of high and low confidence in the model predictions. This can enable forecasters to make statements such as "We're fairly confident that warm, muggy weather will prevail next week" or "A major shift in the weather pattern is likely by the weekend." If you hear a lot of talk about a big weather change coming a week or more in the future, you can be sure that there is plenty of agreement among the various ensemble forecasts.

In closing this section, we are reminded of a saying that is not attributed to Yogi Berra but sounds like something he might have said. And it applies very well to weather prediction: "Forecasting is difficult, especially about the future."

STORY FROM THE TRENCHES

The D-Day Forecast

Early in 1944 during World War II, the German Army occupied Europe. The Allies were already planning for an invasion, an operation so huge that it would take days to put into motion. General Dwight D. Eisenhower wanted a ten-day weather forecast in preparation for D-Day. Although most experts had told Eisenhower that such a forecast was impossible, Henry "Hap" Arnold, Commander of the Army Air Forces, found two groups of meteorologists in the States who were working on long-range forecasting. Hurd Willet headed a group at MIT, while Irving Krick spearheaded an effort at Cal Tech. Arnold put the two groups to work on the D-Day forecast.

Philadelphia-area native (and future television pioneer) Dr. Francis Davis was part of that historic attempt to gain advantage from the weather. Davis, who had graduated from college in 1939 with degrees in physics and math, had enlisted as an aviation cadet in 1941. When the United States needed weather officers for the war effort, Davis had the perfect academic background, so he was sent to MIT to study meteorology. There, Davis worked for Willet on the long-range forecasting project. Davis's assignment was to draw maps of the winds at the 10,000-foot level for various days between 1890 and 1940. The maps were sent to the Pentagon where a database of these historical weather charts, as well as the corresponding surface weather maps, was being compiled. To prepare a ten-day forecast, the current weather map was compared to all the weather maps in the historical database, searching for the best match between present and past. Once a good match, or "analog," was found, the weather that followed that historical day was used as a forecast for the days to follow the current day. (Analog comes from the word "analogy," which means "likeness.") This forecasting technique was appropriately called the "analog method."

Davis was eventually sent to the Pentagon, where he recalls that "we would plot the 10,000-foot map for the day and take it down the hall and put it in a slot in a big metal door. Some time later, out came a list of dates in the past which had maps close to what we had that day." The meteorologists would then check the weather that occurred following those historical "analog" days for guidance in producing the current ten-day forecast. That forecast was then sent to Eisenhower in Europe. Though Davis did not know it at the time, behind those closed doors were some of the first primitive "electronic computing machines," which were helping to look for the best analogs.

Today we know that the analog method fails at pinning down specific weather features that would make such ten-day forecasts useful. No two weather maps are ever exactly alike, and the small differences that exist between any two similar maps at the start of a forecast will grow with time to become a large difference in the details of the weather just several days in the future. The D-Day forecasters, with no previous experience working with analogs, learned this lesson the hard way. "It didn't work out very well," recalls Davis of the most important long-range weather forecast in history. The D-Day invasion was postponed a day, and even then, the weather was still far from ideal. ❖

CHAPTER 3

Winter: December–January–February

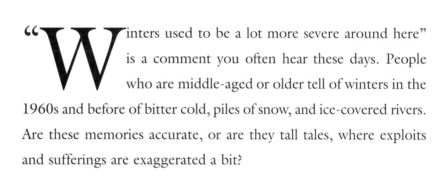

Snowy winter, a plentiful harvest.
—*Benjamin Franklin*

"Winters used to be a lot more severe around here" is a comment you often hear these days. People who are middle-aged or older tell of winters in the 1960s and before of bitter cold, piles of snow, and ice-covered rivers. Are these memories accurate, or are they tall tales, where exploits and sufferings are exaggerated a bit?

Fortunately, plenty of documentation exists about past Philadelphia winters, and not just for the last century. Winters have been documented in this area as far back as the middle of the seventeenth century, at least in a descriptive sense. The picture that emerges shows that big snowstorms and big chills have been part of winter around here for centuries, but on average, winters during the last two decades have, indeed, been milder and less snowy than many of the past ones.

In this chapter, we take a close look at winter in the Delaware Valley, covering averages and extremes in snowfall and temperature. From colonial days to the present, we chronicle the huge snowstorms and the historic cold spells. Were there blizzards in the 1700s and 1800s that rivaled the Blizzard of '96? Did the Delaware River routinely freeze over for months at a time? What was it like at Valley Forge during that historic Revolutionary War winter?

We will also explain why the Philadelphia region as a whole is one of the toughest places in the country to forecast winter weather. One storm brought parts of the area three inches of snow and other parts four feet! We'll discuss how this can happen and why some winter storms produce sleet and freezing rain while others bring all snow or all rain. And we will evaluate the trustworthiness of wind chill as a cold weather indicator.

Finally, a tale of two forecasts gives the behind-the-scenes stories of a television station's weather and news coverage of the biggest snowstorm in modern Philadel-

phia-area history, and a similar account of a big snowstorm that was predicted but never came.

TOUGH FORECASTING ON THE EDGE

Based on astronomical considerations, winter officially begins around December 21 in the Northern Hemisphere. But with all due respect to astronomers, meteorologists tend to think of winter as simply the three coldest months—December, January, and February. Of course, weather conditions are not necessarily bound by either this "meteorological winter" or the astronomical calendar, as cold and snow sometimes arrive in the Delaware Valley in November, and persist even into April.

The Philadelphia area sits between the colder, higher elevations of the Appalachians and Poconos to the north and west and the coastal regions to the south and east, where winter cold is moderated by the ocean. As a result, winter in the Delaware Valley is often spent "on the edge" in terms of both temperature and precipitation. Bitterly cold Arctic air easily filters into northern and western Pennsylvania and northern New Jersey, but its bite is usually restrained by the time it gets south and east of the mountains. During winter storms, the rain-snow line often snakes its way through the area, as rain near the coast becomes ice inland and snow farther north and west. With the Atlantic Ocean only about sixty miles away, the Philadelphia area is positioned for a variety of winter weather, and the potential for significant, moisture-laden storms.

The character of a Delaware Valley winter can vary greatly from year to year. The difference in average temperature between Philadelphia's coldest and warmest winters is more than 15°F. Average winter snowfall in Philadelphia is 20 inches, but the range is from less than an inch to more than five feet. And in nearby cities, average snowfalls are dramatically different. To the north, Allentown averages 31 inches and Wilkes-Barre about 45 inches (with some nearby areas in the Poconos having much more than that); to the southeast Atlantic City averages 16 inches (see Illustration 3.1). Seasonal snowfall in these locations also varies greatly from winter to winter.

Of course, cold and snow often mean significant disruption. Who can forget the ice storms of the winter of 1993–94 or the virtual shutdown of the city in the days after the Blizzard of '96? Nationally, the annual cost of snow removal for streets and highways exceeds $2 billion. For the city of Philadelphia, the cost for plowing and salting averages about $5 million per winter, but the cost exceeded $15 million during the winter of 1995–96. These figures do not include the private clearing of parking lots, driveways, and sidewalks or the expense of repairing pothole-riddled roads. Add on the costs resulting from ice- and snow-related accidents, absences from work and school, airline delays and cancellations, damaged trees and collapsed roofs, and so on, and winter is clearly the major cause of weather-related disruptiveness in the Philadelphia area.

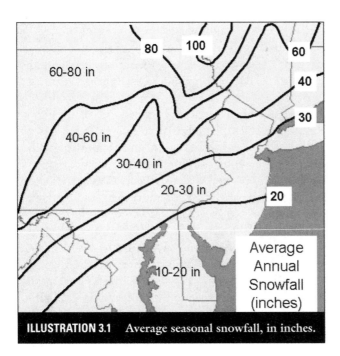

ILLUSTRATION 3.1   Average seasonal snowfall, in inches.

Snow, Sleet, Freezing Rain, or Rain?

In a simple world of weather, there would be just two types of precipitation: rain and snow. When the air temperature near the ground was above 32°F, there would be rain, while temperatures of 32°F or below would produce snow. In the real world, snow has been observed at temperatures above 40°F, while raindrops can reach the ground at temperatures as low as 10°F. The explanation for these oddities could read a little like an Abbott and Costello routine:

Costello: What's up?
Abbott: That's right!
Costello: No, what's up??
Abbott: Exactly!

Meteorologists have to know "what's up," that is, the temperature of the air above the ground. Indeed, the temperature "up there" is critical in determining the form that precipitation takes at the earth's surface.

Most people know that the atmosphere typically gets colder with higher elevation. In the midlatitudes, temperatures are usually below freezing at cloud level. As a result, most precipitation at our latitude starts in the clouds as snow (even during summer). If the temperature of the air below the clouds is 32°F or below all the way to the ground, the snowflakes will survive the fall. Even if a thin layer of air next to the ground is above freezing, snowflakes can reach the surface without completely melting, especially if the snow is falling rapidly and the flakes are big. A complete meltdown, of course, would just result in rain.

Now comes the tricky part. Sometimes the "higher equals colder" rule is broken. When an above-freezing layer of air forms a few thousand feet above the ground, it creates a warm-air sandwich: a layer of relatively mild air wedged between the deep chill at high altitudes and a thinner layer of cold air near the ground (see Illustration 3.2). The result can be sleet or freezing rain. In both cases, the precipitation starts as snow high in the clouds, but as these flakes fall through the warmer layer, they melt into raindrops. If the layer of cold air hugging the ground is cold enough or thick enough, the raindrops can freeze into tiny balls of ice before reaching the surface— that is **sleet** or **ice pellets**. Sleet is often confused with hail, but the two form in very different ways—hail comes from thunderstorms and will be discussed in Chapter 4.

Freezing rain occurs when the melted snowflakes do not have time to refreeze in the air before reaching the ground. Instead, the raindrops freeze on contact with

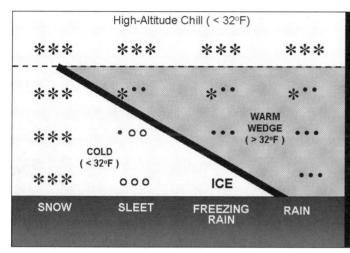

High-Altitude Chill (< 32°F)

COLD (< 32°F)

WARM WEDGE (> 32°F)

ICE

SNOW   SLEET   FREEZING RAIN   RAIN

ILLUSTRATION 3.2   A side view of the atmosphere in a typical "mixed precipitation" winter storm. The right side of the figure might represent the atmosphere above Cape May and the left side the atmosphere over Reading. Near the coast, the wedge of warm air (temperature above 32°F) is thick, so falling snowflakes melt into rain. Farther inland, a thin layer of cold air (temperature below 32°F) remains near the ground, so the rain freezes when it hits the surface, forming freezing rain. Farther northwest (*left in the diagram*), a thick enough layer of cold air remains near the surface to refreeze raindrops or partially melted snowflakes in the air, forming sleet. Finally, well inland, the entire atmosphere is cold enough to support snow.

the cold surface (Illustration 3.2). Sometimes, if the ground is relatively warm prior to the storm, the drops may not freeze on roads; but sidewalks, trees, cars, power lines, and any metal surfaces ice up. If the ground had been cold for a substantial time before the precipitation arrives, even the roads glaze over.

In winter's entire arsenal, freezing rain is the greatest menace. Just a thin coating of ice can produce dangerously slick roads and snap trees and power lines. It is no wonder that freezing rain is notorious for disrupting electrical and communication services. In fact, the practice of burying communication lines grew popular after much of the aboveground telegraph system in New York City was destroyed by freezing rain during a powerful winter storm in 1888.

Why Winter Forecasting Is Tough

Anyone who has lived in an area that gets wintry weather knows how difficult it can be to predict precipitation types and amounts in the cold season. Winter forecasting in some parts of the country is tougher than in others, and the Philadelphia area is as challenging as any.

Part of the difficulty stems from geographical location. The Delaware Valley sits in the heart of the mid-latitudes, so the primary battleground between cold and warm air usually is not too far away in winter. Just 200 miles to the south, in Richmond, Virginia, snow makes up only about 10 percent of the winter precipitation—forecasters in Richmond know that, at least climatologically, snow is relatively uncommon. However, 200 miles to the north of the Delaware Valley in Syracuse, New York, more than 80 percent of winter precipitation falls

as snow—forecasters there are less likely to have to deal with storms complicated by changeovers to sleet, freezing rain, or rain.

Proximity to the Atlantic Ocean also adds to the forecasting challenge, lessening the likelihood of an all-snow winter storm. On average, the closer an area is to the ocean, the less snow it will have. Even in the heart of winter, ocean temperatures off New Jersey and Delaware are typically in the 34–38°F range, and air sitting above the ocean takes on the temperature of the water below. The slightest onshore wind during a storm can bring in just enough relatively mild air to change snow to sleet, freezing rain, or rain, or make it more likely that snow will melt instead of accumulate. One of the main ingredients in most big snowstorms in the Philadelphia area is a very cold air mass already in place before the storm arrives—the colder the air at the start, the more likely that precipitation will stay all snow.

The other major geographical feature that can make for great variability during winter storms is the increase in elevation that occurs as you move north and west from the Coastal Plain onto the Piedmont Plateau. Elevations rise from near sea level along the Delaware River to more than 500 feet in parts of Chester, Montgomery, and Bucks counties. Though the elevation rise is not large in an absolute sense, the extra chill of the slightly higher elevations can make the difference between liquid and frozen precipitation in marginal cases (of which we see many each winter). It is not uncommon for rain to be reported in Millville, New Jersey or in Center City Philadelphia, while it is snowing in Chestnut Hill and Northeast Philadelphia. Also, the slight rise in elevation adds lift to an east or southeast wind, and that extra boost of rising air can mean more clouds and greater precipitation.

Another part of the winter forecasting challenge lies in expectations. Consider that when rain is predicted, meteorologists usually do not specify how much will fall. But when snow is in the forecast, the public expects an amount to be given. In reality, most people cannot tell the difference between, for example, three-tenths and seven-tenths of an inch of rain. Yet assuming a ten-to-one ratio of snow to rain, that range becomes three to seven inches of snow. Everyone will notice that difference, and most people would be upset if seven inches of snow fell on a day when only three was predicted. The fact is, there is simply an inherent expectation for more precise forecasts in winter, but the science of predicting precipitation amounts is not much more precise in winter than at other times of the year. And expectations do not end with the amount of snow. Forecasters are also expected to pin down when the snow will start and end, whether the snow will change to sleet, freezing rain, or rain, and when the temperature will get low enough for everything to stick to the roads.

Even if forecasters get all of those details correct for a particular place, what about neighboring counties? An accurate forecast for the city of Philadelphia may not be on target for the western suburbs, or southern New Jersey. The county-to-county difference is often dramatic. One storm in March 1958 brought only a few inches of snow to the New Jersey shore and central Delaware, while Morgantown, Pennsylvania, just twenty-five miles west of Valley Forge, measured 50 inches (Illustration 3.3a), stranding hundreds of people along the Pennsylvania Turnpike (Illustration 3.3b). This storm was characterized by very large local variability in snow totals even within Philadelphia. The observer for the U.S. Weather Bureau in the city noted that "the increase in snow depth was very marked along Germantown Avenue from Broad and Erie Avenues to Chestnut Hill, a continuous rise in elevation from 95 feet to 420 feet." Though this is an extreme example, it illustrates the fundamental point that just a few miles often makes a world of difference during a winter storm in the Delaware Valley.

The Winter Weather Setup

Low-pressure systems usually get all the attention and all the blame for stormy weather. But a low often works in tandem with an area of high pressure to create clouds and precipitation, especially in winter. This tag-team high is the Rodney Dangerfield of winter storms, getting no respect. But during the winter, high pressure generally means cold, relatively heavy air. Without a strong high in place prior to the approach of a low, there may not be enough cold

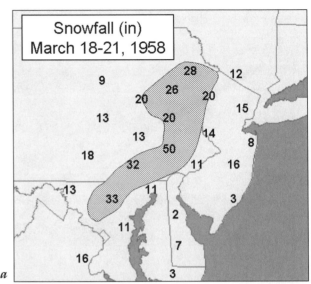

ILLUSTRATION 3.3 *(See plate.)* *(a)* Snowfall totals from the storm of March 19–21, 1958. Areas receiving 20 inches or more are hatched. Morgantown, Pennsylvania recorded 50 inches, while Atlantic City measured only 3 inches (adapted from Kocin and Uccellini, 1990). *(b)* On March 23, 1958, a U.S. Air Force helicopter assists some of the nearly 800 stranded travelers who had been marooned at a rest stop near the Morgantown exit of the Pennsylvania Turnpike about ten miles south of Reading after four feet of snow closed the road (Urban Archives, Temple University, Philadelphia, Pa.).

air to manufacture snow or ice. The cold heavy air also acts as a wedge to lift the relatively warm, moist air that typically streams northward ahead of a low, often producing an extensive area of overrunning clouds and precipitation.

The weather map in the mid-Atlantic can become very complicated when lows approach during winter. Cold air circulated southward by a high in New England or eastern Canada can become trapped between the Appalachian Mountains to the west and the relatively mild air off the coast, a process meteorologists call **cold-air damming**. The result is a shallow layer of cold and heavy high-pressure air that noses down the eastern slopes of the Appalachians. The front that separates the advancing warm air from this stubborn low-level cold air can become very convoluted (see Illustration 3.4). In this "backward *S*" pattern, temperatures in western Pennsylvania with southwest winds can reach 50°F, and southeast winds can push temperatures to similar levels near the Jersey and Delaware shores. By contrast, when low-level winds continue from the northeast in the Philadelphia area and surrounding counties, temperatures hold below freezing.

The situation above the ground is very different, however. A few thousand feet up, above the shallow layer of cold air, warmer southerly winds affect all areas, pushing temperatures well above freezing. In such situations, temperatures might be 30°F at the ground in Philadelphia and 40°F or higher a few thousand feet up (but then colder again above that). Snowflakes falling from high up melt as they pass through this warmer layer, but then the raindrops refreeze in the cold air below or when they hit

ILLUSTRATION 3.5 A tree sags under the weight of an ice storm in January 1994 in Jeffersonville, Montgomery County, Pennsylvania (courtesy of Marie Fee).

ILLUSTRATION 3.6 Ice coats a pine tree in Ogletown, New Castle County, Delaware after an ice storm in January 1994 (courtesy of Kathryn West).

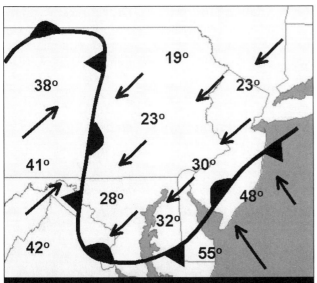

ILLUSTRATION 3.4 A sample "backward *S*" front snakes around colder air trapped east of the mountains and away from the coast. Modern, higher-resolution computer models simulate these cold-air damming situations better than older models, but they still often have difficulty resolving the details.

the frozen ground. Now we have an ice storm. In recent years, the winter of 1993–94 stands out for having several very disruptive ice storms that developed in this classic way (see Illustrations 3.5 and 3.6).

In an ice storm, weather conditions usually change over time. The coastal part of the front often makes progress northwestward, especially during the daylight hours. Temperatures at the ground rise quickly as the front passes, and the ice changes to rain. The last places to see that change are areas north and west of Philadelphia, where the extra hours of sleet or freezing rain lead to the greatest ice accumulations. Predicting the timing of the changeover is a big challenge. Because cold-air damming covers a relatively small geographical area, and occurs in only the very lowest levels of the atmosphere, most computer models do not handle it adequately. In such situations, a forecaster's experience and a sharp eye for subtle changes in the temperature and wind direction often produce the most reliable short-term forecasts.

Depending on the exact location and orientation of the high and of the approaching low, and the persistence of the cold-air damming, winds at the surface in advance of a storm can blow from the north, northeast, east, or even southeast. A north wind will block any ocean influence and keep cold air locked in at low levels. A northeast wind might allow some warming right near the coast, while an east wind will bring warmer air well inland. A strong east or southeast wind can cause a changeover to rain well west of Philadelphia. Without a doubt, the single most important weather variable to follow before and during a winter storm in the Delaware Valley is the wind direction.

WINTER COLD

Though the sun's heating power is on its lowest setting in the Delaware Valley on the first day of astronomical winter in late December, observations show that the lowest average temperatures occur about a month later, in late January. This delay in the time of deepest chill occurs because the continents, the oceans, and the air above them take time to cool down from the warmth of summer and fall. Late January, however, is just the average coldest time—there have been winters when the deepest chill occurred in mid-December, and other win-

ters when the coldest weather did not arrive until late February.

A conventional way to track winter-to-winter variability is to plot average winter temperature over the years, as in Illustration 3.7. The winter of 1976–77 was the coldest on record, with an average temperature of 28.0°F. At the other extreme, the warmest winter on record was 1931–32, with an average temperature of 43.3°F. To put the difference between these two extreme winters into daily terms: from December 1976 through February 1977, the temperature dropped to 20°F or below on fifty-four days, and on thirty-five days the temperature failed to get above 32°F. January 1977 was the coldest month on record, with ice clogging the Delaware River as can be seen in Illustration 3.8. In stark contrast, the lowest temperature from December 1931 through February 1932 was just 23°F, and the temperature went above 32°F every day. In many Philadelphia winters, however, there are generally enough cold days and nights to produce spectacular sights such as the iced-over fountain in Logan Square during the winter of 1924–25 (see Illustration 3.9).

It is clear from Illustration 3.7 that over the last two decades, there has been a general upswing in average winter temperature. Five winters from the 1990s rank in the top twenty warmest winters on record, while none of the twenty coldest winters on record occurred in the 1990s (or the 1980s for that matter). Whether these relatively mild winters in recent decades are just part of the climate's natural variability or a consequence of unnatural factors such as global warming, remains controversial.

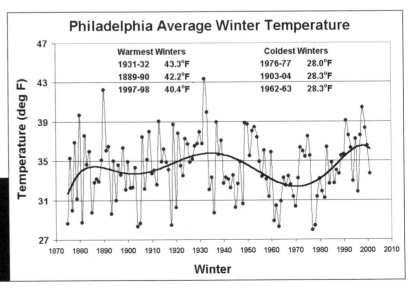

ILLUSTRATION 3.7 Philadelphia's average winter temperature, 1874–75 to 2000–1 (winter is defined as the period December through February). The thicker solid line is a smoothed version of the data intended to suggest longer-term trends.

ILLUSTRATION 3.8   A Coast Guard buoy-tender patrols the Delaware River on January 20, 1977. Ice on the river made the trip hazardous for smaller boats that were usually used by the Coast Guard (Urban Archives, Temple University, Philadelphia, Pennsylvania).

ILLUSTRATION 3.9 *(See plate.)*   The fountain at Logan Square in Philadelphia during the winter of 1924–25 (The Library Company of Philadelphia).

(For more, see Chapter 7.) Regardless of the reason, the numbers speak for themselves: winters overall in the last two decades have been relatively mild.

Even in relatively cold winters, temperatures rarely drop to 0°F in Philadelphia. Officially, there have been forty-five such days since 1874, an average of about one every three years. More than half of these bitterly cold days occurred in January, with most of the rest in February. The ten coldest days are listed in Table 3.1. The record lowest temperature occurred at 8:10 A.M. on the clear Friday morning of February 9, 1934. (February 1934 is the second coldest month on record, behind January 1977.) Illustration 3.10 shows a portion of the official U.S. Weather Bureau observation sheet for that record-setting day. As often occurs with wintertime Arctic air masses, a strong high-pressure system was in control—the pressure that day reached 30.75 inches, a level rarely attained in the Philadelphia area.

There has never been a day in the modern record when temperatures remained below zero all twenty-four hours in Philadelphia. Even on that record-cold day of February 9, 1934, the mercury recovered from −11°F in the morning to reach 10°F late that afternoon. However, there have been nine days since 1874 when single-digit temperatures prevailed all day long. A day such as January 17, 1982, when the official high temperature was 13°F, is also noteworthy. The high occurred at midnight, at the start of the day, but then behind a strong cold front, the temperature fell to −6°F by 8 A.M., was still −2°F at noon, and peaked at 0°F late in the afternoon. Undoubtedly, this was one of the coldest periods from sunrise to sunset in Philadelphia history.

The harshness of winter is compounded when the cold is persistent. In the Delaware Valley, the freezing mark is a reasonable choice for assessing the stubbornness of winter cold. Rarely does the temperature stay below freezing for ten days in a row—the longest streak, fifteen days, has occurred only twice, from late January into early February of 1961 and in February 1979 (see Illustration 3.11). Such persistent cold is intimately tied to the jet stream patterns that dominate the season.

TABLE 3.1   Coldest Days on Record in Philadelphia

Low Temperature (°F)	Date
−11	February 9, 1934
−7	January 17, 1982
−7	January 22, 1984
−6	February 10, 1899
−6	February 11, 1899
−6	January 21, 1985
−5	January 10, 1875
−5	December 30, 1880
−5	January 29, 1963
−5	January 19, 1994

Form No. 1014—Met'1

U. S. DEPARTMENT OF AGRICULTURE, WEATHER BUREAU.
DAILY LOCAL RECORD.

(Station) _Philadelphia, Pa._ , (Day of week) _Friday_ , (Date) _FEB 9 1934_ , 193_

Time (local standard) normal time.	VISIBILITY AT OBSERVATION (SCALE .. TO 0.) A. M.							7					7					P. M.					7						Maximum Temperature.	Minimum temperature.	Mean temperature.	Normal temperature.
	12-1	1-2	2-3	3-4	4-5	5-6	6-7	7-8	8-9	9-10	10-11	11-12	12-1	1-2	2-3	3-4	4-5	5-6	6-7	7-8	8-9	9-10	10-11	11-12		10	−11	0	33
Temperature at end of hour	−4	−5	−5	−6	−7	−6	−8	−9	−7	−2	−1	1	6	7	7	10	9	9	8	8	9	8	7	7					
Nature of precipitation and time of beginning and ending																									
Amounts of precipitation																									
Cloudiness and other conditions						CLEAR																			

8:10 a.m.

Total depth of snow for 24 hours ending __8__ p. m., __0__ inches.
Snow, sleet, and hail on ground __8__ p. m., __1.1__ inches.
Total snowfall, midnight to midnight, __0__ inches.
Precipitation to __8__ a. m., __0__ ins.; to __8__ p. m., __0__ ins.
Total precipitation, midnight to midnight, __0__ inches.
Character of day, _Clear_ Moon phase, _____
Mean daily cloudiness (0-10) __0__

ILLUSTRATION 3.10 The official record of the coldest day in modern Philadelphia, taken from the observation book of the U.S. Weather Bureau. Temperatures are shown at the end of each hour; the record of −11°F occurred between official observation times, at 8:10 A.M., as given in the summary of the day on the right (courtesy of National Weather Service).

Although bitterly cold air masses are constantly being manufactured during winter over far northern latitudes, few reach our area. To tap this big chill and deliver it to the eastern United States for a long stretch of time requires a persistent jet stream flow directly from northern Canada or, in extreme cases, the Arctic. The average wind pattern at around 30,000 feet in mid-January of 1961, in the heart of the record cold stretch that winter, is shown in Illustration 3.12a. The wind arrow passing through Pennsylvania can be traced back nearly to the North Pole. Compare this frigid flow to the relatively mild pattern on January 8, 1998, shown in Illustration 3.12b. The air reaching the Delaware Valley on that day originated over the Pacific and crossed the Gulf of Mexico, and the high temperature in Philadelphia was a record-breaking 69°F. The average temperatures during these two Januarys differed by more than 15°F, due in large part to the persistence of these contrasting upper-air flow patterns.

No discussion of winter temperatures would be complete without mentioning the "January thaw." Though rooted in New England folklore and far from a sure thing, the January thaw is more than just superstition. A slight warming in late January does show up in the data of many cities (particularly in the northeastern U.S.), including New York, Boston, and Philadelphia. Illustration 3.13 shows the average daily high temperature in Philadelphia centered near the time of the January thaw, using the last 100 years of data. There is clearly a bump of a few degrees in the period January 21–26; the low temperature data shows a similar pattern. However, no physical mechanism has ever been identified to explain why a thaw should occur at this time of year. With a careful statisti-

cal analysis, it is possible to prove that the January thaw can be just a random fluctuation—that is, a statistical fluke. Regardless, the January thaw is still somewhat of a meteorological oddity and is so ingrained in the weather psyche of the Northeast that it will undoubtedly remain part of winter lore for decades to come.

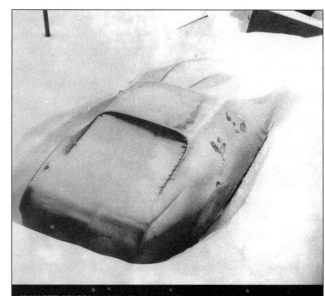

ILLUSTRATION 3.11 A storm dropped 13.9 inches of snow on Philadelphia at the end of a fifteen-day stretch of below-freezing weather in February 1979. The combination of persistent cold and snow left many cars buried, including this one in the Northeast section of Philadelphia (Urban Archives, Temple University, Philadelphia, Pennsylvania).

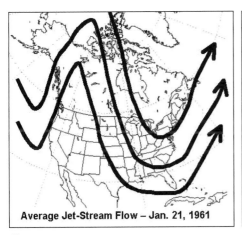

Average Jet-Stream Flow – Jan. 21, 1961

a

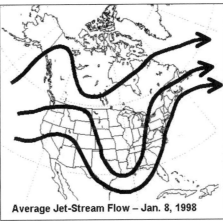

Average Jet-Stream Flow – Jan. 8, 1998

b

ILLUSTRATION 3.12 *(a)* The average jet-stream pattern on January 21, 1961 shows a sharp trough in the East and a pronounced ridge in the West. With this pattern, the air affecting Pennsylvania originated near the North Pole. *(b)* The average jet-stream pattern on January 8, 1998. In this pattern, mild air floods the East.

Wind Chill

During the sultry days of summer, a breeze—whether generated by nature or a fan—brings welcome relief. Air moving past the skin strips away a wafer-thin layer of warm air that normally sheaths the body. Wind also helps increase the evaporation of perspiration from the skin—it takes energy to evaporate water, and some of the energy comes from our body. The overall effect is that wind increases the rate of heat loss, so our skin feels cooler when air moves past.

In winter, the added cooling effect of the wind can cause discomfort that ranges from an annoyance to a serious health hazard. The effect of wind on comfort depends on the individual, varying with factors such as age, physical condition, and level of activity. Nonetheless, the cooling sensation felt on exposed skin as a result of the wind can be estimated by the **wind-chill temperature**, more commonly called the wind-chill factor or simply the wind chill.

The original wind-chill experiments were done in the Antarctic around 1940. Scientists exposed plastic cylinders filled with water of various temperatures to many combinations of wind speed and air temperature and recorded the time it took for the water to freeze. Wind-chill values based on these experiments were used in the United States and Canada for more than half a century. However, because these wind-chill temperatures lacked any physiological basis, there was widespread recognition (particularly in the last decade or so) that revised wind-chill values were needed.

A redesigned wind-chill index was introduced by the National Weather Service in the winter of 2001–2. The new index better incorporates a modern scientific understanding of heat transfer between skin and air in windy weather, backed up by clinical trials. The updated wind chill chart is shown in Illustration 3.14. For most combinations of wind and cold experienced in the Delaware Valley, the new wind-chill values are higher than the old ones by anywhere from 5 to 15°F. For example, in the old system a temperature of 20°F with a wind of 15 mph translated into a wind chill of −5°F; the same combination produces a wind chill of +6°F in the revamped index.

Despite the recent modification, the wind chill does not account for other meteorological factors that can affect the way you feel in cold weather, such as humidity and exposure to direct sunshine. But as a general rule, frostbite can occur on exposed skin in fifteen minutes or

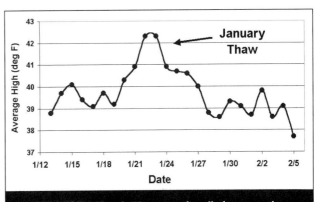

ILLUSTRATION 3.13 A "January thaw" shows up in Philadelphia high-temperature data, peaking on January 22–23.

New Wind Chill Chart
Wind (mph)

Calm	5	10	15	20	25	30	35	40	45	50	55	60
40	36	34	32	30	29	28	28	27	26	26	25	25
35	31	27	25	24	23	22	21	20	19	19	18	17
30	25	21	19	17	16	15	14	13	12	12	11	10
25	19	15	13	11	9	8	7	6	5	4	4	3
20	13	9	6	4	3	1	0	-1	-2	-3	-3	-4
15	7	3	0	-2	-4	-5	-7	-8	-9	-10	-11	-11
10	1	-4	-7	-9	-11	-12	-14	-15	-16	-17	-18	-19
5	-5	-10	-13	-15	-17	-19	-21	-22	-23	-24	-25	-26
0	-11	-16	-19	-22	-24	-26	-27	-29	-30	-31	-32	-33
-5	-16	-22	-26	-29	-31	-33	-34	-36	-37	-38	-39	-40
-10	-22	-28	-32	-35	-37	-39	-41	-43	-44	-45	-46	-48
-15	-28	-35	-39	-42	-44	-46	-48	-50	-51	-52	-54	-55
-20	-34	-41	-45	-48	-51	-53	-55	-57	-58	-60	-61	-62
-25	-40	-47	-51	-55	-58	-60	-62	-64	-65	-67	-68	-69
-30	-46	-53	-58	-61	-64	-67	-69	-71	-72	-74	-75	-76
-35	-52	-59	-64	-68	-71	-73	-76	-78	-79	-81	-82	-84
-40	-57	-66	-71	-74	-78	-80	-82	-84	-86	-88	-89	-91
-45	-63	-72	-77	-81	-84	-87	-89	-91	-93	-95	-97	-98

Temperature (°F) is the left axis.

Frostbite occurs in 15 minutes or less

ILLUSTRATION 3.14 Wind chill chart introduced by the National Weather Service in the winter of 2001-2. For the combinations of wind and cold in the darkened boxes, frostbite occurs in fifteen minutes or less on exposed skin (courtesy of National Weather Service).

sooner when wind-chill values are −20°F or below. So winter weather reports include low wind chills as a warning to the public that the combination of wind and cold is potentially dangerous. The local National Weather Service office issues a **Wind-Chill Advisory** when wind chills are expected to drop to −10°F, and a **Wind-Chill Warning** when wind chills are expected to drop to −25°F. (The threshold for the advisory is slightly lower, −15°F for counties well north and west of Philadelphia.)

Though records of extreme wind chill are not routinely kept, the lowest wind chills in the Philadelphia area have likely been in the −40°F to −50°F range. We reach this conclusion because the winds on the morning of the record lowest temperature of −11°F were steady, out of the north, at 20 to 25 mph, yielding a wind chill of about −40°F. (Using the old wind-chill values, this would have been about −55°F.) Although it is certainly possible that lower wind chills have occurred, this value is likely very close to the limit of winter's harshness around Philadelphia.

WINTER SNOW

In the general west-to-east jet-stream flow that steers weather systems across the United States are several favored storm tracks that can potentially bring snow to the Philadelphia area. One storm track begins on the eastern slopes of the Canadian Rockies, near the province of Alberta. The fast moving, moisture-starved storms that follow this track are called **Alberta Clippers**. Their speedy movement, their path over land, and their relatively dry region of origin mean that clippers usually bring just light accumulations of snow to the Delaware Valley. Only if the center of the storm passes nearby to the south would the accumulation exceed an inch or two. If a clipper passes well to the north, we may see only a few snow or rain showers with the passage of the storm's cold front. In some cases, there may be no precipitation at all—just a blast of cold air in the northwesterly flow behind the storm.

Big snowstorms in the Delaware Valley almost always originate along the Gulf or Southeast coasts and then move northeastward. On that track, easterly winds north of a storm can tap moist air over the relatively warm ocean, and the potential for a big snow greatly increases. That is when East Coast forecasters start using the terms "nor'easter" and, in extreme cases, even "blizzard."

Nor'easter is a general term for a storm that moves up the East Coast bringing heavy snow, ice, or rain (or maybe all three), and often lots of wind as well. Nor'easters are most common from November to April. The name comes from the strong northeast winds that blow ahead of the storm. Benjamin Franklin is credited with recognizing that these storms generally approach from the southwest, even though the winds preceding them blow from the opposite direction. A nor'easter's gale-force winds sometimes approach minimal hurricane strength and often inflict severe and widespread beach erosion and coastal flooding, even if the storm does not bring much snow. (See Chapter 4 for more about this.)

The most severe nor'easters may qualify as blizzards. Though sometimes the term is used anytime it snows and the wind blows hard, a **blizzard**, by definition, must have at least three consecutive hours with winds of 35 mph or higher, and falling or blowing snow reducing the visibility to less than one-quarter of a mile. Extreme cold is not necessary, though it is often part of the package. True blizzards are very rare in the Philadelphia area.

When a significant winter storm approaches, the local National Weather Service office in Mount Holly, New Jersey issues a Winter Storm Watch or Warning. A **Win-**

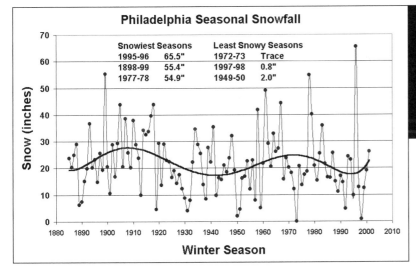

ILLUSTRATION 3.15 Philadelphia's seasonal snowfall, 1884–85 to 2000–1 (a season's snowfall includes October through April). Note the extreme variability from winter to winter. The solid line is a mathematically smoothed version of the data. The long-term average seasonal snowfall is about 20 inches.

ter **Storm Watch** means that heavy snow or ice is possible in the next twelve to thirty-six hours. A **Winter Storm Warning** implies a more immediate and certain threat, indicating that heavy snow or ice is expected within the next six to eighteen hours. The criteria for watches and warnings depend somewhat on location. Five inches of snow is the threshold for the NWS to issue a watch or warning in Delaware, south Jersey, eastern Maryland, and Delaware and Philadelphia counties in Pennsylvania. Farther north, in the rest of New Jersey and eastern Pennsylvania, the threshold is six inches. For ice, the criterion is the same for all counties: if one-quarter inch or more is anticipated, the NWS issues a watch or warning.

In the last several years, Philadelphia-area residents have lived through the snowiest winter on record (65.5 inches, 1995–96) and the second least snowy (0.8 inches, 1997–98), a timely reminder of how variable winters can be from year to year (see Illustration 3.15). It is not unusual for a snowblower to get a workout the first winter you own it, but then sit relatively idle in the garage the next few years.

Table 3.2 lists the five snowiest and five least snowy winter seasons (totals include all snow from October through April). The average snowfall in Philadelphia is about 20 inches, but in one out of every three winters, snowfall has been more than 10 inches from this average. There is no dramatic long-term trend, though, with the notable exception of 1995–96, winter snowfall has generally been close to or below the average since around 1980. It is reasonable to link this tendency for less snow to the relatively mild winters of the last few decades.

Snow and cold are obviously related in that one (snow) requires the other (cold), but there is also a subtle relationship that works the other way. Snow reflects a large percentage of sunlight, explaining why you need sunglasses on a clear day following a big snow. The more sunlight that is reflected back to space, the less the ground and the air heat up. As a result, days with a snow cover tend to be a few degrees cooler than if there were no snow on the ground. With daytime highs suppressed a bit by the snow cover, nighttime temperatures also tend to be lower. In addition, snow is an efficient emitter of

TABLE 3.2 Snowiest and Least Snowy Winters in Philadelphia (in Inches)

Most Snow in a Season (October–April)	
1995–96	65.5
1898–99	55.4
1977–78	54.9
1960–61	49.1
1966–67	44.3
Least Snow in a Season (October–April)	
1972–73	Trace
1997–98	0.8
1949–50	2.0
1930–31	4.1
1918–19	4.5

infrared radiation, so it rapidly loses what little heat it has at night. Snow is also a great insulator because it contains lots of air. As a result, a snow cover tends to prevent heat from the ground from warming the air. The overall effect is that a snow cover helps to preserve itself by feeding back chill to the air, which in turn helps keep the snow from melting.

A snow cover offers obvious recreational opportunities, but it also increases the potential disruptiveness of the winter. Snow and ice that melt during the day often refreeze at night, making sidewalks and roads slippery with "black ice." The average number of days each winter with at least an inch of snow on the ground in Philadelphia is between fifteen and twenty. But just as with snowfall, the range of variation from year to year is huge, from forty-seven days during the winters of 1960–61 and 1995–96 to none during the winters of 1972–73 and 1997–98. Rarely does a snow cover of an inch or more persist longer than a few weeks. The longest stretch on record is thirty consecutive days in January and February of 1905, and there are a few other winters when a snow cover lasted about four weeks. This was not the case following the record-setting blizzard of January 7–8, 1996. Though 16 inches of snow remained on the ground ten days after that storm, another powerful low-pressure system brought enough rain and warm air to melt the rest by January 20, leading to devastating flooding in many areas of the Delaware Valley (see Chapter 4).

Many people have heard the rule of thumb that ten inches of snow melts down into about one inch of water—in other words, snow lovers can dream of an inch of rain as a potential ten-inch snow. This simple rule is a good first estimate, but the exact ratio depends on the air temperature during the snow. The ten-to-one rule works best when temperatures are within a few degrees of 30°F. When the air is slightly warmer, say a few degrees above freezing, snow is often "wet," or water-laden, and the ratio might be closer to five to one. In contrast, at lower temperatures (say, in the 15–20°F range) snow tends to be light and fluffy and the ratio can be closer to twenty to one. That was the case across much of the area during the Blizzard of '96 when temperatures hovered near 20°F during the storm.

Snowfall rates and durations vary, and forecasters use different terms to describe the different types of snowfalls. Steady snows that last for at least a few hours are categorized as light, moderate, or heavy, depending on the horizontal visibility. It is **light snow** if visibility is more than one-half of a mile, **moderate snow** (or just simply "snow") if visibility is between one-quarter and one-half of a mile, and **heavy snow** if visibility is one-quarter of a mile or less. Of the shorter-duration snows that come and go, a **snow squall** is a heavy, brief burst of snow that can put down a quick inch or two. A **snow shower** is a moderate, but again brief, period of snow, which sometimes leaves a light accumulation—perhaps a coating. **Snow flurries** describe a very light and brief snow shower, with no measurable accumulation. Note the common word "brief" in these definitions: squall, shower, and flurry all describe snows that do not last long, on the order of tens of minutes.

The term "snow squall" is rarely heard in forecasts for the Delaware Valley, but it is used frequently to describe the heavy snows that are common around the Great Lakes in late fall and early winter. These **lake-effect snows** form when cold air from Canada flows across the waters of the Great Lakes, which are still relatively warm at that time of year. The air warms, moistens, and rises, often dumping bands of heavy snow on areas just to the south and east of the Lakes, especially where the air coming off the water is given added lift by the terrain (see Illustration 3.16). The Poconos will occasionally get a few inches of lake-effect snow (courtesy of Lakes Erie and Ontario). But a northwest wind downslopes off the Appalachian Mountains into the Delaware Valley, and sinking air tends to dissipate clouds. As a result, it is unusual to see any more than some clouds and a few snow flurries in the Philadelphia area from the lake effect.

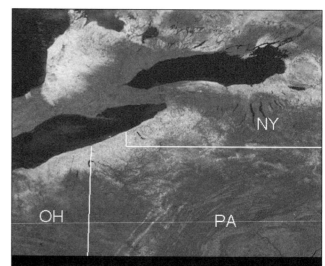

ILLUSTRATION 3.16   A visible satellite image from mid-November shows lake-effect snow on the ground to the south and east of Lakes Erie, Ontario, and Huron (courtesy of NOAA).

ILLUSTRATION 3.17 *(See plate.)* Columbia Avenue in Philadelphia after the 21-inch Christmas Day snowstorm in 1909, the city's largest December snowstorm on record (Library Company of Philadelphia).

two winters, 1957–58 and 1977–78 (see Illustration 3.19), have two. The winter of 1977–78 also is notable for its brutal cold. Coming on the heels of the record setting 1976–77 season, it seemed to indicate a trend. These winters, along with several other very cold winters in the 1960s, had people talking (and some scientists writing) about the coming of the next ice age. The last few decades have certainly silenced that talk.

The blizzard of January 7–8, 1996, with 30.7 inches, leads the list of great Philadelphia snowstorms. This storm was exceptional, not only for the amount of snow that fell but also for the extensive area that received a crippling snow. Parts of nine states, in a swath from southwestern Virginia to Massachusetts, had two feet or more. Illustration 3.20 shows snowfall amounts from the storm. As is typical with East Coast storms, the largest amounts occurred in the higher elevations, with amounts lessening toward the coast.

The topic of snow always raises the question of a white Christmas. How common it is depends on your definition. Although snow has fallen in Philadelphia on Christmas Day in about one out of every four years, about two-thirds of those sightings amounted to only a trace—that is, not even enough to be measured. If a "white Christmas" is defined as at least an inch of snow that falls on Christmas Day, the frequency lessens considerably, to only about one out of every twenty-five years. On the other hand, if we loosen the criteria so that simply having an inch of snow on the ground qualifies, the frequency is about once every eight years. Whatever the definition, Christmas Day snow in Philadelphia is not very common. The city's most spectacular Christmas snow occurred in 1909 with a storm that brought 21 inches, still the third biggest snowstorm on record (see Illustration 3.17). The U.S. Weather Bureau observer that day noted that "moist snow began at 9:10 A.M. The snowfall was a steady, heavy one . . . increasing in amount to 3 A.M. [on December 26]. . . . In places, snow drifts were four to five feet high." Now that's a white Christmas!

Historic Snowstorms of the Modern Record

The Christmas 1909 snow is one of twenty-four snowstorms of 10 inches or more that have occurred in Philadelphia in the period 1884 to 2001 (see the list in Table 3.3). Almost half of them (eleven of twenty-four) occurred in February, with six in January and three each in December and March. Only the winter of 1960–61 has three snowstorms on the list (see Illustration 3.18), while

TABLE 3.3 Biggest Snowstorms (10 Inches or More) in Philadelphia

Date	Total Snow
Jan. 7–8, 1996	30.7 inches
Feb. 11–12, 1983	21.3
Dec. 25–26, 1909	21.0
April 3–4, 1915	19.4
Feb. 12–14, 1899	18.9
Jan. 22–24, 1935	16.7
Feb. 28–March 1, 1941	15.1
Dec. 11–12, 1960	14.6
Feb. 18–19, 1979	14.3
Feb. 5–7, 1978	14.1
Jan. 19–20, 1961	13.2
Jan. 19–20, 1978	13.2
Feb. 15–16, 1958	13.0
Dec. 24–25, 1966	12.7
Feb. 4–5, 1906	12.7
Jan. 28–29, 1922	12.3
March 13–14, 1993	12.0
March 19–21, 1958	11.4
Feb. 17–18, 1900	11.2
Feb. 17, 1902	11.0
Feb. 20–21, 1947	10.6
March 12, 1888	10.5
Jan. 28–29, 1928	10.4
Feb. 3–4, 1961	10.3

ILLUSTRATION 3.18 Thirtieth Street Station shrouded in snow on February 4, 1961, a common scene that winter. There were three snowstorms of ten inches or more between December 11, 1960 and February 5, 1961 (Print and Picture Collection, Free Library of Philadelphia).

board). But because it was so windy during the Blizzard of '96, a different procedure was used at the Philadelphia International Airport. Each hour, the snow that had collected in the rain gauge was melted, and then the snowfall for that hour was estimated based on the air temperature and the amount of meltwater. This alternative procedure is also commonly used to estimate snowfall, but standard guidelines call for taking measurements every six hours—taking them every hour tends to overestimate the snow. Despite the measurement inconsistencies at the airport, other nearby locations reported roughly similar amounts using the snowboard method— for example, 28 inches was measured downtown at The Franklin Institute. So although some uncertainty surrounds the exact amount of snow that fell, the Blizzard of '96 was surely the biggest snowstorm in the official recorded history of Philadelphia.

One other note about the Blizzard of '96: based on the strict definition, the storm was not a blizzard at all, because no official reporting station reached the wind criterion. However, some places that are not official observing sites were undoubtedly subjected to true blizzard conditions. And regardless of what it is called, the storm will long be remembered as a nor'easter of historic proportions.

Prior to the January '96 storm, the biggest snowstorm on record in Philadelphia occurred on February 11–12, 1983. That storm dumped 21.3 inches of snow on Philadelphia, and was nicknamed the "Megalopolitan

(text continues on page 73)

Years after the storm, there is still some controversy surrounding the record-setting total of 30.7 inches at Philadelphia, because of the way the snow was measured. Standard National Weather Service procedure requires that snow be measured every six hours on a flat surface, preferably a sheet of wood painted white (called a snow-

ILLUSTRATION 3.19 A common scene in Philadelphia during the brutal winter of 1977–78; here, cars buried by snow on February 7, 1978 in South Philadelphia (Urban Archives, Temple University, Philadelphia, Pennsylvania).

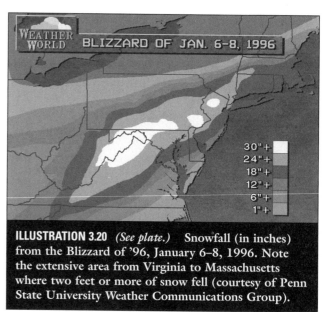

ILLUSTRATION 3.20 *(See plate.)* Snowfall (in inches) from the Blizzard of '96, January 6–8, 1996. Note the extensive area from Virginia to Massachusetts where two feet or more of snow fell (courtesy of Penn State University Weather Communications Group).

January 1996 and March 2001—A Hit and a Miss

These behind-the-scenes stories reflect Glenn's best recollections of how events unfolded at his station, NBC-10. As one of the station's meteorologists, he was, of course, an active participant in the events, and his accounts are intended to give readers a sense of the excitement and stress that accompany weather reporting when a major storm is predicted. Other local stations undoubtedly conducted similar discussions about the storms and how to focus their coverage, but we can only tell the stories we know first-hand.

The Blizzard of January 1996—The Hit

Predicting a major winter storm more than a day or two in advance was almost unheard of a few decades ago. Any meteorologist who tried to do so was criticized immediately as irresponsible. But with the increased accuracy of computer model predictions in the three- to five-day period, public forecasters have become bolder. The March 1993 "Storm of the Century" represents one of the most dramatic successes; meteorologists mentioned the high probability of a major East Coast snowstorm in long-range discussions and forecasts at least five days before the storm arrived. Computer model predictions were very consistent in the days leading up to that storm, giving forecasters a high degree of confidence. The forecasts of the March 1993 storm were so accurate that even the most skeptical forecasters were persuaded that it was at least possible to make accurate forecasts of some big storms several days in advance. Another successful prediction of a huge East Coast snowstorm would come just a few years later, in January 1996.

In the days prior to that storm, Arctic air was approaching the Delaware Valley. A half-inch of chilly rain fell on Tuesday, January 2, and all computer forecasts showed very cold air would follow. High temperatures in the 20s were predicted for Thursday, Friday, and Saturday, with lows in the teens (the actual highs for those days were 24°F, 28°F, and 20°F). While not a record-breaking chill, such cold air gave forecasters confidence that the bulk of the upcoming storm would be snow. On Thursday, January 4, snow was forecast for Sunday, January 7. Predictions of snow that far in advance usually do not draw much attention, but at the National Weather Service, long-range discussions focused on the high probability of a big snowstorm for the mid-Atlantic and Northeast states for the latter half of the upcoming weekend (see Illustration 3.21).

Around the time the NWS issued those discussions, all media outlets in those areas undoubtedly started considering the implications of a big storm. In such cases, especially with a weekend storm, contingency plans need to be discussed well in advance. What about scheduling extra people for the weekend? Reserving hotel rooms? Altering the on-air schedule to add special newscasts to cover the storm? At what point do contingencies become necessities? The answers differ, or come at different times, at the various television stations. At NBC-10, News Director Steve Doerr wanted to know what the "worst case" would be. When the reply was "blizzard," he began looking at weather staffing for the storm. The extended television coverage of a major storm requires as large a weather staff as possible in order to keep up with possible changes in the forecast and to produce the necessary on-air graphics. Glenn was working the morning shift that week, with veteran Herb Clarke doing all the evening shows in place of John Bolaris, who was vacationing in Cancun, Mexico. The decision was made that, if the risk of a big storm was still high the next day, the station would call Bolaris.

When talking about a big storm three days in advance, forecasters use a different strategy than they do when the storm is expected within the next day or so. Over three days, the computer models can change their solutions dramatically, so forecasters try not to commit themselves fully to the "big-storm" idea. While the word "blizzard" was used among station employees, it was not mentioned on air. Once that word is used, there is no way to take it back. Even if the word is dropped the next day as different data comes in, viewers will still remember the *B* word. Instead, the strategy was to highlight the potential for a major storm in the weathercasts and explain the reasons behind that potential.

The case for a big snowstorm changed suddenly early Friday morning. Instead of predicting that the storm would move up along the coast, some of the computer models showed the storm and the heaviest precipitation remaining well south of the Philadelphia

ILLUSTRATION 3.21
Extended forecast discussions from the National Centers for Environmental Prediction (NCEP, an arm of the National Weather Service) and from the local National Weather Service office in Mount Holly three days before the Blizzard of '96 highlight the potential for a major East Coast snowstorm. The NCEP discussion focuses on the agreement among computer models and mentions that the storm may actually be "deeper" (stronger) than predicted by the models. The "Winter Storm Outlook" from the Mount Holly office provides local details for residents of southeastern Pennsylvania, New Jersey, Delaware, and Maryland.

```
EXTENDED FORECAST DISCUSSION FOR 07 JAN THRU 09 JAN 1996
HYDROMETEOROLOGICAL PREDICTION CENTER...NCEP...NWS...WASHINGTON DC
1:15 PM EST THU 04 JAN 1996

...MID ATLANTIC NOR'EASTER AND SNOWSTORM FOR THE WEEKEND... THE MEDIUM RANGE
MODELS ALL CONT TO INDICATE MAJOR EAST COAST CYCLOGENESIS DURING THE FORE-
CAST PERIOD TODAY ...

MANUAL PROGS HAVE ACTUALLY DEEPENED THE SFC PRESSURES OF THE LOW SOMEWHAT
AND THERE IS THE DISTINCT POSSIBILITY THAT THIS LOW WILL BE DEEPER THAN THE
MODELS. VERY COLD AIR IN PLACE OVER THE WARM GULF STREAM WILL YIELD A RIPE
UNSTABLE ENVIRONMENT ... THIS WILL BE THE FIRST SIGNIFICANT SNOWFALL FOR THE
MID ATLC REGION THIS SEASON WITH THE HEAVIEST AMOUNTS ALONG THE COASTAL
PLAIN OF DE/MD/VA.

-------------------------------------------------------------------------


SPECIAL WEATHER STATEMENT - WINTER STORM OUTLOOK
NATIONAL WEATHER SERVICE PHILADELPHIA/MOUNT HOLLY NJ
230 PM EST THU JAN 4 1996

LONG RANGE COMPUTER MODELS SHOW THE POTENTIAL FOR A SNOW STORM FOR THE
PERIOD BEGINNING SATURDAY NIGHT AND ENDING SUNDAY NIGHT.  LOW PRESSURE IS
EXPECTED TO DEVELOP OVER THE SOUTHEAST STATES SATURDAY NIGHT AND THEN INTEN-
SIFY AS IT MOVES NORTHEAST INTO THE ATLANTIC. THE LOW IS FORECAST TO STAY
WELL OFFSHORE...BUT CLOSE ENOUGH TO BRING SNOW TO EASTERN PENNSYLVANIA...NEW
JERSEY...DELAWARE AND THE EASTERN SHORE OF MARYLAND.

THIS POTENTIAL STORM EVEN HAS THE POSSIBILITY OF DROPPING SNOW ALONG COASTAL
REGIONS. SINCE THE LOW SHOULD STAY OUT TO SEA...WARMER ATLANTIC AIR SHOULD
NOT MAKE IT INLAND.

IT IS TOO SOON TO TRY TO FORECAST SNOW AMOUNTS. THE EVENTUAL TRACK OF THE
LOW WILL DETERMINE ANY SNOW TOTALS. HOWEVER...IF THE STORM TRACK IS NOT TOO
FAR OUT TO SEA...SNOWFALL TOTALS COULD BE SIGNIFICANT. STAY TUNED TO YOUR
NOAA WEATHER RADIO OR YOUR LOCAL RADIO/TV STATION FOR LATER UPDATES ON THE
POTENTAL SNOW EVENT.
```

area. Now the question was: "Do we back off the current forecast, or wait until the next computer run comes in before making a change?" This is always a tough call. There have been many instances of the computer models "changing their minds" in one cycle and then going back to the original solution in the next. In many cases, a forecaster's decision comes down to whether there are solid meteorological reasons to change the forecast. Glenn decided to stick with the "big storm" idea, since only the details of the computer model forecast and not the overall weather pattern (which was still favorable for a powerful East Coast storm) had changed. But different forecasters can come to different conclusions from the same computer model guidance. That is exactly what happened in this situation. Instead of agreement among nearly all forecast outlets, some speculated that the new southern track might spare the Delaware Valley from "the big one."

New computer guidance comes in every twelve hours, and the next round would be crucial for deciding which way to go with the forecast. The computer models that started coming in around 11 A.M. EST on Friday suggested that the storm would take the original, more threatening track. It was time to move on staffing issues. NBC-10 called John Bolaris in Cancun, and Glenn read the computer forecast information to him on the phone. John is typical of many "lifers," meteorologists who have had a passion for weather since childhood and whose excitement about a big storm never fades. John cut his vacation short and decided to return to Philadelphia. There are plenty of meteorologists across the country who would make the same decision rather than miss a potentially historic storm.

John recalls: "Flights were already being canceled, since the airlines feared their planes would get stuck in Philly, even though at the time the snow was still some eighteen hours away. My flight from Dallas to Philly got canceled. I was pleading and begging with the ticket agent when a pilot overheard us. He had remembered me from my days doing TV weather in New York City, so he offered to put me on his flight to Newark. I arrived around 1 A.M. Sunday with my luggage still in Dallas, rented a car and drove toward Philadelphia. The snow started to fall on the way down the New Jersey Turnpike. After a quick stop at home to pick up clothes, I arrived in the studio by 8 A.M. and was on the air by 8:30 A.M."

While John was attempting to get back to Philadelphia, it became obvious that a huge snowstorm was coming. By noon on Saturday, the two primary short-range computer models (called the Eta and the NGM) were forecasting 2.16 inches and 1.33 inches, respectively, of liquid precipitation in Philadelphia by 7 A.M. on Monday morning, with temperatures low enough that it would all fall as snow (see Illustration 3.22). Because the air was so cold, the snow would be fluffy, and the ratio of liquid to snow would probably be close to fifteen to one. It was hard to believe, but if the Eta was right, that meant about 30 inches of snow, an amount made even more remarkable given that Philadelphia had never recorded a storm with more than 21.3 inches. On air, there were confident predictions of more than a foot. By Saturday evening, a new computer run had confirmed these incredible amounts, increasing the level of confidence. On the 6 P.M. news, Glenn said: "Even though not a single flake has fallen yet, there is no doubt that we're in for a crippling snowstorm tomorrow."

The station had a dilemma, however. The normal Sunday news schedule was going to be altered by the NFL playoff games. The Eagles/Dallas game would be aired on FOX-29 from 1 to 4 P.M., and NBC-10 would air the AFC playoff game from 4 to 7 P.M., preempting the early evening news. That meant NBC-10 would have no newscasts between 10:30 A.M. (when the morning newscasts ended) and 11 P.M. on the day of what could turn out to be the biggest

snowstorm of the century. Because of the confidence in the blizzard forecast, NBC-10 scheduled a special weather update during halftime of the Eagles game. Although the game would be on another station, there was hope that some of that huge audience would want to know about the storm. Other stations had similar strategies, so there were at least three stations airing special storm updates at halftime. NBC-10 also scheduled a special newscast from 7 to 8 P.M. after the AFC playoff game.

As it turned out, the snow started early Sunday morning, resulting in a change in programming at many stations. For example, the usual schedule at NBC-10 was local news from 6 to 8 A.M., the *Today Show* from 8 to 9 A.M., and then more local news from 9 to 10:30 A.M. Because of the snow, the *Today Show* was preempted (not an easy decision when the network owns the local station). With no break between 8 and 9 A.M., the anchors were left in their chairs for hours. The technical staff, from producers to directors to the studio crews, also continued without a break. There were no scripts. Weather updates were followed by live reports on the storm, followed by interviews. Since it was a Sunday, and the storm was such a threat, the audience for the storm coverage was huge, and the 7 to 8 P.M. special show was one of the highest-rated newscasts in NBC-10's history. John Bolaris's return was of major importance for our coverage, since one person could not possibly work every minute of the storm.

All told, the Blizzard of '96 brought the Delaware Valley to a standstill (see Illustration 3.23). Schools canceled classes for a week. So much snow fell that

```
OUTPUT FROM ETA 12Z JAN 06 96
TTPTTR1R2R3 VVVLI PSDDFF HHT1T3T5
PHL//594548 -3118 343519 19878891
06000803548 -3819 313511 22898991
12000804353 -3423 343511 24909091
18000686657 -1020 290314 27899092
24006848059 01519 260521 31909193
30030968761 09117 180527 36949395
36048979160 10316 110533 39979595
42055989241 10613 010539 40979595
48077988831 09011 980436 32979596

OUTPUT FROM NGM 12Z JAN 06 96
TTPTTR1R2R3 VVVLI PSDDFF HHT1T3T5
PHL//523940 -3019 333519 17888989
06000353772 -2720 333409 21939091
12000345295 -0321 330211 25959091
18000327599 01022 320612 25949090
24000549292 02819 300511 26939092
30002789488 03016 250713 31959195
36042969999 09418 190623 36969197
42045969982 08014 090629 41999399
48044969111 09212 990530 38009599
```

ILLUSTRATION 3.22 Numerical weather forecasts from two different computer models (the "Eta" and the "NGM") produced the morning of January 6, 1996, about twenty-four hours before the snow started. The data has a lot of detail, but notice the numbers that are bolded and underlined in the columns headed by "PTT." These are precipitation forecasts—adding the amounts in each model yields 216, or 2.16 inches in the Eta and 133, or 1.33 inches in the NGM. Considering the dryness of the expected all-snow event, this would imply twenty to thirty inches of snow. For complete instructions on how to decode this forecast data, see Web Sources in the Bibliography.

ILLUSTRATION 3.23 *(See plate.)*   Snow from the Blizzard of '96 piles up in the 2600 block of South Percy Street in Philadelphia (courtesy of Joseph F. Iacuzio).

some had to be dumped in the Schuylkill River, though that practice stopped when the snow began to dam up (see Illustration 3.24). If a storm like that had somehow been a surprise or occurred during the week, the havoc it wreaked would have been much worse. There is no doubt that accurate, consistent forecasts helped reduce the disruptiveness of this historic weather event.

The Storm of March 2001—The Miss

The impressive accuracy of the computer predictions of the March 1993 and January 1996 snowstorms could have inflated meteorologists' confidence and the public's expectations that major East Coast storms could be foreseen many days in advance. During the winter of 2000–1, however, none of the computer models had been particularly accurate in forecasting snow in the eastern United States. The models had forecasted a few nor'easters to dump heavy snow on Delaware, Maryland, and Virginia, but the storms actually intensified farther south, sparing those areas. Other storms that were predicted to stay to the south developed farther north, spreading snow to areas in New England that were supposed to stay dry. In the Philadelphia area, a morning rain on February 5, 2001 had quickly changed to heavy snow around noon, a few hours earlier than most forecasters expected, creating a traffic nightmare. Thousands of cars were stranded for hours during the afternoon and evening on icy highways, preventing snowplows from treating the roads.

This incident heightened meteorologists' awareness of what can happen when hazardous weather is poorly predicted.

The March 1993 and January 1996 blizzards, the January 2000 "surprise" nor'easter (discussed in Chapter 2), and the February 5, 2001 rain-to-snow storm had no direct bearing on how any future storm would turn out. But those events had affected the meteorologists involved in subtle ways. In the practice of an inexact science, staffed by humans, nonscientific factors such as overconfidence or concern about a repeat of a recent poor prediction can affect the judgment calls that go into forecasts and the decisions that result from them. In March 2001, these factors contributed to a forecasting embarrassment from Virginia to Maine involving a major East Coast storm.

Signs of a potentially significant nor'easter began to take shape on Monday, February 26, and Tuesday, February 27. Long-range computer models from the U.S. Navy and from the European Centre for Medium-Range Forecasting (the "European model") were showing a strong low-pressure system just off the East Coast for the upcoming weekend. (The European model is generally the most highly regarded long-range computer model.) A model from the Canadian Meteorological Centre (the "Canadian model") was suggesting a similar solution, although the primary U.S. long-range computer model (known as the MRF, for Medium-Range Forecast model) had been inconsistent from day to day about the potential for a big storm. On Wednesday evening at 10:45,

ILLUSTRATION 3.24   Trucks dump loads of snow into the Schuylkill River on Monday, January 8, 1996 (AP/Wide World Photos).

after seeing the latest computer guidance from the European model, NBC-10 ran a scrolling message, or "crawl," mentioning the "possibility of one of the biggest storms of the past decade." Using a crawl to highlight a weather event several days away is highly unusual, though crawls had been used in the past to promote a big story in the upcoming 11 P.M. news, both for news-related information and for weather. (This crawl would be the main focus of later criticism of the storm coverage.)

On Thursday afternoon, with the computer models still not all in agreement, the forecast discussion from the National Weather Service highlighted a "potentially serious East Coast storm." On Friday morning, the new MRF came out with the "big storm" solution too. Some of the computer models were suggesting an extremely intense and slow-moving storm, one that could cause severe damage at the shore. That afternoon, the forecast discussion from the NWS highlighted a possible "historic blizzard" and compared the situation to some of the worst snowstorms in East Coast history. The local NWS office in Mount Holly, New Jersey issued a Winter Storm Watch for the entire area for the possibility of "one foot or more" of snow. All local television stations were now mentioning the possibility of a major snowstorm.

Just as it did in January 1996, Glenn's station began planning for extensive storm coverage for the upcoming weekend. (Other stations were probably doing so too.) Plans were made for food, hotels, and extra staffing, including special coverage from the shore. The storm was predicted to hit on or near the anniversary of the most destructive nor'easter on record at the Jersey and Delaware shores, the March 1962 nor'easter (see Chapter 4). Also like that one, this storm was expected to last for days, with the potential for increasing water levels with each high tide.

By Saturday morning, the potential storm was within the forecast realm of the short-range computer models, the ones that only go out a few days into the future. One of those models (called the AVN) suggested one to two feet of snow throughout the area. Almost all computer models were consistent in predicting a big storm: 6 to 10 inches of snow for most of the area by Monday morning, not even halfway into the storm! Furthermore, the computer models were suggesting that this storm would move in an unusual way, first heading to the northeast but then looping back briefly to the southwest, prolonging the storm on land. Only the Canadian

model was significantly different, showing the storm moving out to sea more quickly. When many of the computer models produce the same solution regarding a storm's location, duration, and precipitation, forecaster confidence increases. On Saturday morning, the "possibility" of a big storm now seemed more of a "strong likelihood."

Television stations in the Philadelphia area had been leading their newscasts with the big storm potential since Friday, if not earlier. Video of the last big blizzard in January 1996 was aired. Live shots from grocery stores showed the usual run on milk, eggs, and bread. Other live shots from hardware stores highlighted the demand for shovels and salt (many stores were already featuring spring supplies and quickly ran out of salt). People were changing travel plans for Sunday and Monday in anticipation of the storm. Talk of the coming blizzard was everywhere. The Weather Channel's winter weather expert, Paul Kocin, who wrote the definitive book on major East Coast snowstorms, specifically mentioned on air the potential for one to two feet of snow or more in Philadelphia.

Then things changed with the computer model forecast that arrived around noontime on Saturday. The new computer guidance was still indicating a lot of precipitation, but one of the models (the Eta) was showing a layer of relatively warm air at about 4,000 to 5,000 feet above ground that could change the snow to sleet or rain. Was this the first sign of a lesser snowstorm, or just a one-time model fluctuation? Now conflicting information and some doubt had complicated the picture, causing some forecasters to express less confidence in the blizzard scenario.

By early Sunday morning, it was becoming clearer that the warmer air that the Eta predicted would invade much of the area. But how far north and west would it reach? How much snow would fall before the changeover? When would the rain or mixed precipitation change back to snow (the models still suggested this would happen)? Now, instead of just an all-snow storm, it was looking like a three-part storm—Phase 1 would be snow on Sunday, Phase 2 would be a changeover Sunday night to sleet, freezing rain, or rain, and Phase 3 would be a change back to snow on Monday with the potential then for more accumulation. Now forecasts from one media outlet to another were varying a great deal—some lowered predicted snow amounts dramatically, while others concentrated on how much snow would fall in Phase 3. A long-lasting storm might still produce a foot of snow, even with a twelve- to eighteen-hour break in the snowfall. And with the change back to snow expected on Monday, forecasters wondered if there could be a repeat of the February 5 storm, when morning rain turned to snow, creating travel chaos in the afternoon.

By Sunday afternoon new computer guidance still suggested the potential for a lot of snow at the end of the storm. So the National Weather Service issued a Winter Storm Warning for parts of the region not covered in the warning already in effect for counties north and west of Philadelphia. While some area television stations kept their forecasts of much lower snowfall totals, NBC-10 agreed with the NWS assessment of the potentially dangerous Phase 3 and kept amounts high. Snow had been falling through much of the day Sunday, accumulating several inches in the Philadelphia suburbs, but only 1 inch officially at Philadelphia International Airport. The change to a mix of sleet, freezing rain, and rain occurred Sunday evening, and the question of when the mix would change back to all snow was making forecasts for Monday extremely difficult.

As it turned out, the change back to snow did occur on Monday, but the snow was not heavy enough to cause major traffic problems or much accumulation. The strong March sun can warm the ground even when there is a thick cloud cover, so the snow needs to be very heavy to accumulate during daytime, even with temperatures below freezing. The storm intensified farther north and east than predicted, and the looping back toward the southwest was not as extreme as expected, so the heaviest precipitation fell well north and east of the Philadelphia area (see Illustration 3.25). The heavy snow expected at the end of the storm reached back only to Long Island, New York. There, forecasters had drastically lowered snow forecasts, only to see 12 to 18 inches accumulate. But the Philadelphia area experienced only periods of mainly light snow through early Wednesday morning. The official total for the storm would be only 1 inch at the Philadelphia International Airport—the inch that fell on Sunday—with as much as 3 to 5 inches in parts of Chester, Bucks, and Montgomery counties.

In response to the forecast of a foot or more of snow, some airports had closed down or curtailed operations (airlines do not want their planes to be stuck on the ground during a blizzard). Many schools and businesses announced closings for Monday (a few for Tuesday as well). When the heavy snow did not materialize, a lot of people were very unhappy.

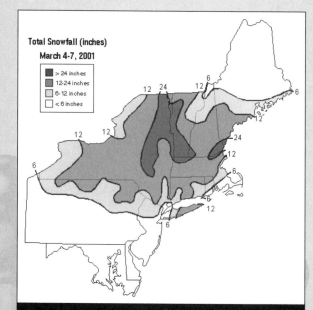

ILLUSTRATION 3.25   The shaded regions received six inches or more of snow from the early March 2001 storm. The heaviest snow remained well north of Philadelphia.

Radio stations and newspapers wanted to know what went wrong. Scenes of the one to two feet of snow in New England were of little consolation to those who braved lines at the grocery stores and lost work time due to the forecasts. The "busted" forecasts generally included the Washington, D.C., Baltimore, Philadelphia, New York City, and Boston areas, making national headlines. Local forecasters, along with those on the Weather Channel, spent hours explaining what happened.

In the Philadelphia area, the bulk of the anger was directed toward NBC-10, and John Bolaris in particular. As one of the best known and most popular weathercasters in the area, John has been the focus of weather talk for years. He had issued the crawl the previous Wednesday, giving the first "heads up" on the potential storm. Criticism of inaccurate forecasts is nothing new to meteorologists; it is an occupational hazard for people who forecast the weather. But some of the accusations focused on "making up" the storm to boost ratings. (Wednesday night was the last night of a ratings period, also known as "sweeps," a four-week span during which viewership for the various stations is closely tabulated and used to set advertising rates for future months.) Such a charge is understandably offensive to any forecaster, since credibility is extremely important for any meteorologist. We know of no meteorologist who would intentionally make a false forecast or exaggerate a storm's intensity just for ratings. Still, in this case, strong denials of those charges did little to appease those who were upset. According to John Bolaris, he received over 1,000 E-mail and phone messages, some containing death threats. A few were supportive, but the overall level of hostility was unmatched in anyone's memory.

In a way, the public's fury at the "busted" storm forecast affirmed the progress in weather forecasting and gave meteorologists a backhanded compliment. People would not be so furious over an inaccurate forecast if they did not already have a high level of trust in weather forecasts. And people would not have that level of trust if meteorologists did not have a pretty good track record of predicting big storms. In this case, that trust was demonstrated in many ways: for example, people took off from work and went to stores for groceries and shovels in preparation; the Pennsylvania Department of Transportation (PENNDOT) was on continuous standby mode for snow removal; and airlines cancelled flights based on forecasts.

What contributed to this forecast debacle (Glenn said on air that "it was my worst forecast in twenty-eight years as a meteorologist")? The confidence built from successful predictions of earlier big snowstorms was obviously a factor. So was the underprediction of snowfall just a month earlier on February 5; like most people, meteorologists do not want to repeat their mistakes. If everyone had gone to work and school on Monday morning, and heavy snow had arrived on Monday afternoon, there would have been legitimate concerns for life and property. That is why the National Weather Service issued Winter Storm Warnings. Should forecasters have showed more uncertainty? Would it have mattered? Perhaps there was too much confidence and too much specificity too soon. Perhaps showing lines at the grocery store and video of the January '96 blizzard helped feed the later fury. (As an aside, events like this raise questions about human behavior. For example, why do so many people buy milk, bread, and eggs when a big storm is predicted? No one starved during the Blizzard of '96, the city's all-time worst snowstorm. Should media take some responsibility for calming the public when a storm is approaching?)

Just like the '96 blizzard-that-was, the February 2001 blizzard-that-wasn't will live in memory for many years—perhaps even decades. ❖

Snowstorm" because of its significant impact in the heavily populated urban corridor of the Northeast. The Big Four metropolitan areas of Washington, Baltimore, Philadelphia, and New York City all received more than 20 inches. The storm began in Philadelphia on a Friday morning and reached its full force that afternoon. Winds of 25 to 35 mph, with occasional gusts over 40 mph, created near-blizzard conditions. By mid-evening Friday, all the interstates were either impassable or closed, the Philadelphia International Airport was shut down, and trains were stopped in their tracks (see Illustration 3.26). There was even thunder during the snowstorm, a very unusual occurrence and testimony to the storm's intensity.

Among the other notable storms in Table 3.3 is the freak early spring snowstorm of April 3–4, 1915, which occurred over Easter weekend. Nineteen inches fell in twelve hours in Philadelphia between 8 A.M. and 8 P.M. on Saturday, April 3. In true April fashion, however, temperatures rebounded to 54°F on Easter Sunday, melting all but 2.5 inches of snow by 8 P.M. that evening. Not since 1841 had this much snow fallen in April in Philadelphia, and no April storm has come close to matching it in the years since.

The "blizzard of '99" of February 12–14, 1899, formed in tandem with one of the greatest outbreaks of Arctic air on record in the eastern United States. The 18.9 inches that fell in Philadelphia, piled on top of nearly a foot from smaller storms the previous week, brought the snow to depths not seen again until January 1996 (see Illustration 3.27). Cape May received 34 inches from the storm, and snow depths there exceeded 40 inches. The snow was compounded by extreme cold: the high temperature was only 10°F in Philadelphia each day from February 11 through 13, with overnight lows of −6°F, 4°F, and 7°F. At Cape May, the ocean was reported to be a mass of ice as far as the eyes could see. The frigid Arctic air even penetrated into Florida, where Tallahassee recorded −2°F, still the all-time record low temperature for the Sunshine State.

The March 1993 "Storm of the Century" does not rank near the top in snowfall in Philadelphia, but the snow was heavy and wet and accompanied by near hurricane-force winds along the coast. More than 100,000 power outages were reported in the Philadelphia area, and extraordinary cold (for March) followed. The storm

ILLUSTRATION 3.26 Traffic is snarled on the Vine Street Expressway on the afternoon of Friday, February 11, 1983, during the second largest snowstorm on record in Philadelphia (AP/Wide World Photos).

Philadelphia, rain the morning of March 11 changed to heavy snow around 11 P.M., eventually accumulating to more than 10 inches as winds blew at 30 to 60 mph and temperatures fell into the teens. At Trenton, 2 inches of rain on March 11 gave way to 21 inches of snow on March 12–13. Farther northeast in New York City, much of the aboveground telegraph system was demolished by a prolonged period of freezing rain, cutting off the city from the outside world for days.

HISTORICAL WINTERS

Winters around Philadelphia have been documented as far back as the mid-seventeenth century, though for the first 100 years or so most observers did not have instruments. Three publications are especially valuable in reconstructing these historical winters. Samuel Hazard, a prominent historian of the early nineteenth century, compiled information about ice cover on the Delaware River from 1681 to 1828, using the effect on navigation as a gauge of winter severity. Charles Peirce's *A Meteorological Account of the Weather in Philadelphia,* written in 1847, gives accounts of each month's weather from 1790 to 1846 in Morrisville, Pennsylvania, about twenty miles northeast of Philadelphia. And American weather historian David Ludlum (see Chapter 1) used both Hazard's and Peirce's work, and the observations of many other co-

set Philadelphia's all-time record low air pressure, 28.43 inches. The intensity of this storm is evident in the classic comma-shape of its cloud pattern, as seen in Illustration 3.28. To its north and west, this superstorm brought an unprecedented swath of blizzard conditions from Alabama into Canada, while on its warm side an intense squall line of tornadic thunderstorms pounded Florida with hurricane-like storm surges. All in all, 100 million people in twenty-six states were adversely affected, and 270 storm-related fatalities were reported.

Finally, the "blizzard of 1888," sometimes called "The Great White Hurricane," paralyzed the East Coast from Chesapeake Bay to Maine and is still considered the most disastrous storm on record at many locations. The storm moved in an unusual way, first heading northeast away from the coast, but then looping back to the southwest, prolonging the impact on land (the kind of track the "non-storm" of March 2001 was expected to take). The storm sank, damaged, or grounded dozens of vessels along the New Jersey coast and in Delaware Bay. In

ILLUSTRATION 3.27 *(See plate.)* The blizzard of February '99 (1899, that is) dumped 18.9 inches of snow on Philadelphia and was followed by days of extreme cold. This scene of South Broad Street was taken February 14, 1899 (Print and Picture Collection, Free Library of Philadelphia).

ILLUSTRATION 3.28 An infrared satellite image of the "Storm of the Century" in March 1993 shows the classic "comma cloud" shape of an intense low-pressure system. The storm's influence stretches from southern Canada through Florida into the Caribbean (courtesy of NOAA).

lonial observers in his two-volume *Early American Winters*. Together, these sources suggest that some colonial winters in the Philadelphia area were more severe than even the harshest winters in the modern record.

Many early settlers of the Delaware Valley came from parts of Europe with relatively mild winters, so the first few bitterly cold seasons came as quite a shock. Two winters became the standards of comparison for colonists who followed:

- The winter of 1697–98 was called "The Terriblest Winter" by settlers. The Delaware River was closed with thick ice for more than three months, and sleighs were able to go from Trenton to Chester. Peter Kalm, a Swedish naturalist who visited Colonial America decades later, wrote that even in the mid-1700s, older settlers used it as a reference with which to compare cold and severe winters.
- The winter of 1704–5 was famous for its deep snows and intense cold. Three feet of snow was reported to have fallen in December, and the Delaware was

frozen over from December 10 to March 10. Of that winter, Philadelphian Isaac Norris wrote: "We have had the deepest snow this winter, that has been known by the longest English liver here. No travelling; all avenues shut; the post has not gone these six weeks; the river fast [frozen]; and the people bring loads over it as they did seven years ago."

Winters of the 1700s

In the first half of the eighteenth century, the winter of 1740–41 became a new benchmark against which future cold winters would be compared. There are no known surviving temperature records from the Delaware Valley, so the winter's severity can only be judged by its effects. The Delaware River was closed to navigation from December 30 until March 24. On January 19, news arrived from Lewes, Delaware, at the mouth of Delaware Bay, that "tis all ice towards the sea as far as the eye can reach." Samuel Hazard reported: "The severity of the winter complained of throughout the country. Cattle dying for want of fodder; many deer found dead in the woods. . . . The winter extremely long and severe."

In the latter half of the eighteenth century, the winters of 1764–65 and 1779–80 stand out as exceptionally severe. During the winter of 1764–65, the Delaware at Philadelphia closed on December 31 and remained frozen until the end of February. Hazard reports that on February 7 "an ox was roasted whole on the river Delaware, which from the novelty of the thing, drew together a great number of people." Though the river was navigable by February 28, a nor'easter struck in late March, described this way in the *Pennsylvania Gazette:* "On Saturday night last there came on here a very severe snow storm, the wind blowing very high, which continued all the next day, when it is believed there fell the greatest quantity of snow that has been known (considering the advanced season) for many years past; it being generally held to be two feet, or two feet and a half, on the level, and in some places deeper." That storm hit while Charles Mason and Jeremiah Dixon were surveying the famous line that would later bear their names. Their journal contained the following entries:

March 21,22,23 snow
March 24 at 9am snow was near 3 feet deep
March 25,26,27 snow so deep we could not proceed

The coldest winter of the eighteenth century in the Delaware Valley (and much of the East) was likely 1779–80, known as "The Hard Winter." According to David Ludlum, this was "the only winter in recorded American history during which the waters surrounding

New York City froze over and remained closed to all navigation for weeks at a time." The Delaware River at Philadelphia froze solid by December 21 and, with ice 16 to 19 inches thick at times, did not open to large ships until March 4. Based on the observations of David Rittenhouse, the temperature in Philadelphia during January 1780 rose above freezing at an observation time only once, and then only for a short time. For comparison, even during January 1977, the coldest month in modern times in Philadelphia, temperatures rose above freezing on ten days. Even allowing for errors in measurement given the relatively crude thermometers in use at that time, January 1780 may have been the coldest month since settlers first arrived in the Delaware Valley.

Three major nor'easters hit between December 28, 1779 and January 7, 1780. Of the second storm on January 3, 1780, which added to the winter's misery, Reverend Henry Muhlenberg at Trappe, Pennsylvania (about ten miles north of Valley Forge) wrote: "Since yesterday afternoon and throughout the night, there was such a snowstorm that the house and yard are so circumvallated [surrounded] that one can scarcely get out or in, and the snow is still falling." The harshness of the winter of 1779–80 prompted the American Philosophical Society to appoint a committee "to make and collect observations on the effects of the severe & long continued cold," one of the first organized attempts to compile weather data in Colonial America (see Illustration 1.3).

Revolutionary War Winters

The infamous "winter at Valley Forge" is probably the most celebrated weather event of the Revolutionary War. But the harshest winter during that war was the previously mentioned "Hard Winter" of 1779–1780, which General Washington's troops spent at Morristown, New Jersey, poorly quartered, poorly clad, and poorly fed compared to the British who occupied New York City just twenty-five miles to the east. Still, no discussion of Philadelphia-area winters would be complete without documenting the winter at Valley Forge, including the story surrounding what is probably the most famous winter scene in American history: the dramatic picture of Washington crossing the ice-clogged Delaware River on the stormy night after Christmas in 1776.

One of Washington's staff provided the following account of the weather at 6 P.M. on Christmas Day 1776: "It is fearfully cold and raw and a snowstorm setting in. The wind is northeast and beats in the face of the men. It will be a terrible night for the soldiers who have no shoes. Some of them have tied old rags around their feet, but I have not heard a man complain." Indeed, observations from North Carolina to New Jersey indicate that the Delaware Valley was on the northern edge of a large nor'easter on December 25–26. In Philadelphia, Phineas Pemberton described December 26 as cold and stormy with northeast winds (see Illustration 3.29). By the morning of December 27, with Washington and his troops back safely on the Pennsylvania side of the Delaware River, Reverend Muhlenberg at Trappe reported that the snow lay "a foot deep and it is bitter cold." The adverse weather undoubtedly contributed to the success of Washington's surprise raid on the Hessians at Trenton, but it also made the crossing and the nine-mile march extremely difficult for the soldiers.

The following winter of 1777–78 was the infamous one at Valley Forge. Observations from Philadelphia and

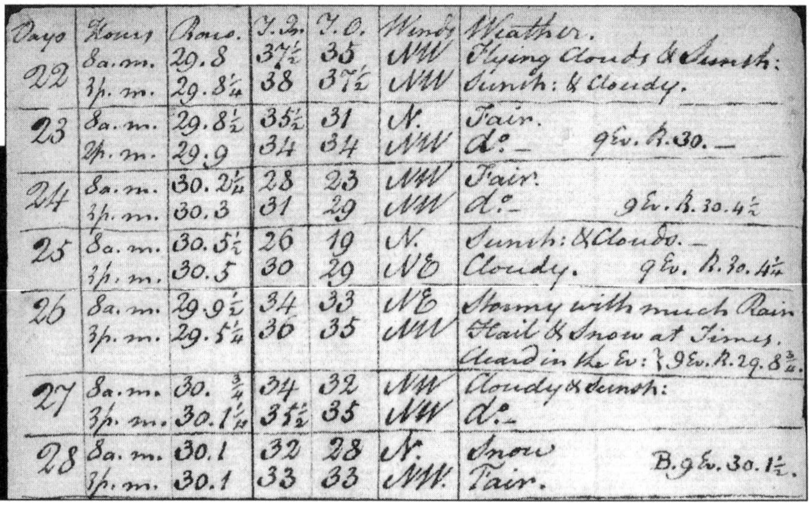

Days	Hours	Baro.	T.In.	T.O.	Winds	Weather.
22	8a.m.	29.8	37½	35	NW	Flying Clouds & Sunsh:
	3p.m.	29.8¼	38	37½	NW	Sunsh: & Cloudy.
23	8a.m.	29.8½	35½	31	N.	Fair.
	2p.m.	29.9	34	34	NW	do.— 9Ev. B.30. —
24	8a.m.	30.2¼	28	23	NW	Fair.
	3p.m.	30.3	31	29	NW	do.— 9Ev. B.30.4½
25	8a.m.	30.5½	26	19	N.	Sunsh: & Cloudy. —
	3p.m.	30.5	30	29	NE	Cloudy. 9Ev. B.30.4¼
26	8a.m.	29.9½	34	33	NE	Stormy with much Rain
	3p.m.	29.5¼	36	35	NW	Hail & Snow at Times. Cleard in the Ev: } 9Ev. B.29.8¾.
27	8a.m.	30.¾	34	32	NW	Cloudy & Sunsh:
	3p.m.	30.1½	35½	35	NW	do.—
28	8a.m.	30.1	32	28	N.	Snow B.9Ev.30.1½.
	3p.m.	30.1	33	33	NW.	Fair.

ILLUSTRATION 3.29 Meteorological observations from December 1776, taken from the weather journal of Phineas Pemberton of Philadelphia. On December 26 at 8 A.M., the morning of the day of Washington's famous crossing of the Delaware, Pemberton observed 34°F and winds from the northeast. His "Weather" for the 26th reads "Stormy with much Rain Hail & Snow at Times. Cleard in the Ev" (American Philosophical Society).

Trappe suggest that it was not a particularly severe winter compared to other Revolutionary War years. But even an average Delaware Valley winter would be enough to cause great suffering to poorly dressed and poorly sheltered soldiers spending winter in the open countryside. The troops arrived at Valley Forge on December 19, amidst "stormy winds and piercing cold." The deepest chill occurred at the end of December and the beginning of March, with temperature in single digits in the city; it was undoubtedly much lower at Valley Forge. But the winter was one of ups and downs, with periods of deep freeze and snow broken by frequent thaws. Despite the occasional moderation, an American general wrote: "One half of our troops are without breeches, shoes, and stockings; and some thousands without blankets. . . . Their shelters were but apologies for dwelling places. . . . The cold winter winds blew through their crevices. . . . For the most part the men lay on the damp earth, padded by a thin coating of scarse [scarce] straw."

Winters of the 1800s

Charles Peirce's *Meteorological Account of the Weather in Philadelphia* includes many specific temperatures (particularly extremes) from 1790 to 1846. Though his observations were taken in Morrisville, across the river from Trenton (where average temperatures are about 2°F lower than in Philadelphia), Peirce's long period of temperature observations is the earliest chance to quantitatively compare colonial winters in this area to those of more recent times.

For example, Peirce reports approximately fifty days on which the temperature fell below 0°F during the fifty-seven years of his observations. (There may have been more; we cannot be sure he mentions them all.) For comparison, during the first fifty-seven years of official temperature measurements in Philadelphia (1874 to 1930), the temperature fell below 0°F on sixteen days (and in the next fifty-seven years, 1931 to 1987, on eighteen days). Part of the difference is Morrisville's somewhat colder climate given its location about twenty miles northeast of Philadelphia. And, undoubtedly, the less frequent extreme cold of the more recent periods is partly the "urban heat island" (see Chapter 2), the artificial heat created by cities. It is also possible that poorly calibrated instruments affected Peirce's observations. Whatever the reasons, his data suggest that Philadelphia has not experienced as many brutally cold days in the last century as it used to.

In the first half of the nineteenth century, two winters featured snows that rival the record-setting winter of 1995–96 in the Delaware Valley. According to David Ludlum, the nor'easter of January 14–16, 1831 was "the greatest snowstorm of the 19th century along the Atlantic seaboard." The snow began in Philadelphia at 8 P.M. on the 14th and continued through the 15th. Southeastern Pennsylvania and southern New Jersey were in the belt of heaviest snow, with estimates of 18 to 30 inches in Mount Holly. In Cape May, New Jersey "snow fell nearly three feet on the level," according to the *Philadelphia Gazette*. In Chester, "The oldest inhabitant cannot remember so severe a snowstorm . . . the mails and stages were much impeded for three days on the road from Philadelphia to Wilmington; but the cross roads, leading westward from the river, are blocked up with snow nearly as high as the top of the fences." In West Chester, it was "one of the most tremendous snow storms ever known in this latitude. . . . It is supposed the snow fell to the average depth of three feet." High winds were reported to have created drifts twenty feet high.

Five years later came another cold and snowy winter. Peirce writes that in January 1836, "There were no less than seven snowstorms. . . . It was generally supposed that during the storm of the 9th and 10th, that nearly three feet fell in this city." In February 1836, "There were eight snowstorms, and it was estimated at the time if the snow had fallen on a level and remained to the end of the last snow, it would have been from eight to ten feet in depth." Even assuming some exaggeration of these amounts, this winter would easily rank at the top of the current list of snowiest Philadelphia winters. Horses and sleds were still crossing the Delaware river on March 5, and as late as March 17 boats had to cut through ice one to two feet thick to reach Chester.

As mentioned earlier, January 1780 may have been the coldest month of the eighteenth century (and possibly in the entire recorded weather history of the Delaware Valley). In the nineteenth century, the strongest candidate for this (dis)honor is January 1857. By that time, the Smithsonian weather observing network was in place, and there were several observations sites in Philadelphia. At the Naval Hospital, the average temperature that month was 20.3°F, while the average at the Pennsylvania Hospital was 22.4°F. For comparison, the average temperature during the coldest month of the twentieth century (January 1977) was 20°F.

One other day in the nineteenth century deserves mention for its extreme cold. On January 8, 1866, the temperature dropped to −12°F in Trenton, −14°F in Haddonfield, −9.5°F at the Pennsylvania Hospital in downtown Philadelphia, and −14°F on Chestnut Street (recall that the modern-day record for cold is −11°F).

ILLUSTRATION 3.30 Ice jam on the Schuylkill River in January 1918. The official U.S. Weather Bureau observation for January 28, 1918 reads: "Heavy ice in both Delaware and Schuylkill rivers. Snow on ground tonight deepest since 1888" (Library Company of Philadelphia).

ILLUSTRATION 3.31 A sight that could hardly be imagined today—ice skating on the Delaware River in early 1918, here at the foot of Buckius Street in Bridesburg. Ironically, Philadelphia's all-time record highest temperature of 106°F would be set later that year in August (Bridesburg Historical Society).

The day was probably the coldest ever known in Philadelphia up to that time. The air pressure at the Pennsylvania Hospital crested at 31.05 inches, while a barometer on Vine Street peaked at 31.10 inches (the modern-day record for high air pressure at Philadelphia is 31.10 inches). Undoubtedly, January 8, 1866 was an extraordinary day in the weather history of the Delaware Valley.

Freezing of the Delaware River

Samuel Hazard's history of ice cover on the Delaware River from 1681 to 1828 indicates that the Delaware froze over much more frequently in the eighteenth and nineteenth centuries than it does in modern times. During thirty of the fifty winters between 1751 and 1800, river ice was severe enough to hamper navigation. Even in the nineteenth century, there were substantially more reports of the river freezing fast than in the twentieth century. During the winter of 1904–5, the weather observer in Philadelphia wrote that "ice extends the full length of the river into the ocean." And both the Delaware and Schuylkill Rivers were ice covered in January 1918 (see Illustrations 3.30 and 3.31). More recently, the Coast Guard closed the river during the winters of 1976–77 and 1977–78 because icebreakers could not keep the river open. Ice was again severe during the winter of 1992–93, but the river did not freeze solid.

Less frequent freezing of the Delaware can certainly result from milder winters, and this is undoubtedly a factor in recent decades. Non-weather issues also contribute to the river's infrequent freeze overs. First, the river is deeper now; the Army Corps of Engineers currently dredges the Delaware to a depth of forty feet from the mouth of Delaware Bay up to about twenty-five miles north of Philadelphia, and to a depth of twenty-five feet from there to Trenton. The deeper the river, the less easily it freezes over. Secondly, the Delaware River is tidal just about as far north as Trenton, and these tidal fluctuations help suppress the formation of ice. The river has, over time, gradually shifted its banks in several places. For example, piers and docks have been excavated several blocks west of the current banks of the river at Philadelphia, indicating that the Delaware River there has shifted to the east since colonial times. Farther south, a careful look at the border between Delaware and New Jersey shows that a small portion of western Salem County, New Jersey is actually part of the state of Delaware. Originally, the state borders were set with the middle of the river as the divide, but the river in that area has since shifted to the west, leaving part of Delaware east of the river. These examples of the meandering of the river suggest that in the past, with a different riverbed, tidal changes might have been smaller, making the river more likely to freeze over.

CHAPTER 4

Spring: March–April–May

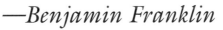

> March windy, and April rainy, makes May
> the pleasantest month of any.
> —*Benjamin Franklin*

More people probably look forward to spring than to any other season, especially if the winter has been long and harsh. Early sunsets, weeks of snow on the ground, and cold spells that kept us indoors can quickly turn a person's thoughts to the return of warmth. It is no wonder that Groundhog Day has become a subject of such fascination. Though Punxsutawney Phil's forecasts cannot be taken seriously, who would not want to start thinking about spring in early February?

But as we saw in the winter chapter, spring is certainly not right around the corner on February 2, in Punxsutawney or in Philadelphia. Indeed, February is the month with the greatest number of big snowstorms. Things start to change dramatically in March, the first month of meteorological spring. By mid-March, daytime temperatures average almost 15°F higher than in early February, and daylight has increased in length by an hour and a half. That is a significant difference.

This rapid transition out of winter's chill opens the door to great extremes of weather. The atmosphere in spring is in turmoil—or, as meteorologists say, there is plenty of "instability." With warm air poised to make its inevitable comeback, but cold air not far to the north, the jet-stream pattern often becomes very wavy. This means that spring low-pressure systems can be slow moving, and very intense. The cold fronts that trail from these lows can spark thunderstorms, some of which can turn violent, even packing tornadoes.

We start this look at meteorological spring by considering how long winter chill can hang on—when were the latest freezes, how deep into spring have snowflakes been sighted, and why do cold fronts sometimes move "backwards" in the spring? Then we explore in detail another

leftover from winter—the nor'easter. These storms were introduced in the winter chapter because they are responsible for our biggest snowfalls. We revisit nor'easters here in detail because the most damaging storm on record to affect coastal New Jersey and Delaware was an early March nor'easter in 1962 that was known more for wind and waves than snow.

Spring is also associated with the return of frequent thunderstorms and the occasional severe weather that they trigger. While there are actually more thunderstorms in summer, the strongest and most violent tend to occur in spring. So we cover thunderstorms in this chapter, along with the hazards that accompany them, including lightning, damaging winds, hail, and tornadoes. We explain what meteorologists look for to identify which thunderstorms might produce severe weather. And we include descriptions of some of the most notable tornadoes in this area from as far back as the 1700s. We also explain the watch and warning system that the National Weather Service uses to alert the public to potential severe thunderstorms and tornadoes. These official advisories need to be communicated to the public, and the quickest way to reach the most people is through television. In this chapter's "Story from the Trenches," we chronicle the tornado that struck Lyons, Pennsylvania in May 1998 on a minute-by-minute basis from the eyes of television meteorologists tracking the storm.

Spring is also a season when snowmelt and heavy rains can combine to produce large-scale river flooding. Though this can also occur in winter, it is most likely in spring, so we discuss river flooding in this chapter. Within that discussion we give an overview of the river systems that drain the Delaware Valley, focusing on the Delaware and Schuylkill Rivers, and talk about some of the more notable large-scale floods that have resulted from heavy rains combined with snowmelt.

Finally, we close this chapter by mentioning some of the notable warm spells that have come in spring, a perfect lead-in to summer in Chapter 5. But we begin this chapter at the other end of spring, with a look at winter's futile attempt to hang on.

FROM WINTER TO SPRING

The transition from winter to spring is rarely smooth in the midlatitudes, and that is certainly true in the Delaware Valley. Most years, spring arrives in fits and starts. Temperatures have reached 80°F in March in Philadelphia only to be followed two weeks later by hard freezes and snow in early April. Sometimes spring seems like the weather equivalent of "two steps forward, one step backward."

A late freeze can disrupt the early growing season and annoy winter-weary motorists tired of scraping ice from their windshields. In the historical record, there are several reports of freezing temperatures in Philadelphia as late in spring as mid-May. A 30°F temperature was ob-

served at the Pennsylvania Hospital in the city on the morning of May 15, 1834. Some of the latest freezes on record in this area probably occurred in 1816, during the infamous "Year Without a Summer" (see Chapter 5 for more details on that year). Charles Peirce, observing in Morrisville between Philadelphia and Trenton, described May 1816 this way:

> The medium temperature of this month was 57°F, and she was really a frosty jade. Her frowns were many, and her smiles few. Northerly winds, with cold frosty nights prevailed, until every green thing was either killed or withered. A melancholy hue appeared to seal the fate of all vegetable life. Buds and small fruit froze upon the trees . . .

Since 1874, when official high and low temperature record keeping began in Philadelphia, the average date of the last freeze has been April 5. But as Illustration 4.1 shows, there is plenty of variation from year to year, with a range from March 11 (in 1953) to May 11 (in 1966). The exceptionally late freeze in 1966, when the temperature dropped to 28°F (the only below-freezing temperature on record in May), was probably influenced by a severe drought which choked the region at that time, because dry air and dry ground promote rapid cooling at night. When interpreting these last-freeze dates, remember that they vary greatly from location to location, with suburban and rural areas tending to experience freezes several weeks later in spring. Also, official temperature readings are taken several feet above the ground, but the air right next to the earth's surface is often several degrees colder (especially on clear, calm nights). So frost can form

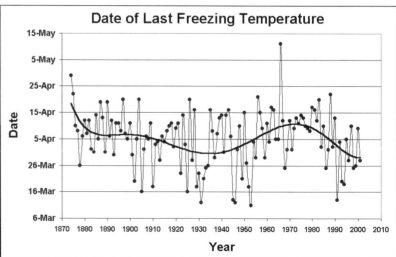

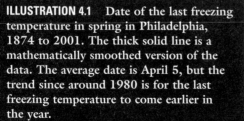

ILLUSTRATION 4.1   Date of the last freezing temperature in spring in Philadelphia, 1874 to 2001. The thick solid line is a mathematically smoothed version of the data. The average date is April 5, but the trend since around 1980 is for the last freezing temperature to come earlier in the year.

on the grass even when official temperature readings are above freezing.

With the sun getting a little higher in the sky each day during spring, truly chilly days are hard to come by, especially as the calendar turns to late April and May. Cold air masses arriving from the Midwest and Great Lakes become more infrequent, and they pack less and less of a chilly punch. But during the spring (and even early summer), there is a source of cool air just to the east and northeast of the Delaware Valley, over the still-chilly waters of the Atlantic Ocean. Sometimes in spring, cold fronts approach from that direction when chilly air literally "backs in" off the water. These wrong-way fronts are called, appropriately, **backdoor cold fronts** (see Illustration 4.2). Temperatures ahead of such a front can reach the 70s or even 80s, but once the front passes, winds swing around to the northeast and cool moist Atlantic air takes over. If it is April or May, and the sky is filled with low clouds, winds are brisk out of the northeast, and temperatures are holding steady in the upper 40s or low 50s in the middle of the afternoon, it is a good bet that a backdoor cold front has recently moved through.

Late-season snow is another sign of lingering cold in spring. Any discussion of such snows for the Philadelphia area must begin with the remarkable 19.4-inch Easter weekend snowfall of April 3–4, 1915, still the fourth largest snowfall in city history. Most of the snow fell on Saturday, April 3 as temperatures hovered near 30°F. But the following day—Easter Sunday—temperatures rebounded to the 50s, and by 8 P.M. only 2.5 inches of snow remained. A storm of similar magnitude occurred even later in the spring in April 1841, according to the weather records of Charles Peirce. That snow accumulated over several days, with 6 inches on April 10, another 10 to 12 inches on April 12, and "a few more inches" on April 13–14.

Snowflake sightings are not that unusual in Philadelphia in April, as snow has been observed in that month in about two-thirds of all years. In most of those Aprils, however, only a trace was observed—not enough to be measured. In the modern record (since 1884), the latest measurable snow in Philadelphia was reported on April 9, in several different years. The most recent was 1996, when 2.4 inches fell, while the most notable was in 1917 when 7 inches accumulated.

A trace of snow has even been reported in Philadelphia in early May. The latest-in-the-season official snowflake sighting occurred on May 8, 1898, with May 6, 1891 a close second. These late snows, though not accumulating, make it easier to believe reports from colo-

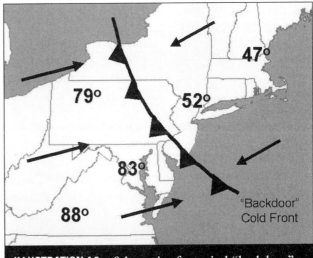

ILLUSTRATION 4.2   Schematic of a typical "backdoor" cold front. Northeast of the front, temperatures are in the 40s and 50s and cool, moist winds blow from the northeast. On the other side of the front, temperatures are in the 70s and 80s with southwest winds. Most computer models do a poor job of predicting the details of backdoor cold fronts.

nial times of measurable snow in the Philadelphia area in early May. An entry from the diary of Jacob Hiltzheimer from May 4, 1774 reads: "The houses this morning are all covered with snow," while in Germantown, "heavy rain on the night of May 3–4 . . . turned to snow toward morning and fell four inches deep." And on May 8, 1803, David Rittenhouse reported a "great north east snow storm" in Philadelphia, accompanied by thunder.

Spring and Instability

Rittenhouse's observation of snow and thunder is certainly unusual. The two typically do not go together, and there is good reason. Thunderstorms usually feed off low-level warm air, and low-level warmth is not favorable for snow. An alternative explanation invokes a buzzword used frequently by television meteorologists to explain severe weather: instability.

Atmospheric instability is a matter of temperature and how it changes with altitude. The earth's surface warms quickly as spring progresses in response to the lengthening daylight and the higher sun angle (as evidence, consider that snow that falls during the daylight in March and April usually does not accumulate on roads unless it is falling very heavily). In turn, the low levels of the air also start to warm. But winter's chill lingers longer at upper levels of the atmosphere, which are far removed from the warming ground. This vertical temperature setup—increasing warmth below stubborn chill—is, in the meteorologist's jargon, "unstable." And this increasing instability helps explain why thunderstorms become more frequent in spring, and also why spring tends to be so windy.

To understand atmospheric stability, think about piling boxes in the back of a van for a trip. Imagine that some boxes are stuffed with old encyclopedias, while others are filled with towels and blankets. Common sense says to put the heavier book-filled boxes on the bottom and the other lighter boxes on top. With that setup—heavier under lighter—the pile is in less danger of tumbling down during the ride.

Now switch over to the atmosphere. Imagine cold air as the heavy boxes and warm air as the light. The air is almost always warmer close to the ground than it is higher up, so there is usually some "tumbling" or overturning going on—currents of rising warmer air and sinking colder air. But in spring, that vertical temperature difference can really become substantial, leading to more frequent and more vigorous currents of rising and sinking air. In essence, the atmosphere is more vertically restless in spring than in other seasons. And rising air, especially rapidly ascending air, leads to clouds, showers, and thunderstorms.

The frequent overturnings in the atmosphere in spring have another consequence. Winds at high altitudes are still pretty strong in spring, with the jet stream frequently overhead. When air from higher up mixes down to near the ground, it brings with it some of that faster momentum. That tends to speed up the winds near the surface, resulting in blustery, gusty spring afternoons, even if the nearest storm system is hundreds or thousands of miles away. Spring's reputation for being good kite-flying weather is well deserved.

As a season of transitions and temperature extremes, spring features some of the most violent weather of the year. Severe thunderstorms and tornadoes are the primary spring weather threat from the eastern slopes of the Rockies to the Ohio and Tennessee River valleys and the Southeast. These severe storms can also threaten the Philadelphia area (to a much lesser extent), but usually not until middle to late spring.

Early in the season, however, nor'easters remain a major weather worry. And these are not just run-down leftovers from winter. Though spring nor'easters are not necessarily big snow producers, their wind, waves, and resulting coastal flooding can still cause considerable damage. In fact, one of the most severe storms to ever affect coastal regions of New Jersey and Delaware in any season was a spring nor'easter in March 1962 that did not have much snow at all. We now take a closer look at nor'easters in general, and specifically that infamous 1962 storm.

NOR'EASTERS

A hurricane can have devastating effects on the coast and its inhabitants. But few hurricanes actually make landfall along the East Coast, and when one does, only a relatively small stretch of coastline surrounding the eye bears the brunt of the storm. On the other hand, the powerful East Coast storms known as nor'easters are much more common and can affect much larger areas. So, even though hurricane winds are stronger, a nor'easter's potential for destruction can be just as great.

Benjamin Franklin was the first to recognize the essential feature of the nor'easter—that surface winds ahead of these storms blow from the northeast, but the storms themselves tend to approach from the southwest. Nor'easters are notorious from the Carolina coast to New England and out into the North Atlantic and have earned the waters off North Carolina the designation as "the graveyard of the Atlantic." They can occur in any month but are most common from November to April. Nearly all of the big snowstorms along the East Coast are nor'easters.

Nor'easters are some of the strongest storms to form anywhere in the country, and part of the reason is the contrast between land and ocean. From about mid-autumn until early spring, the ocean off the East Coast tends to be warmer than the continent. The warm Gulf Stream current, which flows northeastward just off the Carolina coast, enhances the ocean's relative warmth. With cold air entrenched over the continent, there is a built-in temperature contrast between the air over the land and the air over the ocean. So there is a natural tendency for a front to form just offshore, making this region ripe for low-pressure systems to develop and strengthen (see Illustration 4.3).

Meteorologists watch several areas for signs of a brewing nor'easter. Sometimes the seeds of a big storm can be traced back to the Gulf of Mexico, or even farther west, and then the low-pressure center intensifies off the East Coast. At other times, the low forms off the Carolina coast without any prior history. In either case, nor'easters make for tricky forecasting because much of the storm is over the water, where there are few surface observations. A storm that develops quickly and moves

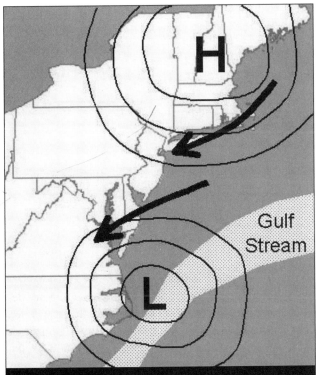

ILLUSTRATION 4.3 The ingredients for a powerful nor'easter. Thin solid lines are isobars (lines of equal pressure). A low-pressure system strengthens off the North Carolina coast, aided by the relatively warm Gulf Stream and the natural temperature contrast between the land and water. A strong high in New England provides cold air and adds to the onshore winds to the north of the developing low.

fast makes for the most difficult forecast, since there is not much time to watch and track it. Sometimes there is a clue that a budding storm is about to intensify—the pressure at or just off the coast will start to fall quickly. A clear wind shift line will develop, with winds from the northeast on land, and from the southeast well offshore. If a trough in the jet stream is approaching from the west to provide upper-level support, watch out—the storm can strengthen explosively, or "bomb out" as meteorologists say.

But you cannot judge the potential destructive power of a nor'easter simply by how low its pressure gets. There are other contributing factors that determine its severity. A partner that is often overlooked is a high-pressure system to the north of the storm. Most of the time, high pressure is associated with fair, relatively calm weather. But when a strong high settles into New England or eastern Canada, and a strong low is developing to the south

along the coast, the high can make things worse. Winds circulating clockwise around the high create a northeast wind at the shore and higher surf, even without a storm. And the greater the high's central pressure, the stronger the onshore winds. As the nor'easter strengthens, the north-south pressure difference increases. The greater the pressure difference, the stronger the winds. And the stronger the winds, the higher the waves and the greater the potential for coastal flooding. The high-pressure area can also act as a block of sorts, helping to slow the nor'easter's northward progress or even causing it to stall, adding to the problems (see Illustration 4.3).

The length of time the wind blows, and the length of ocean over which it blows, also make a big difference in how much damage a nor'easter can inflict. The longer the wind blows from the ocean, the higher the seas will get. Each successive high tide will lead to a higher water level, progressively worsening the coastal flooding. And the larger the distance those winds have been traveling over open ocean toward the shore (known as the **fetch** of the wind), the higher the swells get too.

Finally, the phase of the moon at the time of the storm is crucial. The moon does not cause weather on earth, but it is the primary driver of the tides. Thus the moon can indirectly affect the coastal pounding a particular storm can inflict. Near the time of new and full moons, astronomical tides are slightly higher (the so-called "spring" tides) than in between those times. The same nor'easter that brings a tidal height of seven feet, for example, during a moon's first quarter could lead to a tidal height of nine feet near the full or new moon. And that difference of two feet can mean the difference between minor and severe coastal flooding.

So, the worst damage at the coast occurs with a strong high to the north, a slow-moving storm, and a wind of long duration and long ocean fetch, all of this occurring at or near the time of new or full moon when tides are naturally higher. The most severe nor'easters can generate storm surges—that is, rises above average sea level—of up to fifteen feet, inundating coastal regions.

In 1992, two University of Virginia researchers, Robert Davis and Robert Dolan, developed a scale to rate the intensity of nor'easters (see Table 4.1). They analyzed more than 1,300 storms that occurred between 1942 and 1984. Like the Saffir-Simpson hurricane scale (see Chapter 6), the ratings go from 1 to 5, with Class 5 being the strongest. Though their scale is not commonly used, it does give a way to compare various nor'easters by using the duration and height of the waves produced at the coast.

TABLE 4.1   Dolan/Davis Nor'easter Classification Scale

Class	Average Wave Height	Average Duration	Impact
1 (weak)	6 ft.	8 hr.	Minor beach erosion
2 (moderate)	8 ft.	18 hr.	Some beach erosion and property damage
3 (significant)	11 ft.	34 hr.	Extensive beach erosion, significant dune loss, many structures lost
4 (severe)	16.5 ft.	63 hr.	Severe beach erosion and recession, wider scale of building loss
5 (extreme)	23 ft.	96 hr.	Extreme beach erosion, massive overwash, extensive property damage

About half the storms Davis and Dolan studied fell into the Class 1, or weak, category. Only seven nor'easters, or about one-half of one percent, were Class 5. And only one of those Class-5 storms came close enough to the New Jersey and Delaware coasts to do extreme damage to those shores: the March 1962 nor'easter, also known as the "Ash Wednesday" storm.

The March 1962 Nor'easter

The Class-5 storm of March 1962 did not get a name like a hurricane. Though it hit early in March, this storm is not remembered for its heavy snow. But the area of destruction from this nor'easter covered the entire coastlines of New Jersey and Delaware, and in fact, much of the East Coast from North Carolina to Maine.

The setup for a major nor'easter was classic. An expansive area of high pressure settled from central into southeastern Canada, contributing to the easterly flow along the coast. A low-pressure center developed on March 5 off the Carolinas and strengthened quickly. Wind gusts to 76 mph were reported in Nags Head, North Carolina, a sign that the storm was potent even before it started moving up the coast. And it did not move very quickly, partly because the sprawling high in Canada helped block the storm's movement. In addition, an upper-air cutoff low (see Chapter 2) formed west of the developing surface low. This not only helped intensify the surface storm but also slowed it down. The resulting three-day onslaught of battering east winds and steadily increasing seas occurred near the time of new moon, so tides were relatively high anyway.

The coastal pounding covered 700 miles from North Carolina to Maine, with the New Jersey and Delaware shores right in the middle. The storm's sustained winds of 40 to 70 mph over the open ocean produced waves over thirty feet high. The roiling seas washed a U.S. Navy destroyer, the USS *Monssen*, onshore at Beach Haven Inlet, New Jersey on the southern tip of Long Beach Island (see Illustration 4.4). More than 4,000 buildings were destroyed in New Jersey and nearly 1,500 in Delaware, mostly in shore communities, and nearly every boardwalk was destroyed (see Illustrations 4.5, 4.6, and 4.7). On Long Beach Island alone, five new ocean-to-bay inlets opened up, breaking the island into pieces. Six hundred homes were destroyed on the island, and seven people drowned.

Overall, damage in the United States from the storm was $1 to $3 billion (in current dollars), with twenty-two deaths. Other nor'easters had stronger wind gusts, but none caused such widespread devastation. And these were the days before federally subsidized flood insurance. The people who lost their homes had to rebuild on their own, while they were still paying taxes on a building that no longer existed. The loss of New Jersey beaches in this storm was so extensive that it led to the first large-scale beach replenishment program in the United States.

Away from the coast, on the western fringes of the storm, there was some snow and wind—almost 7 inches fell in Philadelphia on March 6, and northeast winds gusted past 40 mph. But the sun shone on March 7, and most of the snow had melted by the end of the day on March 8. In most places inland, the storm was an inconvenient, but temporary, disruption. It would take years, however, for the shore to recover.

Other Notable Nor'easters

Several other powerful nor'easters that are remembered more for their wind and water power than their snow are

ILLUSTRATION 4.4 *(See plate.)* The destroyer USS *Monssen* lies grounded at Beach Haven Inlet on the southern tip of Long Beach Island after the March '62 nor'easter (Urban Archives, Temple University, Philadelphia, Pennsylvania).

ILLUSTRATION 4.5 Devastation from the March '62 nor'easter, farther south at Avalon, New Jersey (Urban Archives, Temple University, Philadelphia, Pennsylvania).

ILLUSTRATION 4.6 *(See plate.)* Pacific Avenue in Wildwood, New Jersey flooded in the days after the March '62 nor'easter (Urban Archives, Temple University, Philadelphia, Pennsylvania).

described here. Though none of these storms occurred in spring, we include them here because their effects were similar to those of the March '62 storm.

February 4–7, 1920

Although this nor'easter dropped more than a foot of snow and sleet inland, the storm is best remembered for its coastal damage. Four straight days of onshore winds produced a new inlet on the southern end of Long Beach

Island, below Holgate. The new inlet was eventually called Beach Haven Inlet, and today it remains Long Beach Island's southern inlet.

October 26–November 1, 1991: The "Perfect Storm"

The "Perfect Storm" that became the subject of a popular novel and film was a Class-5 nor'easter with a few twists. The storm first showed up on the weather map on

ILLUSTRATION 4.7   The Henlopen Hotel in Rehoboth Beach, Delaware, after the March '62 nor'easter. Note the boardwalk has been washed away (Delaware Department of Transportation/DNREC).

December 10–11, 1992

The jet stream wraps more or less continuously around the globe over the midlatitudes. So a storm that formed many thousands of miles away, even far out in the Pacific Ocean, or the remnants of such a storm, might eventually affect the U.S. East Coast. The nor'easter of December 10–11, 1992, can be traced to just such a Pacific storm—a powerful typhoon named Gay.

A trough in the jet stream that was essentially the ghost of Gay helped intensify this budding nor'easter as it formed over Texas in early December 1992. The storm moved eastward across the Gulf and then turned up the East Coast, intensifying off North Carolina in classic nor'easter style. Wind gusts reached 80 mph in Cape May, with twenty to twenty-five foot waves just offshore. The storm arrived close to the full moon, so tides were relatively high already. Record-setting flooding occurred at Atlantic City and at Reedy Point (where the Chesapeake and Delaware Canal flows into the Delaware River). At Philadelphia, more than 3 inches of rain fell, and tidal flooding was the second worst on record. Farther north, the flooding nearly shut down the New York metropolitan transportation system.

October 26, 1991, developing near Nova Scotia, Canada. Early in its life, it tapped moisture from a weak hurricane named Grace that was dying over the Atlantic Ocean. That was twist number one. Then, in another unusual move, the storm "backed up" towards the East Coast, taking a path opposite the typical west-to-east movement of weather systems (see Illustration 4.8). The center of the low passed within 150 miles of the Delmarva Peninsula, then headed back out to sea.

While all this was happening, a strong high-pressure system developed in southeastern Canada. With the high to the north helping with the easterly flow, severe tidal flooding occurred along the New England and mid-Atlantic coasts. At Atlantic City, tides were comparable to those produced by Hurricane Gloria in 1985. Despite the coastal pounding, the storm did not bring any rain to New Jersey, Pennsylvania, or Delaware. In fact, the weather in Philadelphia during the storm was rather pleasant, with highs in the 60s, though there was a gusty north wind.

At the time of the storm, meteorologists called it the "Halloween Nor'easter of 1991." The term "Perfect Storm" was coined afterward in discussions between Bob Case, who was deputy meteorologist in charge of the NWS Boston forecast office at the time of the storm, and Sebastian Junger, who used the phrase as the title of his book.

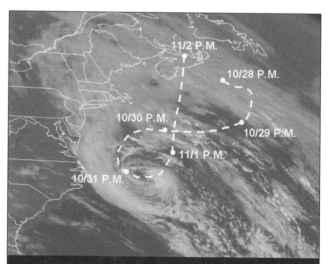

ILLUSTRATION 4.8   The "Perfect Storm" is the swirl of clouds off the coast of North Carolina in this visible satellite image taken the afternoon of October 31, 1991. The dashed line is the bizarre track of the storm. By the time this image was taken, the storm was not a "pure" nor'easter but had evolved into a tropical system, though at the time the National Hurricane Center did not officially designate it as such (courtesy of NOAA).

February 4–6, 1998

Although this nor'easter brought Philadelphia less than a half-inch of rain and just a trace of snow, it devastated many coastal locations, being the worst storm since December 1992.

Coastal New Jersey and Delaware endured damaging winds, severe coastal flooding, extensive beach erosion, and heavy rain. Peak winds reached 81 mph at Cape Henlopen and 73 mph in West Wildwood; at the height of the storm, most tides ran four to five feet above average. The storm struck between spring tides, so the flooding was actually less severe than it might have been. Nonetheless, the peak water level at North Wildwood nearly broke the record set by a devastating hurricane in September 1944 (see Chapter 6).

Twenty-thousand homes and businesses lost power during the storm. Damage was estimated at $17 million in New Jersey and $1.7 million in Delaware. Sussex County in Delaware and Atlantic and Cape May counties in New Jersey were declared federal disaster areas. Boardwalks at Rehoboth and South Bethany in Delaware were badly damaged. On the New Jersey shore, beach erosion left ten-foot cliffs in Avalon; about 75 percent of the sand on Brigantine Beach was carried away; Margate's beach was described as "destroyed." Farther north, at Sandy Hook, about 80 percent of newly replenished sand was lost as several hundred feet of beach disappeared.

THUNDERSTORMS

As the spring calendar turns from March into April and May, the threat from nor'easters diminishes, but the risk of thunderstorms in the Philadelphia area rapidly increases. And, thus, so does the potential for severe weather—strong winds, flash flooding, and occasionally even hail and tornadoes.

In a way, every thunderstorm is potentially dangerous, since every thunderstorm brings lightning. But to be officially classified as "severe" by the National Weather Service, a thunderstorm must produce wind gusts of at least 57.5 mph (50 knots), hail at least three-quarters of an inch in diameter (about the size of a penny), or a tornado. One of the responsibilities of the National Weather Service is to warn the public when thunderstorms approach severe levels.

Severe Weather Watches and Warnings

In the early 1900s, U.S. Weather Bureau forecasts occasionally mentioned the potential for severe weather, but official policy prohibited the use of the word *tornado*. The conventional wisdom was that a tornado forecast would cause panic among the public and do more harm than good. The ban on the word *tornado* was lifted in 1938, but very few forecasts mentioned tornadoes even in the 1940s. The first successful tornado forecast in the U.S. was issued on March 25, 1948 by two weather officers at Tinker Air Force Base in Oklahoma. In 1952, the U.S. Weather Bureau organized a unit to focus solely on severe storms. By the end of the 1950s, this branch was given authority to issue severe weather and tornado "bulletins," which were then used by forecasters in local Weather Bureau offices.

Today, the responsibility for warning the public about severe thunderstorms and tornadoes lies with different parts of the National Weather Service. The Storm Prediction Center (SPC) in Norman, Oklahoma has nationwide responsibility for the first two stages of the process. Each morning, SPC issues a severe weather outlook for the whole country. This outlook outlines those areas where severe thunderstorms might develop in the day ahead, describing regions as having a slight, moderate, or high risk. The rarely used "high risk" category is restricted to days when SPC expects a large outbreak of severe weather, including tornadoes, over a wide area. When SPC includes our area in their severe weather outlook, we are almost always in the "slight risk" category. This suggests only a chance of a few severe thunderstorms.

When it becomes clear that some thunderstorms are likely to reach severe levels over the next few hours, with strong winds and hail the primary threat, SPC issues a **severe thunderstorm watch**. This means that conditions are favorable for severe thunderstorms to form anywhere in the watch area. If SPC believes the primary threat will be tornadoes, they will instead issue a **tornado watch**.

The watch area, sometimes called a watch "box," is a rectangular-shaped region that typically might be about the size of Pennsylvania. With such a large area, there is no implication that any particular spot in the box will get severe weather—it is just a possibility. Typically, a watch box is in effect for four to six hours. When the Delaware Valley is in a box, it is usually in the warm humid air ahead of a potent cold front (see Illustration 4.9). In this case, the threat of severe weather at the "beginning" of the box (that is, the side closest to the cold front) will come before the threat on the opposite side, even though these areas are technically in the same box. If the threat of severe weather ends in part of a watch box before the box is scheduled to expire, SPC will reduce the area covered by the box or cancel the watch entirely.

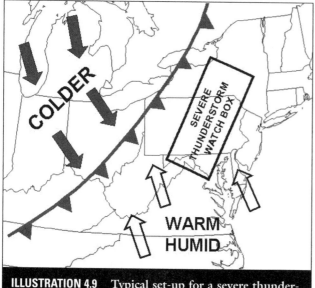

ILLUSTRATION 4.9   Typical set-up for a severe thunderstorm watch area in the warm sector ahead of a cold front. Note that the box is about the size of the state of Pennsylvania and is oriented the same way as the cold front.

Once severe weather is actually occurring or is imminent, the responsibility for alerting the public shifts to the local National Weather Service offices. Thus, for the Delaware Valley, **severe thunderstorm warnings** and **tornado warnings** come from the NWS office in Mount Holly, New Jersey. These warnings can be based on indications from Doppler radar that the thunderstorm may be severe, or on reports of actual severe weather from the public or trained storm spotters. There are more than 2,000 storm spotters within the thirty-four counties served by the Mount Holly forecast office.

The implementation of Doppler radar and increased experience in interpreting the images it produces have led to a higher probability of detecting severe weather before it hits. For example, up to the early 1980s the average lead time for tornado warnings was "negative"; that is, many tornadoes had already been on the ground for several minutes producing damage before the NWS issued a tornado warning. Now the average nationwide lead time for tornado warnings is about twelve minutes and steadily increasing.

Types of Thunderstorms

By the numbers, the thunderstorm capital of the country is the Tampa Bay area in Florida, where the average is more than 90 thunderstorm days per year (based on the period 1950–94). But if the potential to produce large hail and powerful tornadoes is the measure, the Plains states are the clear winner. The Philadelphia area averages about twenty-five days per year with a thunderstorm, and most of the thunderstorms are relatively unremarkable. Though summer is prime thunderstorm season, some of the most severe storms in the area occur in middle to late spring as strong cold fronts encounter increasingly warmer and more humid air. For that reason, we cover thunderstorms in detail in this chapter.

The run-of-the-mill thunderstorm that simply feeds off the heat and humidity of the day is called an **air-mass thunderstorm**. To create such storms, blobs of the warmest, most moist air rise in **updrafts**, typically building by late afternoon into towering cumulonimbus clouds. Air-mass thunderstorms are usually widely separated, bubbling up here and there but not everywhere. If the radar image looks like measles, you are probably looking at air-mass thunderstorms. They might last thirty minutes or so and are rarely severe in the official sense.

Air-mass thunderstorms usually do not last very long because they limit themselves. As raindrops fall, they drag cool air down with them. This creates a downward movement of air within the thunderstorm called a **downdraft**. The downdraft competes with the updraft and eventually overwhelms it, causing the thunderstorm to gradually die.

A thunderstorm has a better chance of living longer and becoming severe if the downdraft and updraft stay separate. **Vertical wind shear** usually accomplishes this—that is, the wind speed or the wind direction must change significantly through the depth of the thunderstorm. For example, the winds at 3,000 feet might be from the south at 20 mph, while 10,000 feet up winds may be from the west at 60 mph. With this change of wind with height, thunderstorms tend to "tilt" a little (see Illustration 4.10). This tilting can separate the updraft and downdraft, allowing the thunderstorm to persist longer and raising the potential for severe weather. When meteorologists study upper-air weather charts in potential severe weather situations, one of the key indicators they are looking for is vertical wind shear.

Vertical wind shear also increases the likelihood that a thunderstorm will produce a tornado. To see how, place a pencil across the palm of one hand. With the palm of the other hand, rub across the pencil so that it rolls. This rubbing action is analogous to vertical wind shear: your top hand represents faster winds at higher altitudes, while your bottom hand represents slower winds closer to the ground. In the atmosphere, such wind shear can produce

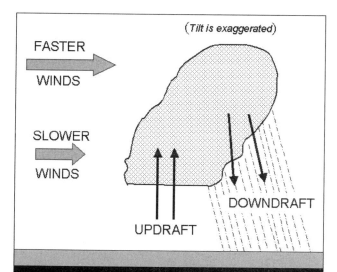

ILLUSTRATION 4.10   Schematic view of a severe thunderstorm. The faster winds aloft cause the cloud to tilt (the tilt is exaggerated here), so the downdraft occurs in a different part of the thunderstorm than the updraft. This allows the storm to persist and also favors severe weather.

form, supercells require not only significant vertical wind shear and exceptionally moist low-level air but also a layer of dry air above the low-level humidity. At first thought, dry air might seem completely the opposite of what a growing thunderstorm needs. And, indeed, a layer of dry air can initially put a "lid" on rising warm, humid air from below, preventing smaller thunderstorms from developing. Over time, however, heat and humidity build below the layer of dry air. Eventually, if this pent-up energy can break through the cap of dry air, supercell thunderstorms can explode to great heights. It is a lot like the explosion of hot humid air that occurs when the lid on a pot of boiling water is removed.

In an average year, ten or so squall lines might move through the Delaware Valley, bringing quick bursts of strong, sometimes damaging winds and small hail. However, the layer of dry air that temporarily "puts a lid on" the low-level heat and humidity is usually missing, so thunderstorms rarely grow to supercell proportions in this area. But if they do, watch out. When a tornado has touched down in the Delaware Valley, it was usually a supercell that spawned it.

a horizontal roll of air inside a thunderstorm cloud. In some situations, the thunderstorm's updraft can tilt the horizontal roll so that it becomes vertical. If the rotating circulation drops from the base of the thunderstorm, a **funnel cloud** is born. And if the funnel cloud's circulation connects with the ground, it officially becomes a **tornado**.

Even if vertical wind shear is present, a long-lasting thunderstorm usually needs some extra mechanism to keep the air rising, beyond the heat and humidity that drives air-mass thunderstorms. In the Delaware Valley, strong cold fronts often provide that additional lift. Sometimes, a chain of powerful thunderstorms called a **squall line** forms in the warm humid air ahead of such a cold front. A satellite view of one of the most intense squall lines ever to affect Pennsylvania is shown in Illustration 4.11. On May 31, 1985, this army of thunderstorms spawned the greatest tornado outbreak on record in the state—twenty-one twisters caused sixty-five deaths in a six-hour period. One of the tornadoes in extreme western Pennsylvania had winds estimated at 300 mph.

The largest, most intense thunderstorms are called **supercells**. They can persist for hours and have high tornado-producing potential. Several of these monster thunderstorms make up the squall line in Illustration 4.11. To

ILLUSTRATION 4.11   A visible satellite image at 7 P.M. EST on May 31, 1985 shows one of the most severe squall lines on record to affect Pennsylvania. The squall line is not solid, but rather composed of several huge supercell thunderstorms. Twenty-one tornadoes formed in Pennsylvania as this line of storms crossed the state (courtesy of NOAA).

Lightning

Tornadoes and hurricanes usually get more attention, but lightning causes more casualties in the United States than any other storm-related phenomenon except floods. Lightning is responsible, on average, for about 80 deaths and 300 injuries each year in the U.S. During the period 1959–94, more than 500 Pennsylvanians were injured and more than 100 killed by lightning, ranking the state third in injuries (behind Florida and Michigan) and tenth in fatalities among all states.

Ancient societies thought lightning bolts were the acts of angry gods casting their wrath upon mortals below. Benjamin Franklin approached the problem more scientifically in 1752 with his famous kite experiment. Franklin found that lightning was a large, visible discharge of electricity—a spark in the air. His experiment demonstrated that electrified clouds, and therefore lightning, are similar to static electricity. Anyone who has walked across a carpet and then created a shocking spark by touching a doorknob has evidence of how lightning works.

Today, we can explain the electrical event that Franklin observed. To do so requires zooming into the tiny spatial scale of the **atoms** that make up all matter. Atoms are made of even tinier particles, some that carry electrical charge. The important ones for making lightning are **electrons**, and they carry a negative charge. When electrons are transferred from object to object, imbalances in electrical charge build up—these imbalances are essential for a discharge of lightning. Atoms with an excess of electrons carry a negative charge, while atoms short on electrons are positively charged. You may have observed the process of "charge separation" at work in a clothes dryer. As clothes tumble and rub against each other, excess electrons can build up on some pieces of clothing. Then, when shirts, pants, and socks are pulled from the dryer, tiny discharges of static electricity occur, reducing the charge imbalance.

One theory explains charge imbalances in thunderstorms as the result of contact between different-sized precipitation particles. For example, most cumulonimbus clouds contain both tiny lightweight snow crystals and heavier balls of ice commonly known as hail. (We do not experience hail and snow from most thunderstorms, however, because they melt on the way down.) When snow crystals and hailstones collide in the cloud, the heavier hail gains electrons while the snow crystals lose them. The lighter crystals, now positively charged, tend to gather high in the cloud. The heavier hailstones, with their extra electrons and negative charge, congregate lower. This imbalance in charge sets the stage for lightning.

Most lightning flashes do not strike the earth's surface—it is estimated that only 10–20 percent of all strokes are cloud-to-ground (see Illustration 4.12). Most lightning bolts stay within a particular cloud, while a very small percentage go from one cloud to another or from the cloud to the air. In any case, a bolt heats the air instantaneously to temperatures near 50,000°F. In turn, the air surrounding the bolt expands explosively, producing a sound wave that moves away from the bolt in all directions. You hear that sound wave as thunder.

Light travels so fast (about 670 million mph) that you see the lightning essentially when it happens. But sound travels much slower—about 600 mph, or about one mile in five seconds. As a result, you can estimate the distance to a lightning bolt by counting the number of seconds between seeing the bolt and hearing the thunder. For every five seconds, the lightning is about one mile away. This "flash-to-bang" rule also explains why thunder sometimes lasts a few seconds. All parts of a lightning bolt

ILLUSTRATION 4.12   Lightning bolts over Connie Mack Stadium on June 30, 1959. We hope the game was stopped well before these bolts appeared (Urban Archives, Temple University, Philadelphia, Pennsylvania).

produce sound, but not all parts of the bolt will necessarily be the same distance away from you. When you hear
a continuous rumble, you are hearing thunder from various parts of the same bolt that lie different distances away.

All lightning produces thunder, but the sound usually
cannot be heard beyond fifteen miles or so from the bolt
because of the dampening effect that the atmosphere has
on sound. Distant lightning that seems to have no thunder is sometimes called "heat lightning," suggesting that
the bolt somehow results from the heat. Not so. "Heat
lightning" is simply lightning from a thunderstorm that
is too far away for the thunder to be heard.

No place is absolutely safe from lightning, but some
places are safer than others. A sturdy closed building is
the best shelter, especially if a lightning rod is mounted
on top. The rod's elevated position increases the likelihood that it will be struck instead of other parts of the
structure. The lightning rod is connected to an insulated
wire that channels the electrical discharge to a long metal
rod that is driven far into the ground. Fully enclosed
metal vehicles such as cars and buses with the windows
rolled up also provide good protection from lightning,
but not because of the rubber tires. The metal "skin"
of the vehicle acts as a conductor of electricity to the
ground. As long as you are not in contact with metal
inside the car, you are relatively safe.

The majority of lightning injuries and deaths occur
outdoors (see Illustration 4.13) in open fields, under
trees, on tractors or mowers, or on golf courses. As far
as gauging when lightning is a threat, the "30–30 rule"
is useful: if 30 seconds or fewer pass between seeing lightning and hearing its thunder, the lightning is close
enough to be a hazard. And you should wait thirty minutes after seeing the last lightning flash before resuming
your normal outdoor activities. More than half of all fatal
lightning strikes occur *after* the thunderstorm has passed.

If you are about to be struck by lightning, you may
experience a tingling sensation on your skin or feel your
hair stand on end. If this happens, immediately squat low
to the ground on the balls of your feet. This position
minimizes height and also contact with the ground. The
latter is important because electricity from a nearby lightning strike can be conducted through the earth to you—
so the more of your body touching the ground, the
greater your risk. This kind of indirect lightning strike
happens more often than you might think. In 1998 in
Ocean County, New Jersey, one person was killed and five
others injured when lightning struck a nearby tree and
traveled through the saturated ground to the tent where
they were sleeping. The next year an almost identical inci-

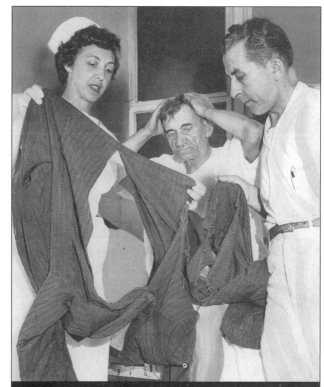

ILLUSTRATION 4.13   This man was struck by lightning
while seeking shelter under a tree on East River Drive,
near the Girard Avenue Bridge, on the afternoon of
June 13, 1959. His coat and shirt were split in the
back and the left leg of his trousers was ripped to
shreds. He was treated at Hahnemann Hospital for
burns on his back and left leg. Like most lightning
victims, he survived (Urban Archives, Temple University, Philadelphia, Pennsylvania).

dent occurred in Burlington County, resulting in one
death and six injuries.

On average, thunderstorms produce about 25,000,000
cloud-to-ground lightning strikes per year in the conti-
nental United States. This estimate comes from the National Lightning Detection Network (NLDN), which was
established in 1989. The NLDN consists of over 100
ground-based sensing stations that pinpoint the location
of a cloud-to-ground lightning bolt within seconds of
a strike.

Illustration 4.14 shows NLDN data for our area from
1989 to 1999. The areas shaded light and dark blue
received the least lightning, about two strikes per square
mile each year, on average. The distribution of these areas
of relatively low lightning frequency is easy to explain.
The higher elevations of northern New Jersey and the

Poconos get fewer thunderstorms because they are generally cooler and less humid. And the relatively cool waters off the Jersey coast suppress thunderstorm development during the spring and summer. The areas in yellow and orange received the most lightning, an annual average of more than five strikes per square mile. There

is no simple explanation for the locations of these areas of maximum lightning, though it is tempting to suggest that the excess heat of the Philadelphia metropolitan area contributes to the maximum that stretches from the city northeastward into New Jersey. More years of data are probably required to reach any trustworthy conclusions.

A squall line of particularly strong thunderstorms that moved through the Delaware Valley the night of September 14, 2000, gives a sense of how much lightning a potent group of thunderstorms can produce. The storms moved into the western Philadelphia suburbs about 11 P.M. and reached Philadelphia and Wilmington around midnight. They then moved into central and southern New Jersey and further south into Delaware after 1 A.M., where they lingered until around 6 A.M. Data from the NLDN indicated 10,380 lightning strikes across the area between 9 P.M. and 6 A.M., an average of more than one strike per square mile in just that one night.

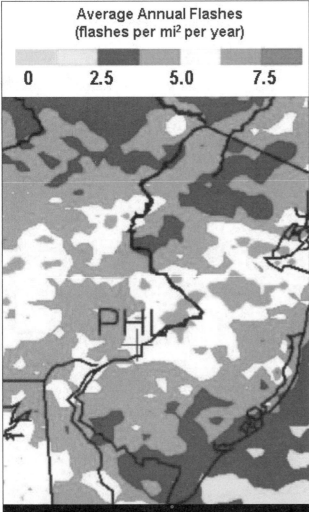

ILLUSTRATION 4.14 *(See plate.)* Annual average lightning density in Philadelphia and surrounding areas as detected by the NLDN, for the period 1989–99 (Philadelphia is marked with "PHL"). The least lightning strikes, represented by light and dark blue, are offshore and in the cooler, higher elevations of northern New Jersey and northeastern Pennsylvania. Orange represents the most lightning strikes, more than five strikes per square mile per year. Because of imperfections in the NLDN, the actual number of lighting strikes is probably twice what is shown here (The Lightning Project, Texas A&M University).

Thunderstorm Winds

The 50-knot (57.5 mph) wind threshold for a thunderstorm to be officially "severe" is based on the wind's potential to cause damage. Winds this strong can knock down or uproot trees and do considerable structural damage to some buildings. Lesser winds are inconvenient, but rarely cause substantial damage.

With regard to thunderstorm winds, meteorologists make a distinction between straight-line winds and winds that show a circulation. As the name implies, straight-line winds blow consistently from one direction, at least in a particular local area. Trees knocked down by straight-line winds all fall like dominoes in the same direction. In contrast, trees blown down by a tornado form a twisted pattern when viewed from a distance or from an airplane.

Strong straight-line winds can usually be traced back to the downdraft of the thunderstorm. When cooler downdraft air strikes the ground, it spreads out in all directions like tap water from a faucet hitting the sink below. A location near the "splashdown" will experience a wind blowing consistently from the same direction (this wind is the "cool before the storm" you sometimes feel as a thunderstorm approaches). A concentrated straight-line burst of winds associated with a thunderstorm downdraft is called a **downburst**. The most intense downbursts are known as **microbursts**, and their straight-line winds can reach 100 mph. When thunderstorm winds cause extensive damage in the Philadelphia area, the culprit is usually a downburst or microburst and not a tornado.

ILLUSTRATION 4.15 *(See plate.)* Damage from a probable microburst on May 24, 1933 at 26th and Harrison Streets in Camden. Note that all trees are blown down in the same direction, strong evidence that this damage was not caused by a tornado. Officials estimated damage at $1 million (1933 dollars) in Camden County alone (Urban Archives, Temple University, Philadelphia, Pennsylvania).

Before Hurricane Hazel in 1954, the strongest winds officially recorded in Philadelphia occurred on May 24, 1933, associated with what was probably a microburst (though they were not called that at the time). Winds surpassed 80 mph in parts of the area and officially reached 68 mph in Philadelphia. The winds caused widespread damage (see Illustration 4.15) and were reported to have swayed the tall PSFS Building in Philadelphia (which had just become the nation's first modern skyscraper when it opened the previous year). The official U.S. Weather Bureau observation that day reads:

> This was an unusually destructive storm. When it spread over the city the dense black clouds shut off practically all sunlight. . . . All records for high wind velocities in Philadelphia were broken. Inspection tours . . . revealed several hundred large trees that were uprooted or broken off. Considerable damage was done to telephone, telegraph, and electric wires. There was no evidence of tornado action in any part of the city.

Hail

Hail is often confused with sleet, but the two types of precipitation form in completely different ways and are predominant at different times of the year. Sleet is a cold-season phenomenon that forms when raindrops or partially melted snowflakes freeze into tiny ice pellets as they fall through a layer of very cold air (see Chapter 3). Ice pellets do not grow large—they are uniformly tiny in size.

Hailstones, in contrast, form in thunderstorms, so hail occurs primarily in the warm season. Hailstones begin as frozen raindrops or small, snow-like particles called **graupel**. If these particles are suspended high in the cold reaches of a thunderstorm, new layers of ice will form on them. Though their weight means that hailstones tend to fall, speedy thunderstorm updrafts can loft them back up. The faster the updrafts, the larger the hailstones they can suspend, so the size of a hailstone is proportional to the speed of the updraft. In powerful supercell thunderstorms in the Plains, updrafts can exceed 50 mph, and hailstones sometimes grow to the size of baseballs and even softballs.

Hail that large is relatively unknown in the Philadelphia area, where hailstones usually do not get much bigger than peas and rarely grow to the three-quarter-inch size that officially qualifies a thunderstorm as severe. How unusual is hail that big? As examples, consider that during the period from 1981 to 2000, three-quarter-inch hail was observed on just eleven days in Camden County, New Jersey, and on only nine days in both New Castle County, Delaware and Philadelphia County. And on each of those occasions, only a tiny fraction of the county was actually affected. The largest hail reported in the last decade in the Philadelphia area fell on May 29, 1995, when baseball- and tennis ball-sized hail (2.5 inches in diameter) caused $3 million in damage to greenhouses in Chester County, Pennsylvania and damaged vehicles along Interstate 295 in Burlington County, New Jersey (see Illustration 4.16). On rare occasions, small hail has coated the ground in downtown Philadelphia, as happened in the middle of the afternoon on May 1, 1953 (see Illustration 4.17).

One truly exceptional hailstorm struck in colonial times at Burlington, New Jersey on May 25, 1758. It

ILLUSTRATION 4.16   These hailstones, ranging in size from one inch to two and a half inches across, fell in Parker Ford, Pennsylvania in northern Chester County, just south of Pottstown, on May 29, 1995. The ruler in the foreground is six inches long (courtesy of the Ames property, Parker Ford, Pennsylvania).

caused a "scene of devastation" with crops destroyed and gardens ruined. One eyewitness wrote:

> . . . yesterday, I was a spectator of one of the most extraordinary storms of hail and rain as perhaps has been seen in America. . . . At first there came a little rain but was soon followed by some large stones of hail. . . . It seemed as if the whole body of the clouds were falling. . . . The street appeared as another Delaware, full of floating ice. . . . I tho't in the midst of it,

of Noah's flood. . . . The thickest of it continued about 15 or 20 minutes more . . . in some places the hail had drifted 6 inches thick, some of which remained on the ground till night, notwithstanding it continued warm after the storm.

Flash Flooding

Flash floods are a thunderstorm's greatest killer, accounting for an average of 140 fatalities in the United States each year. Water is deceptively heavy, and people often underestimate the power of moving water. A cubic yard of water weighs about 1,600 pounds, and the force of only six inches of swiftly moving water can knock people off their feet. Cars can easily be swept away in just two feet of moving water. Never drive into water of unknown depth! You may not get out—many flooding deaths occur in cars that are carried away by water.

In the Philadelphia area, rainfall rates in the core of a thunderstorm might typically be one to three inches per hour. When such a thunderstorm becomes stationary or moves very slowly for a few hours, or when thunderstorm after thunderstorm moves over the same area like cars on a train track, localized flash flooding becomes a concern. Whatever the cause, the adjective "flash" implies a rapid rise of streams, creeks, and small rivers, sometimes in a matter of minutes. Most flash flooding events occur during summer, when the atmosphere is most loaded with moisture. In Chapter 5 we discuss some of the notable thunderstorm-induced flash floods that have struck the Philadelphia area over the years.

ILLUSTRATION 4.17   Marble-sized hail covers the 700 block of South Broad Street in Philadelphia during the late afternoon of May 1, 1953 (Urban Archives, Temple University, Philadelphia, Pennsylvania).

TABLE 4.2   Fujita Scale of Wind Speed and Damage

Rating	Wind Speed (mph)	Damage Class	Character of Damage
F0	40–72	Light	Some damage to chimneys; branches break off trees; crops heavily damaged; some shallow-rooted trees uprooted
F1	73–112	Moderate	Mobile homes overturned; trees uprooted; shingles peeled off; windows broken; moving autos pushed off road
F2	113–57	Considerable	Roofs torn off houses; weak structures and outbuildings destroyed; mobile homes demolished; autos blown off highways
F3	158–206	Severe	Walls torn off well-constructed frame houses; most trees in a forest uprooted, snapped, or leveled; trains overturned; autos lifted off ground and rolled some distance
F4	207–60	Devastating	Well-constructed frame homes leveled; cars thrown some distance; trees debarked by flying debris; heavy missiles generated
F5	261–318	Incredible	Strong frame houses lifted off foundations and disintegrated; steel-reinforced concrete structures badly damaged; automobile-sized missiles fly distances of 100 meters (about 330 feet)

TORNADOES

No other country experiences as many tornadoes as the United States does. During the 1990s, an average of 1,200 twisters a year were reported on U.S. soil. In fact, the most tornado-prone region of the world is a corridor known as **tornado alley** that stretches from eastern and northern Texas across Oklahoma and Kansas into Nebraska.

The ingredients for making twisters typically start to come together in late winter over the Gulf states, as warm humid air that had been banished to more southern latitudes for the winter tries to surge back northward. This warmth and moisture inevitably meets advancing colder, drier air along cold fronts, and strong thunderstorms often erupt. Warm humid air expands farther north in early spring, and tornado frequency really picks up in April, peaking nationwide in May and June. Since thunderstorms favor the warmest part of the day, tornadoes tend to as well. In fact, about 80 percent of all twisters occur between noon and midnight. The likelihood of a thunderstorm producing a tornado increases dramatically if considerable vertical wind shear is present.

Meteorologists rank tornado intensity and damage potential on the **Fujita scale**, or F-scale (see Table 4.2).

It was developed around 1970 by Dr. Theodore Fujita, the world's most renowned tornado expert. On the F-scale, the weakest tornadoes are rated F0 and the strongest F5. About 85 percent of all reported tornadoes are F0 or F1; less than 1 percent make F4 or F5 status, but these cause most of the deaths, injuries, and damage. Almost all Delaware Valley tornadoes are relatively weak, rated F0 or F1. A few twisters rated as high as F3 have touched down in southeastern Pennsylvania and parts of New Jersey, but no tornadoes rated F4 or F5 have been recorded within 100 miles of Philadelphia. The primary reason we see so few strong tornadoes is that this area lacks a nearby source of mid-level dry air, which is typically necessary to generate supercell thunderstorms.

Direct wind speed measurements are not available for most tornadoes, so almost all F-scale ratings are estimated after the fact, based on a survey of the damage. However, in May 1999, a portable Doppler radar being used by University of Oklahoma researchers measured a wind speed of 318 mph in an F5 tornado near Oklahoma City, a world record for top wind speed. Tornadoes such as this F5 produce almost total destruction in their paths, with the only complete protection being underground. In any tornado, the airborne missiles of debris sent flying by the wind are the greatest hazard. So the first rule of tornado safety is to get out of the way of the debris by seeking shelter in a basement or the lowest floor of a house, away from windows.

Reliable tornado record keeping in the United States began in the 1950s. In that decade, the average number of observed tornadoes was around 450 per year. In the 1990s, the average had increased to about 1,200 per year. This remarkable (and at first glance, scary) rise in reported tornadoes almost surely is the result of changes in the ways that we are making observations, not of changes in the weather. The increase is entirely in the F0 and F1 tornadoes—a sure sign that many weaker tornadoes simply escaped detection in the past. Now, with Doppler radar helping meteorologists to better identify potential tornado-producing thunderstorms, and with a larger population that is more spread out and more weather savvy, we are better observers. We do not miss as many of the weaker tornadoes anymore. But with people living and working in more places than ever before, tornadoes that previously would have struck open farmland or forest now have a better chance of going through populated areas.

Illustration 4.18 shows the total number of reported tornadoes for each year in Pennsylvania since 1950 (top graph), and also the number of strong tornadoes per year (those rated F2 or higher, bottom graph). As is the case nationally, there is an obvious increase over the years in the total number, but there is no clear trend in the frequency of strong tornadoes. The same is true for New Jersey and Delaware, though these states get fewer tornadoes overall. In the 1990s, Pennsylvania averaged twenty-two tornadoes per year, Delaware two, and New Jersey four; Pennsylvania's larger area is the main reason for its relatively higher number compared to its neighbors. However, these averages hide a tremendous year-to-year variability. For example, fifty-seven tornadoes touched down in Pennsylvania in 1998, with eleven of them rated F2 or higher. One year later, in 1999, just nine tornadoes were reported, all rated either F0 or F1. One of those weak twisters, an F0, touched down briefly in Marconi Plaza in South Philadelphia on January 19. Though this tornado was only twenty yards wide and on the ground for just two-tenths of a mile, it injured eighteen people because it touched down in a highly populated area.

That January tornado was quite the exception, because if there is such a thing as tornado season in Philadelphia and surrounding areas, it is the period from late May through the end of July (see Illustration 4.19). The most tornadoes have occurred in July, but twisters rated F2 or higher (the few there are) peak in late May and early June when potent cold fronts and squall lines are more common. Most tornadoes in the Delaware Valley are F0 or F1 and are short lived, lasting only a few minutes. Most have narrow tracks, on the order of ten to perhaps a hundred yards wide, and are on the ground for only a mile or two. Because tornadoes are unusual in this area and typically do not last very long, photographs of local tornadoes are very rare. Illustration 4.20 shows an F1 tornado that moved eastward through Somers Point and Linwood, New Jersey (just southwest of Atlantic City) on July 5, 2001. The tornado caused scattered damage along a path about two miles long and seventy-five yards wide.

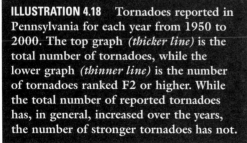

ILLUSTRATION 4.18 Tornadoes reported in Pennsylvania for each year from 1950 to 2000. The top graph (*thicker line*) is the total number of tornadoes, while the lower graph (*thinner line*) is the number of tornadoes ranked F2 or higher. While the total number of reported tornadoes has, in general, increased over the years, the number of stronger tornadoes has not.

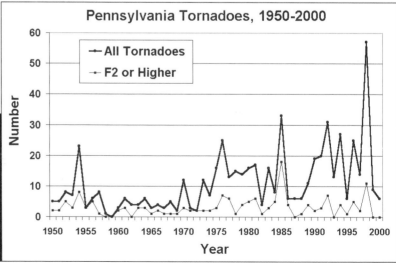

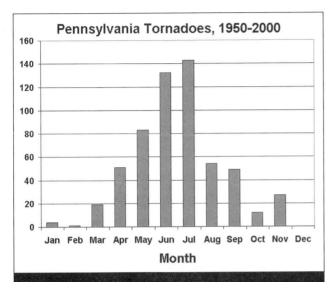

Pennsylvania Tornadoes, 1950-2000

ILLUSTRATION 4.19 Pennsylvania tornadoes by month for the period 1950 to 2000. Although spring is the primary tornado season in the most tornado-prone states of the Midwest, June and July are the most common months for tornadoes in Pennsylvania, New Jersey, and Delaware.

Though many tornadoes cause some damage, injuries and particularly fatalities from tornadoes are uncommon in the Delaware Valley—the risk of being injured by a tornado in this area is very, very small. During the period 1951 to 2000, there have been eight tornado fatalities and approximately 250 injuries reported in the thirty-four-county area served by the Mount Holly NWS forecast office. Most of the injuries and deaths resulted from a small number of the stronger tornadoes.

Illustration 4.21 shows the touchdown site of all reported tornadoes in Philadelphia and surrounding areas during the period 1950–2000. Though there may seem to be a lot, remember that this map shows fifty years' worth of tornadoes. With such a relatively small number of tornadoes, it is hard to pin down any definite patterns, but there is some correlation between population and tornado sightings. Notice that there are no reported tornadoes in New Jersey in the relatively unpopulated stretch of the Pine Barrens from central Atlantic Country northward through the eastern half of Burlington County into western Ocean and Monmouth counties. On the other hand, there tend to be concentrations of tornado reports in more populated areas, such as around the larger cities and near the shore.

Figure 4.21 does not include **waterspouts**, which are tornadoes that form or move over water. Generally a few are sighted each year off the Jersey and Delaware coasts. Because weather tends to move from west to east, most of these waterspouts stay offshore, but occasionally one will come inland and cause damage. One of the strongest in recent years formed just southeast of Beach Haven, New Jersey on August 20, 1999. The waterspout moved onshore at Holgate on Long Beach Island, briefly reached F2 intensity with winds estimated at 120 mph, then moved back into Little Egg Harbor and dissipated. Damage on the island was estimated at $4.2 million, with dozens of homes, buildings, and vehicles damaged.

Significant Philadelphia-area Tornadoes Prior to 1900

From the way that early settlers of the New World described tornadoes, it is obvious that they did not know quite what to make of them. Whirlwind and tempest, along with hurricane and spout, were used loosely and interchangeably by colonists to describe these localized twisting storms. Tornado did not come into common use until the mid-1800s, and even for decades after that there was confusion in terminology (even among scientists).

ILLUSTRATION 4.20 *(See plate.)* An F1 tornado that moved through Somers Point and Linwood in Atlantic County, New Jersey, just after 6 P.M. on July 5, 2001. The photograph was taken from mid span of the Longport/Somers Point bridge, looking west (courtesy Ronald Fallon).

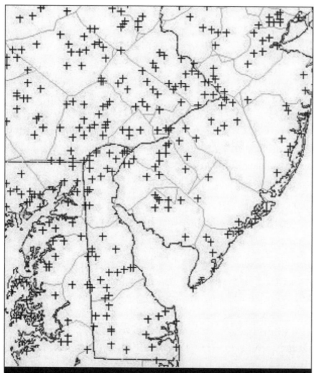

ILLUSTRATION 4.21   The touchdown sites of all reported tornadoes in the Philadelphia area and surrounding counties from 1950 to 2000 are marked with crosses (+). It is probably not a coincidence that tornadoes are least frequently reported in the least populated areas.

August 1724

The first recorded tornado in the Philadelphia area touched down somewhere near Valley Forge and moved east, closely paralleling the present-day Pennsylvania Turnpike through what are now Montgomery and Bucks counties. We have the following description from Philadelphia's *American Weekly Mercury* newspaper:

> . . . about the hour of 12 . . . there began a most terrible and surprizing Whirl-wind. . . . At Plymouth the whole roof was pulled off a big barn and carried out in the lot; a woman's skirt sailed seven or eight miles through the air. . . . It took up almost all the apple trees in the orchard by the root and carried them some distance. People were in danger of being carried off right in their houses . . .

August 1755

While riding horseback in Maryland, Benjamin Franklin chased a storm and later wrote an account of the now-famous incident:

> . . . we saw, in the vale below us, a small whirlwind beginning in the road, and showing itself by the dust it raised. . . . When it passed by us, its smaller part near the ground appeared no bigger than a common barrel; but, widening upwards, it seemed, at forty or fifty feet high, to be twenty or thirty feet in diameter . . . I tried to break this little whirlwind, by striking my whip frequently through it, but without any effect.

What Franklin describes is an encounter with what we now call a **dust devil**. These tornado cousins do not descend from severe thunderstorms as true twisters do. Rather, dust devils are small whirls that form when strong heating by the sun promotes fast, rising currents of air over dry landscapes. Just as wind blowing past a corner of a building can cause a whirl of leaves to form, wind blowing past hills can impart a spin to currents of rising air, generating a dust devil. They are usually short lived and harmless, with winds generally no more than 40 to 50 mph.

June 19, 1835

A violent tornado moved east through the heart of New Brunswick, New Jersey on a track about seventeen miles long and several hundred yards wide. Because of the tornado's location, close to both New York City and Philadelphia, for the first time in U.S. history a tornado's aftermath was extensively studied by many of the leading scientists of the day. The list included James Espy, The Franklin Institute's meteorologist, and Professor Alexander D. Bache, another prominent Philadelphia scientist. Bache's detailed drawings of the damage caused by this twister (shown in Illustration 4.22) may be the first tornado survey ever published.

This tornado caused five fatalities and approximately forty injuries and destroyed or badly damaged about 120 buildings in New Brunswick alone. The descriptions of destruction and reports of people and heavy objects being carried by the wind indicate that this was a very strong tornado, perhaps an F3. A tornado of this magnitude taking this track today would cause a major disaster.

August 3, 1885

The headline in the *Philadelphia Inquirer* the next day blared:

WHIRLING WINDS.
A TORNADO VISITS PHILADELPHIA
AND CAMDEN.

Loss of Life and Great Destruction of Property
in the Two Cities and on the River

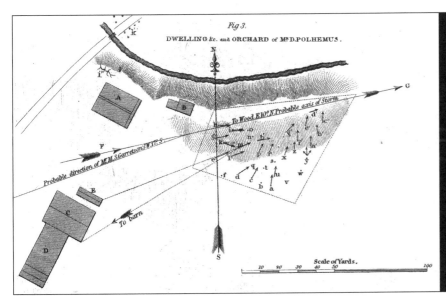

Fig 3.

DWELLING &c. and ORCHARD of Mr D. POLHEMUS.

To Wood E.10°.N Probable axis of Storm

Probable direction of Mr M.S.Garretson: W.3°.5.

To barn

Scale of Yards.
10 20 30 40 50   100

ILLUSTRATION 4.22   Portion of a survey of the New Brunswick tornado of 1835, drawn by Alexander Bache, prominent Philadelphia scientist and Professor of Natural Philosophy and Chemistry at the University of Pennsylvania. This may be the first tornado survey ever published. The arrow from *F* to *G* shows the tornado's path across the property of a Mr. D. Polhemus. The letters *A, B, C, D,* and *E* mark buildings; lowercase letters indicate trees knocked over by the wind, with the small arrows showing the direction the trees fell (from *Transactions of the American Philosophical Society*, n.s., 5 [1837]: 419).

The story continued:

> . . . yesterday the vicinity of Philadelphia was visited by the phenomenon of a tornado of terrific force . . . leaving in its track a series of calamities and destruction such as were never known in the neighborhood of Philadelphia to occur from the violence of the wind. . . . The rush of the cyclone was estimated to be 500 feet wide. Its appearance was that of a dense black cloud revolving at a terrific rate . . .

The large tornado moved quickly from the vicinity of Greenwich Point, near the current Walt Whitman Bridge, northeast across the Delaware River to Camden. It then headed north along the river streets of West Camden, then crossed the river again to the Port Richmond area of North Philadelphia. In five minutes, houses were unroofed, huge trees were wrenched up by the roots and thrown in every direction, and hundreds of telegraph and telephone wires were downed. Overall, damage was estimated at $500,000 (1885 dollars). Six people were killed and more than one hundred injured. This tornado was likely at least an F2. Of course, a repeat of this twister today could be catastrophic.

Significant Recent Philadelphia-Area Tornadoes

Since reliable record keeping began in 1950, about a half-dozen F3 tornadoes have occurred in southeastern Pennsylvania, all north or west of Philadelphia. The most recent was the Lyons (Berks County) tornado in May

1998. Only four tornadoes rated F3 have touched down since 1950 in the entire state of New Jersey, while no tornadoes that strong have been reported in Delaware in that period (though one F2 that touched down in Kent County on July 21, 1983 caused two fatalities and nine injuries).

July 27, 1994—Limerick, Montgomery County, Pennsylvania

This F3 was the last tornado to cause a fatality in the Delaware Valley. It touched down shortly before midnight in Limerick, just east of Pottstown, and spun through a development known as "The Hamlet," causing three fatalities and twenty-five injuries. Four houses in the development were leveled and sixteen were damaged beyond repair. A savings bond from one of the houses was found the next day fifty miles away. Damage in this development alone was $2 million, and total damage in Limerick was estimated at $5 million. This tornado exemplifies how we are becoming more susceptible to severe weather simply by our expansion into previously undeveloped territory. This development was only two years old.

May 31, 1998—Lyons, Berks County, Pennsylvania

An F3 tornado devastated the north side of the borough of Lyons, about fifteen miles northeast of Reading. With a path approximately 120 yards wide and eight miles long, the tornado injured seven people (five in Lyons) and damaged or destroyed about forty homes (see Illustrations

ILLUSTRATION 4.23 Damage from the Lyons (Berks County) tornado on May 31, 1998. After an assessment of the damage by National Weather Service officials, the tornado was classified F3 on the Fujita scale (courtesy of National Weather Service).

ILLUSTRATION 4.24 More damage from the Lyons tornado. Such extreme tornado damage is rare in eastern Pennsylvania, New Jersey, and Delaware (courtesy of National Weather Service).

4.23 and 4.24). At one farm the tornado toppled a huge silo; at another, the remains of a large metal storage bin rested on top of the house. The tornado damage led to a presidential declaration of a disaster area for Berks County.

The Lyons tornado was one of twenty-four twisters in Pennsylvania that day. A few hours after the Lyons tornado, an F2 tornado touched down near Willow Grove in southeast Montgomery County and crossed into extreme northeast Philadelphia County. All told, the tornado's path length was a little more than five and a half miles and its width was about 200 yards. There were no fatalities or injuries, but property damage was estimated at $1.8 million.

RIVER FLOODING

Flooding can occur at any time of the year and can result from a variety of different weather situations. Regardless of the cause, the local National Weather Service office in Mount Holly, New Jersey is responsible for issuing all flood watches and warnings for our area. In this chapter, we have already mentioned flash flooding resulting from the torrential downpours of slow-moving thunderstorms. This flooding is usually very localized and confined to areas that border smaller streams and creeks.

Flooding on main-stem rivers usually results from rain over a longer period of time and over a larger area. The heavy rain from tropical systems has caused some of the worst river flooding in our region's history; in fact, almost every record flood level along the Delaware and Schuylkill Rivers followed either Hurricane Agnes or Hurricane Diane (see Chapter 6). But heavy rains from intense nontropical low-pressure systems can occur any time of year. And when these rains combine with snowmelt during winter or early spring, severe flooding along the two rivers and their tributaries can result.

The Delaware River forms the entire eastern border of Pennsylvania. Its main stream begins in extreme northeastern Pennsylvania, near Hancock, New York, and the river eventually empties into Delaware Bay. Overall, the Delaware River drains approximately 13,000 square miles of New York, New Jersey, eastern Pennsylvania, and Delaware (see Illustration 4.25). Eight counties in Pennsylvania, including Bucks, Delaware, Montgomery, and Philadelphia, and three counties in New Jersey, including Cumberland and Salem, fall entirely within the Delaware River basin. Nine counties in Pennsylvania, eleven in New Jersey, eight in New York, and all three in Delaware are partially in the basin. Because of this extensive drainage basin, the drought and water supply situation in the Philadelphia area cannot be evaluated solely based on local conditions.

The Delaware's largest tributary is the Schuylkill River, which flows southeast through Reading, Pottstown, and Norristown before emptying into the Delaware

The Lyons Tornado of May 1998

Although television viewers are accustomed to program interruptions for news and weather alerts, they probably are not aware of the behind-the-scenes deliberations that result in a written message or announcement delivered by a news anchor or weathercaster. What kind of change in weather conditions justifies an interruption of regular programming? What if the program is a one-time event with a huge audience?

On May 31, 1998, the station where Glenn worked was televising game six of the NBA finals in what was, at the time, thought to be Michael Jordan's last season. The Sunday night prime-time broadcast was expected to have a tremendous following, even people who were not true basketball fans.

In Philadelphia, weather forecasters' attention focused on the changing local weather situation because the entire area was under a tornado watch. At NBC-10, the main question was, "What if a tornado warning is issued during the game?" Ordinarily, we would immediately break into programming with a special report. But television stations hesitate to break into special-event coverage, and they are even more reluctant to break into commercials (any extraneous words or pictures during a commercial mean that a station loses all the revenue from the ad, requiring a free "make good" at a later time). However, if a tornado was on the ground and heading quickly for a big population center, some stations would even interrupt commercials. (While working in Atlanta, Glenn Schwartz once broke into a commercial to report on a tornado sighting: neither management nor the sales department complained.) Regarding special-event coverage, meteorologist Mace Michaels learned a hard lesson in his first weekend at WFLA-TV in Tampa, Florida in June 2000. A waterspout in Tampa Bay persuaded him to break into programming with a ninety-second live Doppler radar update. The report interrupted the final two putts of Tiger Woods's record-breaking finish at the U.S. Open. Thousands of angry calls flooded the station.

At our station we had good reason to think that interrupting the NBA finals could cause a huge uproar. While the relatively few people in the tornado's path would be grateful for the warning, many others would object. Loudly. Some tornado warnings even turn out to be "false alarms"—indications from Doppler radar that do not result in a confirmed tornado. Even before the 6 P.M. news, our weather team and management formulated a plan to deal with the potentially volatile situation. During the news, we would emphasize that the area was in a tornado watch box to heighten the public's awareness of the threat. The basketball game would not be interrupted unless a confirmed tornado was on the ground and headed for a large population center. Any warnings issued during the game would be relayed via a "crawl" (the horizontal scroll of words at the bottom of the screen).

At a television station, tornado warnings require quick decisions and an immediate method of getting the information on air, because tornadoes can come and go within a matter of minutes. Every major station in the Philadelphia area has an automated system that is capable of relaying a crawl literally seconds after a warning is issued. These systems can be set to function without human input, but in that mode only the basics can be relayed to the public: for example, "A tornado warning is in effect for Delaware County until 8:15 P.M." The only decision required about the immediate airing of a warning is whether to wait until after a commercial break. Decisions on giving more specifics or doing a "squeeze-back" (the program stays on, but is placed in a small box while the live update airs in another small box) or a full programming interruption are made by the meteorologist or the news manager on duty.

The weather was quiet through the first half of the game. Our weather team was prepared for a special report at the half, but because there was nothing urgent to relay, only the crawls appeared to remind viewers of the potential for severe weather. But shortly after halftime, things changed quickly.

8:37 P.M. The National Weather Service issued a tornado warning for Berks County, about fifty miles northwest of Philadelphia. The NWS Doppler radar indicated the strong possibility of a tornado.

8:38 P.M. A crawl giving the warning went on the air, with Berks County highlighted on an accompanying map. Undoubtedly, other local television stations aired the warning as well.

8:50 P.M. In an extraordinarily helpful development, the NWS repeated the warning with added details. This time the warning stated that a trained storm spotter had sighted a tornado that was moving toward Lyons, a borough of about 550 people some fifteen miles northeast of Reading.

8:53 P.M. We added the message "A confirmed tornado is on the ground . . . and moving toward Lyons" to the basic warning information that went on the air earlier. This is an option on most weather warning computers, but it takes a minute or two for the meteorologist to type the details in manually.

8:58 P.M. Five minutes later, the crawl ran again because there was actually a tornado on the ground, justification enough to ward off any complaints after the event.

9:00 P.M. An F3 tornado with winds up to 200 mph hit Lyons, causing tremendous damage (see Illustrations 4.23 and 4.24).

According to the NWS post-storm report, "The north side of the borough was destroyed . . . about 25 homes were heavily damaged, several collapsed. Miraculously, only five people were injured." NWS interviews revealed that many residents had been watching the NBA finals and saw the warning. The specific information about the danger to Lyons caused some people to take cover in their basements and kept them from harm, despite the property damage.

The severe weather warning system worked perfectly in the case of the Lyons tornado. National Weather Service meteorologists issued timely warnings, relying on Doppler radar and storm spotters; television and radio stations relayed the warnings and enabled people in the path of the storm to take action that may have saved their lives.

And no one had to miss a play from Michael Jordan's "last game." ❖

at Philadelphia. Narrower and shallower than the Delaware, the Schuylkill River floods more readily and more frequently. Flooding on the Delaware River is relatively rare, especially south of Trenton where the river widens and deepens. In addition, the river is tidal below Trenton, and the daily tidal fluctuations often have more influence than heavy rains. A typical variation between high and low tide is on the order of five or six feet at Philadelphia.

Historical documents refer to floods on the Delaware River as far back as the spring of 1692, when early settlers of Trenton were driven from their homes by waters rising to the second story. Another great flood of the Delaware occurred in January 1841, when the river rose to a record high level between Easton and Trenton. In the twentieth century, snowmelt combined with two bouts of heavy rain produced the "St. Patrick's Day Flood" of March 1936. Rainfall ranged between five and ten inches in a week over much of the northern Delaware River watershed, concentrated where the snowpack contained about the same amount of water. The worst damage occurred north of Trenton (see Illustration 4.26), although 200 people were left homeless in Trenton and some of the main streets of Camden flooded around high tide.

In recent years, major flooding occurred on both the Schuylkill and Delaware Rivers in late January 1996,

ILLUSTRATION 4.25 The Delaware River watershed. Because the Delaware River is fed by reservoirs in New York State and drains parts of northeastern Pennsylvania and southern New York, water conditions in those areas are an important factor in flooding and droughts in the Philadelphia area (Delaware River Basin Commission).

ILLUSTRATION 4.26   An aerial view of Easton, Pennsylvania, about fifty miles north of Philadelphia, during the St. Patrick's Day Flood of March 1936. The Lehigh River, which feeds into the Delaware River at Easton, flooded a section of the city (Urban Archives, Temple University, Philadelphia, Pennsylvania).

almost two weeks after the Blizzard of '96. That storm had dumped between twenty and thirty inches of snow all across the basins of the Delaware and Schuylkill Rivers on January 7–8. Ten days later, one to two feet still remained on the ground in most places from Philadelphia northward. Melted down, this snowpack contained between three and five inches of water. On January 18–19, temperatures soared past 60°F with a moist southerly flow ahead of a cold front. The snow melted rapidly, and then heavy rain of one to three inches fell in a six to eight hour period with the arrival of the front. By January 20, most of the snowpack was obliterated, particularly south of Allentown and Trenton.

Flash flooding of almost every small stream and significant roadway began on the afternoon of January 19. Though the rain ended that evening, larger streams and rivers continued to rise from the runoff, and major

flooding continued through January 21. All gauges along the Delaware and Schuylkill Rivers exceeded flood stage, with the Delaware at Trenton reaching its highest crest since the floods of August 1955. In Philadelphia, flooding along the Schuylkill River put Main Street in Manayunk under two to three feet of water. At Fairmount Park, the river crested more than two feet above flood stage, statistically considered a once-in-fifty-year occurrence.

LOOKING AHEAD: GLIMPSES OF SUMMER IN SPRING

We began this chapter with examples of how spring provides occasional reminders of winter. As a season of transition, spring also sometimes affords glimpses ahead to summer.

The temperature has never reached 90°F in Philadelphia in March, and even days above 80°F are rare in that month, with one occurring every three or four years, on average. March 1990 featured five straight days of 80°F warmth from March 12 to March 16. But in typical spring fashion, cold weather returned late in the month, with more than two inches of snow on both March 24 and April 7. Then, emphasizing spring's fickleness, a three-day period of 90°F heat arrived in late April.

In the official records, the earliest-in-the-year 90°F reading in Philadelphia occurred on April 7, 1929. The 94°F reading on April 18, 1976 is the highest temperature ever measured in April. Philadelphia has never officially recorded a 100°F day in April or May, the closest being 97°F on May 30–31, 1991. The earliest 100°F readings occurred on June 5–6, 1925, considered summer in the meteorological calendar. And if you are looking for more heat, read on. Summer is up next.

CHAPTER 5

Summer: June–July–August

> It's a sure sign of summer if the chair gets up when you do.
> —*Walter Winchell*

Summer is a season of maximums in the Delaware Valley—in temperature, hours of daylight, air pollution, humidity, and coastal population.

Hot weather is a staple of Philadelphia summers. In this chapter, we will identify the all-time hottest days, months, and summers, as well as discuss how the city's characteristic row homes can make the heat worse and endanger lives. We will explain why high humidity combined with excessive heat makes people feel so uncomfortable and increases health risks, and how the combination of heat and humidity led to the largest weather-related loss of life in Philadelphia history. By way of contrast, we will also present an amazing account from nearly two hundred years ago about a "year without a summer."

Hit-and-miss thunderstorms are also a staple of Philadelphia summers. Though they tend to be less violent than their spring counterparts, these thunderstorms usually move more slowly, which means meteorologists must constantly be on the lookout for flash flooding from torrential downpours. But paradoxically, summer's intense sunshine can also dry out the ground quickly, so if thunderstorms are infrequent, dry spells and even drought can develop. We will look at the makings of flash floods and droughts and review some of the worst in Philadelphia-area history.

Summer is also pollution season. Decades ago, the pollution problem in the Philadelphia area was much worse, and much more visible (and we have a picture to prove it). These days, ozone smog raises the most concern, locally and nationally, because too much ozone in the lower atmosphere endangers human health and the environment. On the other hand, ozone at high altitudes is absolutely

necessary for life on earth. What is the difference between "good" and "bad" ozone? What is the "ozone hole," and does it affect us here? In this chapter, we will introduce you to ozone, the most misunderstood gas in the atmosphere.

For most people in the Delaware Valley, summer would not be complete without a trip to the shore. The weather and climate of the coast, and the natural processes that move air, water, and sand up and down the shore are part of our concern. What long-term good (if any) are structures such as groins and jetties and processes such as beach replenishment? We report on what coastal experts say about these issues.

Hurricane season peaks in autumn, so we leave a detailed discussion of tropical storms and hurricanes for the next chapter. But Agnes, a tropical system that occurred in early summer, led to the worst natural disaster in Pennsylvania history. We include in this chapter a "Story from the Trenches" in which Philadelphia television weather pioneer Herb Clarke recalls the week in June 1972 that Agnes dominated the news.

HEAT AND HUMIDITY

A heat wave rather than a hurricane, tornado, blizzard, or flood resulted in the highest death toll in modern Philadelphia weather history. In July 1993, 118 people died from heat-related causes.

In this area, the National Weather Service defines a **heat wave** as a period of at least three straight days when

ILLUSTRATION 5.1   Staying cool in the fountain in front of the Philadelphia Museum of Art during a heat wave in August 1975 (Urban Archives, Temple University, Philadelphia, Pennsylvania).

temperatures reach 90°F. Many of our summers have several heat waves, and many heat spells are accompanied by high humidity. Few summers, though, have a heat wave that is intense or long-lasting enough to cause any more than isolated fatalities.

Rarely do heat waves in our area create dramatic pictures for the evening news or for the front pages of newspapers. Instead, we are more likely to see folks lining up at the local water-ice stand or children cooling off in backyard and public pools (see Illustration 5.1). In a typical heat wave, the sky is usually clear except for an increasing haze, and few, if any, thunderstorms show up on Doppler radar. Winds are usually light. Nothing much changes from day to day. The heat wave of July 1993 began that way. But as the stretch of heat and humidity persisted, the weather made grim headlines as reports of heat-related deaths started accumulating. What made that heat wave so different?

A typical heat wave in the Delaware Valley starts with an area of high pressure in the western Atlantic called the "Bermuda high" by meteorologists. If this high is situated just right, a persistent southeast wind flow sets up, pushing warmer water from the offshore current, the Gulf Stream, closer to the coast. With a pattern like this occurring in July 1993, ocean temperatures off southern New Jersey rose to almost 80°F, about as warm as that water ever gets. Winds off an exceptionally warm ocean are a key factor in increasing the humidity, and thus raising the potential danger of an extended period of heat.

With the humid air established, the threat is enhanced if the Bermuda high builds westward closer to the East Coast, causing winds to weaken. This is what happened during the July 1993 heat wave. The air grows humid and stagnant as high pressure brings day after day of sunshine. Pollutants build up because there is no rain to clean out the air or wind to mix it around.

Daytime high temperatures tend to get the most attention during a heat wave, especially if they approach 100°F. And, indeed, temperatures reached triple digits on three days during July 1993 (see Table 5.1). But in terms of a heat wave's potential danger, overnight lows are at least as important as daytime highs, and that was certainly true in July 1993. One effect of high humidity is that it prevents temperatures from decreasing much after dark. When temperatures stay relatively high at night, homes without air conditioning do not cool off much. When sunshine returns the next day, inside temperatures rise to even higher levels. This is especially true in urban areas of Philadelphia where tightly packed brick row homes with red brick walls and tar roofs, and often with poor airflow, can become the equivalent of brick ovens. This cycle of gradually increasing indoor temperatures was a critical factor in many heat-related deaths in July 1993. Overnight low temperatures were near or above 75°F for ten consecutive nights, with three of those nights at least 80°F (see Table 5.1). The combination of hot days and very warm nights made the difference. Also, the temperatures in Table 5.1 were the "official" readings at the Philadelphia International Airport—undoubtedly, temperatures were several degrees higher (both during the day and at night) in more urbanized areas.

TABLE 5.1 Heat Wave of July 1993 (Temp. in °F)			
	High Temp.	Low Temp.	Maximum Dew Point
July 4	95	74	73
July 5	96	75	72
July 6	95	74	75
July 7	98	77	76
July 8	100	79	77
July 9	100	80	75
July 10	101	81	76
July 11	96	81	73
July 12	94	76	75
July 13	96	76	70

One of the health effects of high humidity is that the body must work harder to prevent overheating. Perspiration is less likely to evaporate when the air is laden with water vapor, as it is on really muggy days. The evaporation of sweat is a natural cooling mechanism because some of the energy needed to evaporate sweat comes from the body. Less evaporation means less cooling and more water left on the skin, and both make you feel more uncomfortable and increase your risk of overheating. The elderly are at greatest risk because as people age their body cooling systems become less efficient at dissipating energy.

As explained in Chapter 2, relative humidity is the most commonly reported measure of atmospheric moisture, but it does not accurately convey how humid the air feels. Typically, relative humidity is highest in the morning, when the temperature is lowest. And on most afternoons during the July 1993 heat wave, the relative humidity averaged a not-so-uncomfortable-sounding 45 percent. Instead of relative humidity, meteorologists use the dew point to describe how humid the air feels. The dew point is straightforward to interpret—the higher it is, the more humid you feel. Once the dew point gets above 65°F, just about everyone agrees that it is humid, and with a dew point above 70°F the air feels tropical. Dew points of 75°F or higher are rare in the Delaware Valley, reached on only the muggiest of days. The highest dew point on each day of the July 1993 heat wave is also given in Table 5.1; you can see how humidity compounded the discomfort of that period.

To quantify how the humidity of the air affects your comfort and the stress that the weather puts on your body, the National Weather Service reports the **heat index**, also known as the **apparent temperature** (see Table 5.2). The heat index is intended to give a sense of what the air temperature "feels like" when the effect of humidity is taken into account. As an example, a temperature of 100°F, coupled with a dew point of 75°F, results in a heat index of 115°F. This is high enough to make heat stroke, cramps, or heat exhaustion likely for anyone exposed to it for a prolonged period. (See Table 5.2 for descriptions of these heat-related ailments.) Heat indices rarely reach that extreme level in the Philadelphia area but approached it during the July 1993 heat wave.

The heat index only tells part of the story, however. Factors other than heat and humidity also affect how you feel in summer (and other times of year). These include the color, fabric, and thickness of your clothes, whether the wind is blowing, and whether you are in direct sunlight. The heat index does not take any of these into

TABLE 5.2 The Heat Index and Various Heat-Related Illnesses

Temperature (°F)	Dew Point (°F)				
	60	65	70	75	80
110	112	115	120	126	135
105	107	111	115	121	129
100	102	105	109	115	123
95	96	99	103	108	115
90	91	94	97	100	107
85	86	88	90	93	99
80	81	82	86	87	91

Most heat-related illnesses can be prevented by drinking lots of fluids and being sensible about exertion in hot, humid weather. Here are the most common heat-related ailments:

Heat cramps. Severe muscle spasms that typically begin suddenly in the hands, calves, or feet. Heat cramps are caused by excessive loss of fluids and salts resulting from heavy sweating.

Heat exhaustion. Heat exhaustion also results from excessive loss of fluids from heavy sweating. Heat exhaustion occurs more readily on a dry day, when the rate of evaporation of sweat is fast. Symptoms of heat exhaustion include fatigue, weakness, a drenching sweat, and often cold, clammy skin.

Heat stroke. Also known as sunstroke, heat stroke can be life threatening. It can occur on humid days when evaporation of sweat is reduced. A person can't sweat enough to lower body temperatures, which can soar to dangerous levels, particularly during strenuous exercise. A person with heat stroke may be delirious or even unconscious. Heat stroke requires immediate, intensive treatment to reduce the body's temperature.

account, even though you will certainly feel more comfortable on a hot, muggy day if there is a breeze and you are in the shade wearing thin, loose-fitting clothes. Thus, the heat index should be used to gauge, in a general sense, whether the combination of heat and humidity poses a danger to health. But remember that other factors, some that you can control, also influence the way you feel on hot, humid days.

Just as the National Weather Service issues watches and warnings for severe weather, they also issue notices for health-threatening heat and humidity. Excessive Heat Watches and Warnings are based on both temperature and humidity, so the heat index comes into play. In the Philadelphia area, the National Weather Service issues an Excessive Heat Warning when the heat index is expected to reach 100°F for at least three hours a day on two straight days (a watch is issued if these conditions are just a possibility). These watches and warnings are especially relevant in highly urbanized areas, where buildings are tightly packed and brick walls and tar roofs tend to absorb heat efficiently.

After the 1993 heat wave, the NWS worked with Dr. Laurence Kalkstein of the University of Delaware, an expert on the health effects of weather and climate, to upgrade its excessive heat advisory system. The goal was to better predict when hot, steamy weather would become a significant health concern. Kalkstein's methods not only factor in temperature and humidity, but also cloud cover, wind, and the expected duration of the heat spell. He correlated these variables with information about heat-related deaths in the Philadelphia area. The result was a computer program that helps in deciding whether to issue an Excessive Heat Watch or Warning for an upcoming bout of oppressive weather.

This procedure came on-line in 1995, and none too soon. That summer turned out to be the hottest on record in Philadelphia. And July 1995 featured a three-day period of super-muggy weather that may very well have been the most humid of the century in this area—the average dew point on July 15, 1995 was 80°F. Though seventy-two heat-related fatalities were reported in Philadelphia during the summer of 1995, the city health department estimated that about three hundred deaths were averted thanks to the revamped excessive heat warning system. Kalkstein has extended his research and devised a new "Heat-Stress Index" that was used on an experimental basis during the summer of 2001 in Philadelphia and also in New Orleans.

The lesson here is that warnings about excessive heat and humidity should be taken just as seriously as any other kind of weather advisory. Though tornadoes, floods, and severe thunderstorms are more dramatic and may strike quicker, it is high heat and humidity that wears us down and potentially poses the biggest weather-related health hazard.

The Hottest of the Hot

Hot, of course, is a relative term. What is "hot" depends, at the very least, on who you are and where you are. For example, it might take a temperature of 100°F for life-long residents of Dallas, Texas to feel hot, but 90°F is a reasonable threshold around the Delaware Valley.

On average, the temperature gets up to 90°F on twenty-five to thirty days a year in Philadelphia. The num-ber of 90°F days is one way to gauge a summer's heat. When you look at lists of the summers with the highest average temperature, those with the most 90°F days, and those with the longest stretches of 90°F weather, there is a lot of overlap (see Table 5.3). The summer of 1988 had the longest streak of 90°F weather (eighteen days in a row), and also the second-most 90°F days (forty-nine). The hottest summer, 1995, had a seventeen-day stretch of 90°F weather and also forty-nine days of temperatures of 90°F or above. Illustration 5.2 shows the variability of average summer temperature in Philadelphia since official records began in 1874. The four hottest summers on record, and five of the top eight, all occurred in the 1990s.

TABLE 5.3   Three Ways to Measure Summer Heat

Hottest Summers by Average Temperature

Year	Average Temperature (°F)
1995	78.5
1994	78.3
1993	78.2
1991	77.9
1900	77.1
1973	77.1
1988	77.1
1999	77.1

Hottest Summers by Longest Stretches of 90+°F Days

Length	Dates
18 days	July 29 to Aug. 15, 1988
17 days	July 20 to Aug. 5, 1995
13 days	Aug. 24 to Sept. 5, 1953
12 days	July 23 to Aug. 3, 1999
12 days	July 12 to July 23, 1952
12 days	June 25 to July 6, 1901

Hottest Summers by Most 90+°F Days

Year	Number of 90+°F Days
1991	53
1988	49
1995	49
1943	42
1983	41
1993	41

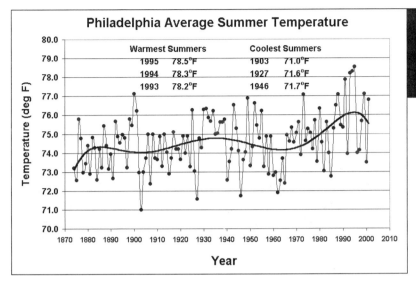

Philadelphia Average Summer Temperature

Warmest Summers		Coolest Summers	
1995	78.5°F	1903	71.0°F
1994	78.3°F	1927	71.6°F
1993	78.2°F	1946	71.7°F

ILLUSTRATION 5.2 Philadelphia's average summer temperatures, 1874 to 2001. The thicker line is a mathematically smoothed version of the data. The four warmest summers on record occurred in the 1990s.

Whether this tendency for relatively hot summers over the last fifteen years or so is just natural variability or related to other factors (such as global warming) is still a matter of debate (see Chapter 7).

The sun is highest in the sky in late June, but July is usually the hottest summer month. Though 90°F readings have been observed in Philadelphia as early as April 7 and as late as October 10, the probability of reaching 90°F on any given day is greatest from late June to early August, peaking at around 35 percent in mid-July. This hottest part of the summer is sometimes called the Dog Days, a term with an astronomical origin. Other than the sun, the brightest star in the sky is Sirius, which the ancient Egyptians called the Dog Star. In midsummer, Sirius rises and sets at about the same time as the sun. The ancients thought that Sirius's energy added to the sun's to help create heat waves. What they did not know was that Sirius is more than half a million times farther away than the sun, so its contribution to heating the Earth is insignificant. Thus, the Dog Star—or any other star for that matter—cannot be blamed for helping to bolster summer heat on earth.

Ninety degrees may be hot, but the threshold for extreme summer heat in the Philadelphia area has to be 100°F. Between 1874 and 2001, there have been just fifty-seven days when the temperature reached or exceeded the hundred-degree mark, an average of only about one every two years. By far, the most 100°F days have occurred in July (thirty-five days), with June (ten days) and August (nine days) distant runners-up. The hottest of the hot are listed in Table 5.4. Philadelphia's all-time record high temperature is 106°F, set on August

7, 1918 (see Illustration 5.3). On that day, the morning low was just 84°F. Under blazing sun, the temperature reached 100°F by 11 A.M. and stayed above 100°F until 6 P.M. Wilmington also set its all-time record high that day, one-upping Philadelphia by reaching 107°F.

The all-time high temperature records for Pennsylvania, New Jersey, and Delaware are all close to 110°F. Two of these three state records were set on July 10, 1936, a day still considered the hottest of the twentieth century in the mid-Atlantic region. The Pennsylvania record of 111°F was established in Phoenixville, a few miles northwest of Valley Forge in Chester County, while the New Jersey record of 110°F was set in Runyon, about twenty-five miles northeast of Trenton. The all-time Delaware state record is 110°F, set at Millsboro, just south of Georgetown, six years earlier on July 21, 1930.

TABLE 5.4 Hottest Days on Record in Philadelphia

Date	High Temperature (°F)
August 7, 1918	106
July 10, 1936	104
July 3, 1966	104
July 2, 1901	103
August 6, 1918	103
July 21, 1930	103
July 9, 1936	103
July 4, 1966	103
July 15, 1995	103

ILLUSTRATION 5.3 The official U.S. Weather Bureau observation form for August 7, 1918, the hottest day on record in Philadelphia. Temperatures are listed every hour, with a description of the sky condition below the temperatures. On the right is the summary for the day, showing that the "Maximum temperature" of 106°F occurred at 3:45 P.M. (courtesy of National Weather Service).

Cool Summers

Of course, some summers are hotter than others. However, the difference between Philadelphia's hottest summer (1995, average temperature 78.5°F) and coldest summer (1903, average temperature 71°F) is much less than the difference between the highest and lowest average temperature of any other season. This relatively small range results from the sun's relatively high position in the sky during summer, which means that daylights are long; sunshine simply controls temperatures to a greater extent in summer than in any other season. Even in that "coolest" summer in 1903, the temperature hit 90°F on ten days in July.

A summer that truly stands out as "unusually" cool occurred in the early nineteenth century, decades before organized systems of weather observing were in place. Yet there is good reason to trust the observations that are available from what was called the "Year Without a Summer."

In April 1815, a volcano in Indonesia named Mount Tambora exploded violently in what was probably the largest volcanic eruption in at least the last few centuries. An estimated thirty-six cubic miles of ash and dust spewed into the air, enough material to cover all of Philadelphia with a quarter-mile of debris. Just as important as the amount of debris was the explosiveness of the eruption; material was thrown very high into the atmosphere, by some estimates more than twenty-five miles. When material gets that high into the air, it is not removed easily by natural processes such as rain or settling to the ground.

Although it seems unlikely that an eruption halfway across the world would affect the weather in the eastern United States, high-altitude winds eventually spread the volcanic debris around the globe. The effect was to reflect a small percentage of sunlight back into space for a while after the eruption, which in turn produced a "global cooling." (A similar but smaller cooling occurred after Mount Pinatubo in the Philippines blew its top in 1991.) In North America and Europe alike, the following year, 1816, became known as the Year Without a Summer.

The following accounts of Charles Peirce from Morrisville, near Trenton, describe the summer of 1816. His observations demonstrate what a truly exceptional period that was:

June 1816: The medium temperature of this month was only 64, and it was the coldest month of June we ever remember; there were . . . severe frosts on several mornings. . . . Every green herb was killed, and vegetables of every description very much injured. . . . From six to ten inches of snow fell in various parts of Vermont; three inches in the interior of New York; and several inches in the interior of New Hampshire and Maine.

July 1816: The average temperature of this month was only 68, and it was a month of melancholy forebodings. . . . It seemed as if the sun had lost its warm and cheering influences. One frosty night was succeeded by another, and thin ice formed in many exposed situations in the country. Indian corn was chilled and withered. . . . Northerly winds prevailed a great part of the month; and when the wind changed to the west, and produced a pleasant day, it was a subject of congratulation by all.

August 1816: The medium temperature of this month was only 66 and such a cheerless, desponding, melancholy summer month, the oldest inhabitant never, perhaps, experienced. This poor month entered upon its duties so perfectly chilled, as to be unable to raise one warm, foggy morning, or cheerful sunny day. . . . Every green thing was destroyed, not only in this country but in Europe. Newspapers received from England said, "It will ever be remembered by the present generation, that the year 1816 was a year in which there was no summer."

To get a sense of the relative chill of the summer of 1816 in the Philadelphia area, consider that the average temperatures that Peirce reported would make June, July, and August of 1816 the coldest on record in Philadelphia for those months. His observations give an overall average summer temperature of 66°F, a full 5°F lower than the average during the coldest summer in the modern record in Philadelphia! Even considering that Peirce's observations were made about twenty-five miles northeast of Philadelphia (which means his readings would be about 2°F lower anyway), and also allowing for some instrument error, the summer of 1816 was very likely the coldest summer in this area (and much of the northeastern United States) since Europeans arrived in the New World.

SUMMER PRECIPITATION: DRENCHING RAIN AND DROUGHT

The heat and humidity so characteristic of a Delaware Valley summer also help make the period from June through August the wettest three-month stretch of the year. There is simply more fuel around—more of the sun's energy to warm the air and help make it buoyant, and more moisture in the air for thunderstorms to work with. But the heat of summer can also produce a very different result: strong sunshine means rapid evaporation, so the ground dries out fast. If thunderstorms become more "miss" than "hit," a deficit of moisture can quickly develop.

Thunderstorms and Flash Flooding

Heavy rain can occur any time of year in the Philadelphia area. But, by far, summer is the season when the atmosphere is most loaded with moisture. If we consider just those days on which at least two inches of liquid precipitation fell in Philadelphia (207 such days in the period 1872 to 2001), and categorize them by the season in which they occurred, we find that summer has almost as many (102 days) as the other three seasons combined (105 days).

Some of these super-soggy summer days result from tropical systems, such as the remnants of Hurricane Agnes in June 1972. (See the "Story from the Trenches" in this chapter.) But for the most part, the heavy rains come from individual gutter-gushing thunderstorms that, more often than not, develop in the heat and humidity of a summer afternoon or evening. Their torrential rains can cause localized flash flooding, overwhelming the drainage systems of streets, underpasses, and other low-lying areas, and causing water levels in creeks and small rivers to rise rapidly. Typically, this kind of localized flash flood occurs several times each summer in the Philadelphia area.

Most of the year, thunderstorms, steered by speedy winds a few miles above the ground, move along quickly enough so that flooding is not a problem. But those high-altitude winds are typically much weaker in summer and, at times, nearly calm. When this happens, thunderstorms can sit over the same spot for hours. Even if the steering winds are not that lazy, flooding can still occur if the winds blow parallel to a line of storms. When that happens, one thunderstorm after another passes over the same location like railroad cars in a train passing over a track. Appropriately, meteorologists call this process **training**.

In the Philadelphia area, it is not unusual for a thunderstorm in the moisture-laden air of summer to put down one to three inches of rain in an hour. So if thunderstorms are stationary or training, huge amounts of rain can fall, with devastating results. A classic example of flash flooding occurred on July 30, 2000, in Upper Southampton Township in Bucks County, Pennsylvania. Doppler radar and some observers at the ground estimated about eight inches of rain in three hours. Local creeks rose rapidly. Neshaminy Creek, near Langhorne, Pennsylvania, which typically runs about a foot deep, rose above flood stage to 12 feet in about two hours (see Illustration 5.4), causing tremendous damage to nearby homes. Three hours later, the water had fallen below flood stage, and by the next afternoon the creek had almost returned to its original level.

One of the worst flash floods in Philadelphia-area history occurred some 150 years ago. On August 5, 1843, a spectacular cloudburst over Delaware County turned the small creeks and streams entering the Delaware River into raging torrents. Rainfall that was estimated at as much as 16 inches in three hours caused flash flooding that destroyed thirty-two bridges and resulted in at least nineteen deaths. Of the flood, the Delaware County historian wrote: "The great freshet [sudden stream rise] of the 5th of August . . . may be regarded as one of the most extraordinary events that have occurred with the limits of our county since it was first settled by Europeans."

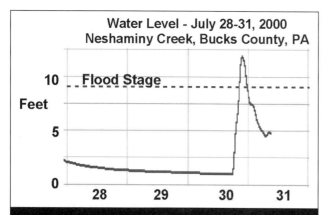

ILLUSTRATION 5.4 Flash flooding is illustrated by the rapid rise of Neshaminy Creek after extreme localized rainfall on July 30, 2000 in Bucks County, Pennsylvania. The rise in water level would be smaller and slower in the wider and deeper channel of a river (courtesy U.S. Geological Survey).

Tornadoes were also reported that day in Delaware County and farther east near the Schuylkill River, where thirty coal barges were sunk.

When extreme flooding (or any type of extreme weather) occurs, there is a temptation to categorize it as a once-every-so-many-years event. A common example is the "hundred-year flood," implying that a flood of that magnitude has a 1 percent chance of occurring at any given point in any given year. This kind of description can be misleading, for several reasons.

Such categorizations are based on historical data, and the only local measuring station with a fairly reliable long-term record is Philadelphia (and that observing station has moved over the years). The standard determined from the Philadelphia data does not necessarily extrapolate well to other nearby locations, where the topography and amount of urbanization might be different. Increased urbanization typically means that less and less rain is necessary to cause a "hundred-year" flood. Plus, with a larger population living in more places and with better weather monitoring systems, it is more likely now than in the past that extreme weather events will be reported, making them seem more common. And even if a rainfall at a particular location truly qualifies as a hundred-year storm, that is no guarantee it will not happen again in the next ninety-nine years. An event that has a 1 in 100 chance of occurring at any given time can, theoretically, happen more than once in a much shorter period of time—it is over the long haul that the once-per-century average would pan out.

Drought

If any season is burdened with pressure to perform, it has to be summer. It is the time of year when the atmosphere is most laden with moisture, and thus, on average, it is the wettest of all the seasons. If the expected precipitation does not materialize during summer, the shortfall can often be felt for months or longer. Still worse, summer's intense sunshine can quickly evaporate moisture from the ground, much faster than in any other season. If thunderstorms are infrequent, and the preceding months did not build up a surplus of rain, drought can easily develop in summer or in the months that follow.

During summer, heat and dryness can feed off each other to worsen the situation. Imagine sunshine beating down on the land. Some of the sun's energy goes toward heating the ground and thus warming the air above. But some energy goes toward evaporating water and thus drying out the ground. As the land dries out, less of the sun's energy is needed for evaporation, so more goes toward heating. The more the ground is heated, the warmer the air gets; and the warmer the air gets, the more the ground dries out. And so on. Only significant rains can break the vicious cycle.

The term **drought** is usually reserved for a period of moisture deficit that combines several features: it must be long lasting and widespread in area, and it must have a significant economic impact, perhaps on crops or water supplies. One of the earliest mentions of drought in Colonial America dates from 1762. Of the weather in the Philadelphia area, Charles Peirce writes:

> The severest drought ever experienced in America was in the summer of 1762. Scarcely a sprinkle of rain fell for nearly four months, from May to September. Vegetables of every description perished.

A letter from a Mr. John Watts of New York City, dated August 23, 1762, reads in part:

> All necessarys for Life both for Man and Beast are astonishingly dear & scarce, owing in a great Measure to the most severe Drouth that ever was known in this part of the World. How food will be provided for Cattle during a long dreadful Winter, God alone knows.

Because the Delaware River is tidal up to about Trenton, drought brings an additional impact to the Delaware Valley: salt-laced water can shift farther upstream during a drought because of reduced stream flow. As the saltier water moves upriver, it can increase corrosion control and water treatment costs for industries and public water suppliers. The upstream advance and retreat of saltier water

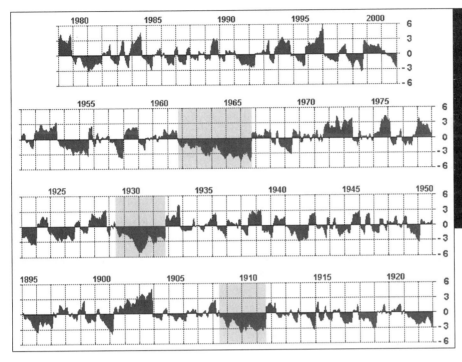

ILLUSTRATION 5.5 The Palmer Drought Severity Index (PDSI) from 1895 to 2001 for southeastern Pennsylvania. Positive values indicate a long-term moisture surplus and negative values a long-term moisture deficit. Values less than −3 indicate severe drought. The time periods corresponding to three of the longest and most serious droughts are shaded: 1961–66, 1929–32, and 1908–11 (courtesy of National Climatic Data Center).

on the Delaware River is usually monitored in terms of the "salt line." You cannot see the salt line—its location is defined based on drinking water quality standards established by the government. The salt line's average location varies over the course of the year, from several miles south of the Delaware Memorial Bridge in April to a few miles north of the Delaware–Pennsylvania state line in October; but during droughts it can shift many miles upriver.

Droughts do not develop overnight. Although dryness can quickly worsen during summer, the seeds of drought are often sown months before. A good example is the drought during the summer of 1999. It can be traced back to the second half of 1998 when less than half of the average precipitation fell in the Delaware Valley, building up a deficit of nearly a foot of water. The period January to May 1999 had near-average rainfall, but then the driest June–July period on record followed. Less than 2.5 inches of rain fell in Philadelphia during those two months (the average is about eight inches). During the drought, the salt line on the Delaware River migrated nearly to the Philadelphia International Airport, about fifteen miles north of its average location at that time of year. More frequent rains returned in the second half of August, and the short drought ended in mid-September when Hurricane Floyd brought between 6 and 12 inches of rain. That scenario—tropical moisture contributing to a drought's demise—often plays out in our area.

At the simplest level, drought is a matter of supply and demand—runoff and evaporation claiming significantly more water than precipitation delivers over long time periods. But evaporation and runoff depend on many other variables, such as temperature, soil moisture, and the presence of a snow pack. So to more accurately assess drought, it is useful to compute a "drought index" that assimilates a wide variety of data on rainfall, snowpack, temperature, stream flow, and other water-supply indicators into one number that describes the big picture.

One index that is widely used in the United States is the Palmer Drought Severity Index (PDSI). The PDSI can be used to compare different locations and different times, so it is useful for placing conditions of extreme dryness (and wetness) in a historical perspective. Illustration 5.5 shows the PDSI for southeastern Pennsylvania from 1895 to 2001. Positive values of the PDSI indicate a moisture surplus, negative values a moisture deficit. A PDSI value of −3 indicates severe drought, while +3 indicates a very moist spell. As with other measures of climate, the PDSI shows tremendous variability over time.

Based on the PDSI, the drought during the summer of 1999 was nothing compared to the worst droughts of the last century. The period from 1908 to 1911 was exceptionally dry, reducing the Schuylkill River to a trickle at times (see Illustration 5.6). Philadelphia's driest year,

ILLUSTRATION 5.6 *(See plate.)* The Schuylkill River at Flat Rock Dam in the summer of 1909 in the midst of a severe drought (Library Company of Philadelphia).

1922, was at the center of a three-year drought from 1921 to 1923. And the period from July 1929 to September 1932 saw several months of extreme drought, including the lowest PDSI value ever in the Philadelphia area in January 1931. That drought was peaking just as the infamous Dust Bowl years in Kansas, Oklahoma, and Texas were beginning.

In terms of duration and overall intensity, however, the most severe drought on record in this area started in 1961 and persisted through the summer of 1966. If there was a "Dust Bowl" in the Philadelphia area in modern times, that was it. The dominant upper-air pattern during the drought was a familiar one for dryness in the northeastern United States—a ridge over the Midwest and a trough off the East Coast. With this pattern, a dry northwesterly flow dominated Pennsylvania, New Jersey, and Delaware much of the time, limiting the flow of moist tropical air from the south. As a result, low-pressure systems had less moisture to work with, fewer thunderstorms formed during the summer, and tropical systems tended to stay away from the area. From January 1963 to August 1966, a deficit of nearly 40 inches of liquid precipitation built up—essentially, that is a year's worth of rain and snow that did not fall. During that forty-four-month period, only ten months had above-average precipitation. The second and third driest years in Philadelphia history are 1965 (29.34 inches) and 1964 (29.88 inches). The summers were especially dry, with the months of May, June, July, and August accounting for 70 percent of the precipitation deficit. October 1963 has the dubious distinction of

being tied for the driest month ever, with only 0.09 inches of rain.

The sparseness of precipitation severely reduced the flow rate on the Delaware River in 1964 and 1965 (see Illustration 5.7). With the downstream flow reduced, there were serious concerns that the salt line might reach the Torresdale intake plant, where Philadelphia took in about half its water from the Delaware River, and contaminate the city's water supplies. In October 1965, the salt line crept north to just past the Benjamin Franklin Bridge, still the farthest upstream advance on record, but remained about ten miles south of the Torresdale plant. The limited water flow of the river created great conflict among government leaders from urban areas of New Jersey, New York, Delaware, and Pennsylvania over the equitable allocation of the Delaware basin waters. Mediating the dispute was the first big test for the newly created Delaware River Basin Commission, which today still manages the river system. Local relief from the drought came in the form of more than a foot of rain in Philadelphia spread over a five-week period in the fall of 1966. By the end of the winter of 1966–67, the reservoirs of the Delaware (which are located in the Catskill Mountains of New York State) were back to near-normal levels.

ILLUSTRATION 5.7 *(See plate.)* The Delaware River at Trenton on August 30, 1963. The average flow rate of the river at Trenton in late August is about 4,000 cubic feet per second (ft³/s); the average this day was 1,930 ft³/s. Stream flow would fall to 1,300 ft³/s in October of 1963, and to the all-time daily average low of 1,240 ft³/s on July 10, 1965 (Delaware River Basin Commission).

AIR POLLUTION

Air pollution is not exclusively a summer problem. But one of the most troublesome pollutants today is ground-level ozone, which forms only in sunny, warm weather. Thus, air pollution season in the Philadelphia area is primarily from May to September. Air pollution is also not exclusively a city problem, though pollution emissions and thus pollution levels in the air tend to be highest in urban areas.

Air pollution can be visible or invisible, smelly or odorless, and it can be produced indoors or outdoors; it can take the form of gases, liquid drops, or solid particles. Some air pollution is natural, belched from volcanoes or produced in forest fires. But, by far, human activities produce the most air pollution, with transportation, power plants, industry, and consumer uses (such as home heating) as the primary sources. Some of the principal air pollutants are:

- *Ozone:* The main ingredient of smog (a term created by combining "smoke" and "fog"). Ozone is not emitted directly into the air, but rather forms when other air pollutants chemically react on sunny hot days.
- *Particulates:* Solid or liquid particles suspended in air that can sometimes be seen as soot or smoke. Particulates come mainly from motor vehicles and power plants.
- *Nitrogen Dioxide:* One of a group of gases called nitrogen oxides that form when fuel is burned. The main sources are motor vehicles and electric utilities.
- *Sulfur Dioxide:* Produced when fuel containing sulfur is burned and during other industrial processes.

Today, the most publicized local, regional, and national pollution concern is ozone's adverse affect on health. However, there is increasing evidence that particulates are a serious health threat as well, especially in urban areas. In large cities such as Philadelphia, researchers have found that on days when particulate levels jump but remain within federal standards, there are more hospitalizations and deaths of elderly people with cardiac and lung disease. In addition, hundreds of other chemicals enter the air each day, primarily from factory smokestacks and vehicle exhausts. Some of these toxic air pollutants, such as benzene, are known human carcinogens.

Besides being a health hazard, air pollution has other detrimental effects. Pollution increases haze and reduces visibility. In addition, nitrogen dioxide and sulfur dioxide react with water vapor in the air, creating acidic clouds and **acid rain**. In turn, air pollution and acid rain speed up the corrosion of metals and materials, contributing to the decay of buildings and statues. That decay is particularly noticeable in urban areas. A 1987 University of Delaware study of the deterioration rates of tombstones in the Philadelphia region found that the rate was ten times higher in the city compared to the nearby countryside, with sulfur dioxide the main pollutant responsible. The rate of decay greatly accelerated between 1930 and 1960, a period when Philadelphia was one of the most polluted urban centers in the United States. Here is a brief look at the history of the air pollution problem, both locally and nationally.

A Short History of Air Quality

Air pollution is far from a new problem, although the pollutants of most concern have changed over the years. Today, on a national level, the Environmental Protection Agency (EPA) oversees the protection of public health and the environment from the adverse effects of air pollution. At the state level, agencies such as Pennsylvania's Department of Environmental Protection and Delaware's Department of Natural Resources and Environmental Control assist in this responsibility. In Philadelphia, the Division of Air Management Services—a branch of the Department of Public Health—operates air-monitoring stations across the city and works with industry to help regulate and reduce air pollution emissions.

Air pollution problems both locally and nationally can be traced to the Industrial Revolution, which really took hold in the United States after the Civil War. Over time, air quality declined as homes and coal-burning industries belched smoke, soot, and other pollutants into the air. By 1940, the air over some large cities was so dirty that automobile headlights had to be turned on during the day. The first major documented air pollution disaster in the United States occurred in October 1948 in the small community of Donora, Pennsylvania, which sits in a river valley about twenty miles south of Pittsburgh. Over a five-day period, twenty people died and thousands became ill as a pall of industrial pollution became trapped over the city.

Another major air pollution episode occurred in urban regions of the Northeast from November 10 to November 22, 1953, contributing to the deaths of at least 175 people in the New York City area from cardiac and respiratory distress. Smog in Philadelphia was so bad that some people wore masks over their mouth and nose, even indoors (see Illustration 5.8). During the episode, the Philadelphia weather observer reported "smoke" every hour for twelve straight days. During some afternoons, visibility dropped to only three-quarters of a mile because

ILLUSTRATION 5.8 *(See plate.)* Smog was so bad in downtown Philadelphia in mid-November 1953 that some workers at the *Inquirer* building wore masks *indoors* (Urban Archives, Temple University, Philadelphia, Pennsylvania).

of the pollution. And these observations were taken at the airport, not downtown.

With air pollution episodes such as these as motivation, Congress passed the Air Pollution Control Act in 1955, the first federal legislation to address air quality. That bill acknowledged pollution as a national concern and granted $5 million annually for research, but did nothing to regulate emissions. The Clean Air Act of 1963 was the first legislation to set emissions standards for stationary sources such as power plants and steel mills, while a follow-up amendment in 1965 established standards for automobile emissions. But legislation could not keep up with industrialization, and the 1970 Clean Air Act was needed to establish new standards for air quality and emissions and to increase funds for research. Twenty years later, the Clean Air Act of 1990 not only addressed air quality standards and motor vehicle emissions but also alternative fuels, toxic air pollutants, acid rain, and stratospheric ozone depletion.

Air pollution laws have been on the books in Philadelphia at least as far back as the early 1900s. The introduction to the city's "1904 Smoke Ordinance" gives insight into the pollution issues at that time:

> An Ordinance To Regulate The Emission Of Smoke From Chimneys, Stacks, Flues Or Open Spaces Within The City Of Philadelphia, Providing A Color Scale For The Measurement Of The Degree Of Darkness Of Such Smoke, Making It Unlawful To Permit The Escape Of Smoke Of A Certain Degree Of Darkness And Providing A Penalty For The Violation Of This Ordinance.

Despite such ordinances, pollution emissions in Philadelphia increased rapidly in the first half of the twentieth century, driven by a rapidly growing population, expanding industry, and increasing use of automobiles. Starting in the mid-1950s, however, a switch from coal to cleaner-burning oil and natural gas resulted in major reductions in the emissions of sulfur dioxide, nitrogen dioxide, and particulates. From the mid-1960s to the mid-1970s, emissions of smoke and particulates from trash incineration were reduced as well. Meanwhile, the Philadelphia area was evolving from a diversified industrial base to a more service-oriented community.

These efforts transformed the city's air quality. In 1966, the first year that comprehensive emissions data are available, a total of 1,115,000 tons of particulates, sulfur oxides, nitrogen oxides, carbon monoxide, and other pollutants was emitted into Philadelphia's skies. By 1985, emissions had dropped to 363,000 tons, a decrease of nearly 70 percent. Almost one-half of the reduction in emissions resulted from the switch from coal to low-sulfur oil and gas. Another one-third resulted from controls on the sources of air pollution. The remainder came from industrial shutdowns, cutbacks, and energy conservation.

At the present time, the only pollutant that exceeds federal standards on a regular basis in the Philadelphia area and surrounding counties is ground-level ozone. For that reason, and because ozone is also at the center of another high-profile environmental issue (the "ozone hole"), it deserves a closer look.

Ground-level Smog: The "Bad" Ozone

Ozone is probably the most misunderstood gas in the atmosphere. Often you will hear that there are two types—"good" ozone and "bad" ozone. In reality, ozone is ozone—a colorless, poisonous gas that is actually a relative of oxygen. Ozone's good/bad reputation arises from its two dramatically different effects, depending on where the ozone resides in the atmosphere. The EPA has a catchy rhyme that summarizes ozone's double life: "Ozone is good up high, but bad nearby."

"Up high" means primarily between ten and thirty miles above the earth's surface, in the layer of the atmosphere called the **stratosphere**. At such high altitudes, ozone protects life on earth by absorbing some of the sun's ultraviolet radiation, preventing those harmful rays from reaching the surface. This is ozone's "good" side. But ozone can also form at ground level on hot sunny days as the product of chemical reactions involving other pollutants. Down here, where we can breathe it, ozone is a noxious pollutant and the primary ingredient in smog.

For ground-level ozone to form, other pollutants must be present first. One group of these pollutants, the nitrogen oxides, comes mainly from motor vehicle exhaust and power plants. Nitrogen dioxide is a member of this group. The other pollutants are chemicals called **volatile organic compounds**—VOCs for short. These are vapors from fuels, solvents, paints, and other industrial and consumer products—basically, a wide variety of chemicals that evaporate easily. In any urban area, dozens (if not hundreds) of different VOCs are in the air in small amounts. If you have ever smelled the aromas from gasoline, paint, and other consumer products, you have smelled VOCs.

The ozone-producing reactions between nitrogen oxides and VOCs require sunshine and hot weather, so ground-level ozone is mainly a warm-season phenomenon—May to September in the Philadelphia area. Ozone concentrations usually peak in late afternoon, when temperatures are highest and nitrogen oxides and VOCs have had time to accumulate in the air. And ozone pollution tends to be a bigger problem when winds are light and the polluted air is not dispersing. (However, ozone-polluted air can blow in from other areas, so even strong winds do not guarantee low ozone concentrations.)

As for ozone's adverse health effects—breathing air with elevated ozone levels can irritate and inflame the respiratory system. Much like the way the sun causes sunburn on the skin, ozone can "burn" cells in the lung's airways, potentially leading to permanent damage. According to the EPA, ozone smog is linked to 10 to 20 percent of all summertime respiratory-related hospital admissions in the northeastern United States. Children are especially vulnerable to ozone pollution because their respiratory systems are not completely developed. Asthma can also be aggravated by ozone pollution. The rate of death from asthma in the United States is three times greater now than it was just twenty years ago. Children make up 25 percent of the entire population but account for 40 percent of asthma cases.

Being active outdoors, especially during the summer, also increases your susceptibility to ozone pollution, regardless of age. Physical exertion causes you to breathe faster and more vigorously, so ozone-polluted air can penetrate deeper into the lungs. Even healthy adults who are exercising outside can experience reduction in lung function on days when ozone pollution is high.

The increased concern about the unhealthy effects of ozone pollution has led to a greater emphasis on publicizing and predicting the threat. Next-day forecasts of ozone levels for the Philadelphia area (and many other cities nationwide) are issued from May to September, and increasing numbers of media outlets relay the forecasts. To simplify the reporting, the EPA defines air quality as good, moderate, moderately unhealthy, or unhealthy, and assigns each category a color and a corresponding health notice (see Table 5.5). When ozone concentrations reach the Code-Orange and particularly Code-Red categories, people are advised to take protective action.

The "Ozone Action" program helps raise public awareness in Code-Red situations. In the Philadelphia

Table 5.5 Air Quality Index		
Category	Color	Cautionary Statements
Good	Green	None
Moderate	Yellow	Unusually sensitive people should consider limiting prolonged outdoor exertion
Moderately unhealthy	Orange	Susceptible people should limit prolonged or heavy outdoor exertion
Unhealthy	Red	Susceptible people should avoid, and everyone else should limit, prolonged or heavy outdoor exertion

area, this program is managed by the Delaware Valley Regional Planning Commission (DVRPC). When the forecast calls for ozone levels to reach the unhealthy range (Code Red) the next day, the DVRPC declares an Ozone Action Day. This declaration is a health notice, aimed particularly at those most susceptible to ozone's harmful effects. But it is also a call for everyone to try to reduce the build-up of ozone. As part of the Ozone Action program, more than 150 companies and organizations in the Delaware Valley have voluntarily agreed to help the DVRPC get the word out on Ozone Action Days by encouraging their employees to reduce ozone levels.

The key to reducing ground-level ozone is to cut the emissions of nitrogen oxides and VOCs, the pollutants necessary for ozone to form. Nitrogen oxides come mainly from motor vehicles and power plants, so limiting driving and conserving energy are good strategies when an Ozone Action Day is declared (they are good ideas all the time, really). You can reduce VOCs by properly using and sealing cleaners, paints, and other chemicals. When refueling, try not to spill any gas, and always tighten the gas cap.

Other local initiatives aimed at reducing the amount of nitrogen oxides and VOCs in the air include:

- Vehicle emission inspections. These are required in Philadelphia, Delaware, Chester, Montgomery, and Bucks counties in southeastern Pennsylvania, and all of New Jersey and Delaware.
- Gasoline recovery systems to reduce spills and escaping vapors. These are the flexible covers you see on the nozzles of gas pumps.
- Use of reformulated gasoline in areas with the worst smog pollution, including southeastern Pennsylvania and all of New Jersey and Delaware. This type of gas does not evaporate as easily and produces about 15 percent fewer VOCs when burned.

To some extent, there has been progress in combating ozone smog pollution. Overall, ground-level ozone concentrations have decreased since the late 1970s, both nationally and in the Delaware Valley. Nonetheless, tens of millions of people still live in areas that do not meet the national health standards for ground-level ozone. These "non-attainment" areas include Philadelphia, Montgomery, Bucks, Chester, and Delaware counties in southeastern Pennsylvania; Kent and New Castle counties in Delaware; and nearly all of New Jersey and northeastern and central Maryland.

The ground-level ozone pollution problem will not go away easily—not in the Delaware Valley or elsewhere.

Many of the products and services we use on a daily basis produce the very pollutants that make ozone formation possible. And because the number of hot, sunny days varies so much from year to year, the weather will always be a wild card in determining how frequently ground-level ozone reaches unhealthy levels in any given summer.

Stratospheric Ozone: The "Good" Ozone

The build-up of ground-level ozone pollution to unhealthy levels is not a natural process. Without human-made pollutants, it would not occur. But ozone does form naturally in the stratosphere, ten to thirty miles up. There, far from the surface, ozone poses no threat to human health. In fact, without this **ozone layer**, life as we know it could not exist. This protective shield of high-altitude ozone absorbs the sun's most damaging ultraviolet (UV) radiation, preventing this radiation from reaching the ground where it can lead to increases in skin cancer, cataracts, and immune system disorders. So any threat to this "good" stratospheric ozone is bad news.

Here is where humans again enter the ozone story. In 1928, researchers at General Motors developed a new group of gases called **chlorofluorocarbons** (CFCs for short). Originally used only as refrigerants, CFCs proved very versatile over the years—among other applications, they have been used as propellants in aerosol cans and in the production of foams such as polyurethane. Over time, a very tiny amount of CFC gas seeped into the air—current concentrations amount to only about one CFC molecule in every one billion air molecules.

But sometimes a little can go a long way. In 1974, chemists Sherwood Rowland and Mario Molina of the University of California at Irvine discovered that CFCs pose a threat to stratospheric ozone. Once in the air, CFCs do not flush out easily—they stay long enough to spread worldwide and for some of the gas to make its way up into the stratosphere. There, through a series of chemical reactions, chlorine from the CFCs destroys ozone. Rowland recognized the potential seriousness of the problem when he told his wife, "My work is going very well, but it looks like the end of the world."

The discovery made headlines worldwide (and eventually earned Rowland and Molina the Nobel Prize in Chemistry). In the years that followed, CFC releases into the atmosphere actually declined. The United States even banned CFC use in aerosol spray cans. But by then it was too late. There was already evidence that too much CFC gas had accumulated in the air.

Since the late 1950s, British scientists had been measuring stratospheric ozone above the Antarctic. They noticed that each year during late winter and early spring (September and October in the Southern Hemisphere) ozone concentrations would mysteriously decline, only to rebound in the months afterward. By the mid-1970s, however, the amount of ozone that disappeared was increasing, and the recovery was not as great. They had discovered the **ozone hole**, a yearly thinning in the stratospheric ozone above Antarctica. (Many scientists prefer the term "ozone depletion area" simply because there is not really a "hole" in the atmosphere but rather just a decrease in one gas.) In any case, the depletion has progressively worsened so that during some Septembers and Octobers in the 1990s, more than half the ozone in the Antarctic stratosphere disappeared. In September 2000, the ozone depletion area was the most extensive ever observed, three times larger in area than the United States (see Illustration 5.9). Research has shown that chlorine from CFCs and other human-made chemicals is the culprit in the ozone depletion.

But why over Antarctica? And is such extensive ozone depletion occurring anywhere else, closer to home? After all, the Antarctic is a long way from the Delaware Valley.

The Antarctic stratosphere is unique in a few ways. First, it is colder there in late winter than anywhere else in the stratosphere at any time of year. At temperatures of −80°F or below, special clouds called **polar stratospheric clouds** form. The chemical make-up of these clouds speeds up the reactions that destroy ozone. Second, the isolation of the air over Antarctica is a factor. During winter, a jet stream called the **polar vortex** develops around the continent. This ring of air acts like a fence encircling the continent, essentially preventing air over Antarctica from mixing with air from other latitudes, which have more ozone.

Now to the other pole. An Arctic counterpart to the Antarctic ozone hole has not been observed, mainly because the air in the Arctic stratosphere is not as cold as in the Antarctic. However, some thinning of Arctic ozone does occur in late winter and early spring, and stratospheric ozone concentrations are now about 25 percent lower during those months than in the mid-1970s. Although this does not approach the ozone loss in the Antarctic, it is potentially a more serious health concern because most of the world's population lives in the midlatitudes of the Northern Hemisphere. Overall, since 1979, the total amount of stratospheric ozone over the midlatitudes (and that includes the Delaware Valley) has decreased between 3 and 6 percent. This translates into

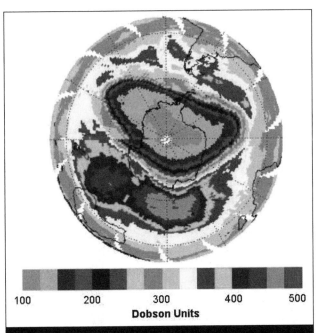

100   200   300   400   500
Dobson Units

ILLUSTRATION 5.9 *(See plate.)* Looking "down" on the ozone hole on October 1, 2000. The South Pole is in the middle. The region of depleted ozone levels is the area of pink and gray roughly centered over Antarctica. "Dobson Units" measure ozone concentrations; values from 300 to 350, shown here as yellow or brownish-yellow, are about average (courtesy of NASA).

a 4 to 7 percent increase in the amount of ultraviolet radiation reaching the ground—not a desirable trend given the link between overexposure to ultraviolet radiation and increased incidences of skin cancer.

Fortunately, there is good reason to believe that the trend will not continue indefinitely, thanks to considerable international cooperation aimed at reducing releases of CFCs. In 1987, forty-three nations signed the Montreal Protocol, an agreement that required a 50 percent cut in CFC production and use by 1999. Several amendments quickened the phasing-out process so that by 1996, CFCs were no longer being produced in developed countries. As a result, the total amount of ozone-depleting chemicals in the lower atmosphere peaked in the mid-1990s and is now slowly declining. Higher up in the stratosphere, where CFCs are harder to remove, the peak is expected to occur early in the twenty-first century. Without a doubt, the Montreal Protocol is working.

To say that this was a case of a potentially catastrophic atmospheric cancer being caught before it spread too far is not an overstatement. If the damaging effects of CFCs on ozone had not been recognized in the 1970s, and swift action had not been taken, at least half the ozone above our heads in the midlatitudes of the Northern Hemisphere would probably be gone today. That is ten times the actual loss. Still, the ozone depletion problem is far from solved. Because CFCs have a long atmospheric lifetime, the ozone hole will likely be an annual occurrence for decades to come. And there are other human-made, ozone-eating chemicals besides CFCs that are currently being monitored and regulated. But if there are no surprises, and the provisions of the current ozone treaties are followed, the ozone layer worldwide should slowly recover. It might, however, take most of the twenty-first century.

One positive side effect of the ozone-depletion issue is a heightened public awareness of the potential danger of overexposure to the sun's rays. By now people know to protect their skin when out in the sun for extended periods of time, especially in the middle of the day when the sun is highest in the sky. Sunscreen does for individuals essentially what the ozone layer does for the entire planet—it prevents the sun's most damaging ultraviolet rays from getting through. Still, many people tend to disregard the need for sunscreen at home, even though most of us log the most sun exposure while gardening, jogging, walking the dog, or doing other routine activities. One place where public awareness about skin cancer risks is very much on display is at the local pool, where the aroma of sunscreen usually fills the air. Philadelphia-area residents looking for a bigger "pool" need only head east to the beaches of New Jersey and Delaware. And that is how we close this chapter—with a trip to the shore!

THE SHORE

Yes, summer in the Philadelphia region brings heat, humidity, and occasional air-quality problems. But one of the advantages of living in this area is the easy access to the shore. The word *shore* is not commonly used elsewhere in the country, so a statement such as "I'm going to the coast (or the beach) for the weekend" immediately identifies the speaker as a transplant to this area. People from our area "go down the shore" for weekends, vacations, or the entire summer.

Some people have speculated that the fascination with the ocean goes back to the comfort of the womb. Whatever the reason, millions of Americans who visit the shore as children develop a life-long love of the sun, the sand, and the water. The primary shore season begins on Memorial Day weekend and lasts until Labor Day, coinciding with the hottest and most humid time of year. Many of us anticipate our escape from the sweltering cities months before the sea-

son. Drawn by the more comfortable weather, the fun at the beach, and the freedom from work or school stress, we accumulate great memories and return year after year.

The Jersey shore was the first of all United States ocean beaches to be developed extensively. According to Professor Orrin Pilkey, a noted coastal geologist at Duke University, the first advertisement for beachfront tourist accommodations appeared in 1802 in the Philadelphia *Aurora* magazine, proclaiming the "beauty of the shore and quality of the lodging in Cape May."

As meteorologists, we think that a major reason for the appeal of the shore lies with the wind.

Sea Breezes

In summer, the prevailing wind direction in Philadelphia and surrounding counties is from the southwest. But that is not the case close to the shore. From late spring into summer, the ocean exerts a profound influence on the wind almost daily in the form of the **sea breeze**. This localized wind forms because of the vastly different rates at which land and sea heat up. Land warms much faster than water, so when the sun comes out, air over the land quickly becomes warmer than air over the nearby ocean. The warmer air over the land rises, and in response, cooler air from over the water moves inland. This creates an onshore wind—the sea breeze. Unless the overall weather pattern favors a strong offshore wind in the morning, the onshore sea breeze usually takes over by midday.

Here is a typical sequence of daily wind shifts at the shore during the late spring and summer: at night and in the early morning, the land is often cooler than the ocean—the reverse of the situation during the day. So there is a slight tendency for the air over the water to rise and a light breeze to blow offshore—a **land breeze**. But as the ground starts heating up, the wind becomes calm, or nearly so. With more heating, a wind from the southeast develops, starting a temperature drop at the beach as cooler air moves ashore. As the southeast wind increases, the cooler air moves even farther inland, sometimes tens of miles. On rare occasions, the sea breeze can even reach the Delaware River. The leading edge of the cooler ocean air can even act as a sort of "mini" cold front, lifting the warmer air over the land and forming a line of clouds and sometimes even a few showers (see Illustration 5.10).

The cooling touch of the sea breeze is most dramatic in May and June, when ocean temperatures are still relatively low, in the 50s and 60s. On a day when temperatures reach the 80s farther inland, shore areas can be

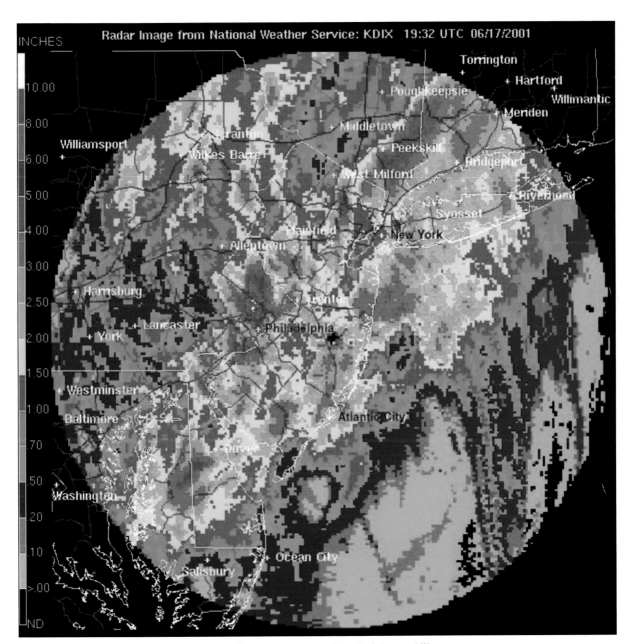

Plate 1.20   Doppler-radar estimated rainfall from remnants of Tropical Storm Allison

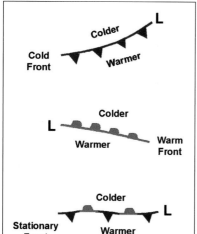

Plate 2.6   Three basic types of fronts

Plate 2.11   Topographic map (elevation in feet)

Plate 2.18   ENIAC, developed in the 1940s at the University of Pennsylvania

Plate 3.3b   A helicopter assists stranded Pennsylvania Turnpike travelers after a March 1958 snowstorm

Plate 3.9   The fountain at Logan Square, winter of 1924–25

Plate 3.17   Columbia Avenue in Philadelphia after the 1909 Christmas snowstorm

Plate 3.20   Snowfall (in inches) from the Blizzard of '96

Plate 3.23
Blizzard of '96 in
South Philadelphia

Plate 3.27   After the blizzard of February 1899 in Philadelphia

Plate 4.4 A destroyer is grounded on Long Beach Island after the March '62 nor'easter

Plate 4.6   Wildwood, New Jersey, flooded by the March '62 nor'easter

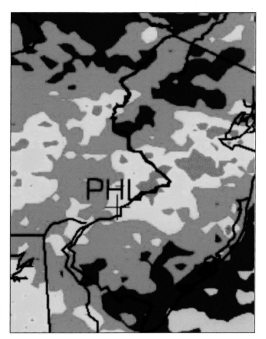

Plate 4.14   Average annual lightning density

Average Annual Flashes
(flashes per mi² per year)

0   2.5   5.0   7.5

Plate 4.15   Damage in Camden from a probable microburst in May 1933

Plate 4.20   An F1 tornado in Atlantic County, New Jersey, on July 5, 2001

Plate 5.6   The Schuylkill River during a severe drought in 1909

Plate 5.7   The Delaware River at Trenton, New Jersey, during a drought in 1963

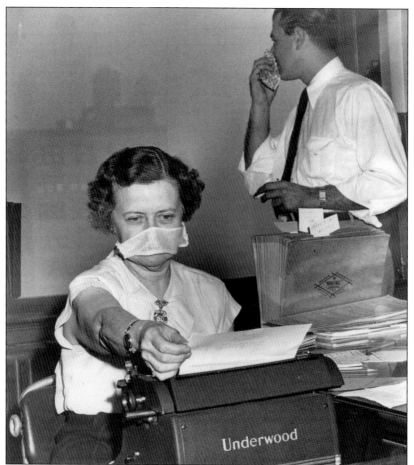

Plate 5.8   Masks were needed indoors because of smog in downtown Philadelphia in November 1953

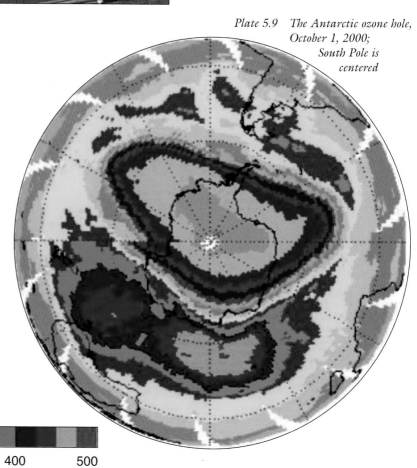

Plate 5.9   The Antarctic ozone hole, October 1, 2000; South Pole is centered

100   200   300   400   500

Dobson Units

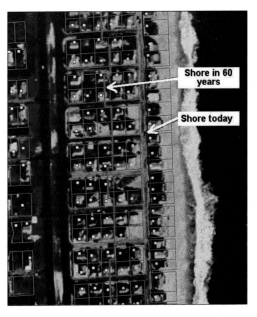

Plate 5.14 Aerial view of South Bethany, Delaware, showing projected shore erosion

Shore in 60 years

Shore today

Plate 6.10 Margate City, New Jersey, after the September 1944 hurricane

Plate 6.12   Boats damaged in Philadelphia by Hurricane Hazel in October 1954

Plate 6.13   Flooding in Philadelphia from Hurricane Connie, August 13, 1955

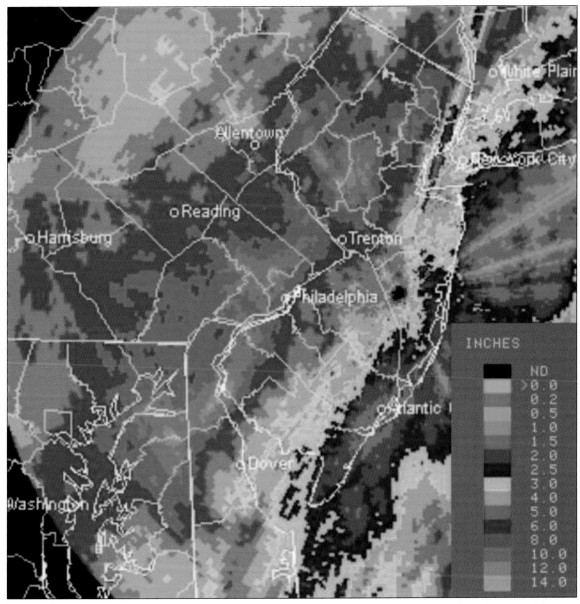

INCHES

	ND
	>0.0
	0.2
	0.5
	1.0
	1.5
	2.0
	2.5
	3.0
	4.0
	5.0
	6.0
	8.0
	10.0
	12.0
	14.0

Plate 6.18   Doppler-radar estimated rainfall from Hurricane Floyd

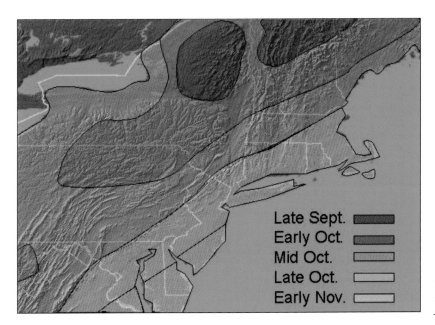

Late Sept.
Early Oct.
Mid Oct.
Late Oct.
Early Nov.

*Plate 6.24
Average peak times
for fall foliage*

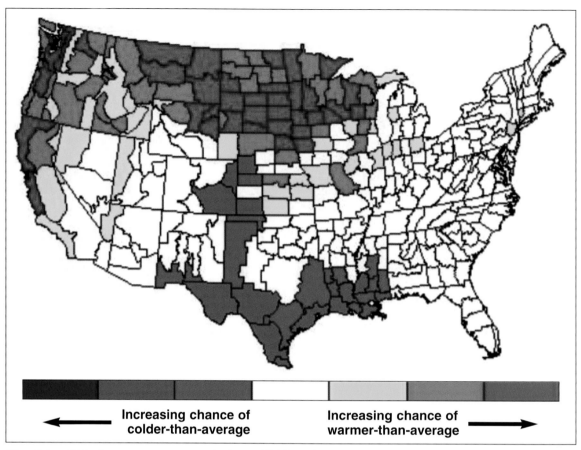

Plate 7.4a   Risk of temperature extremes during an El Niño winter

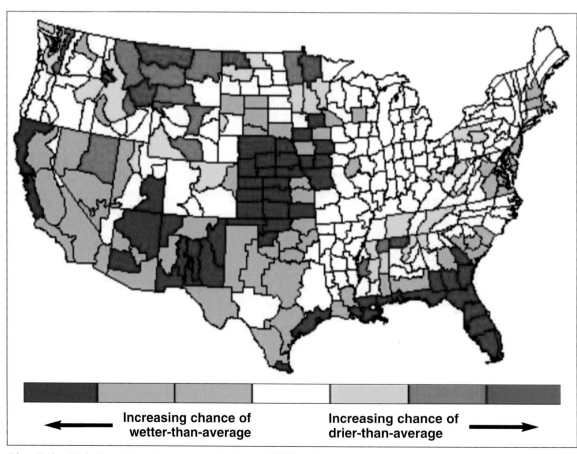

Plate 7.4b   Risk of precipitation extremes during an El Niño winter

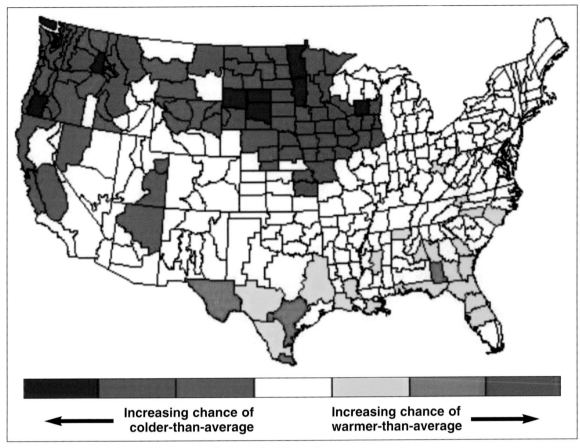

Plate 7.5a Risk of temperature extremes during a La Niña winter

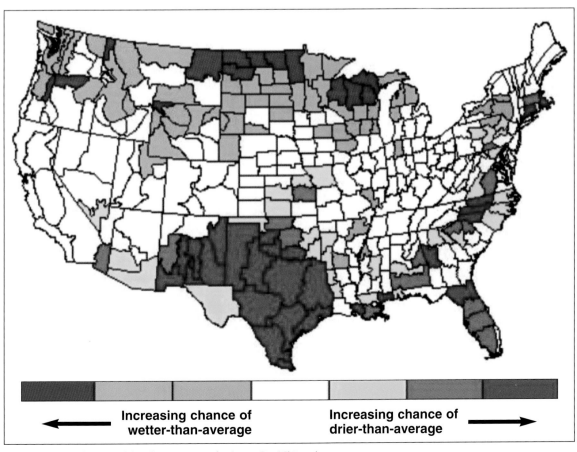

Plate 7.5b Risk of precipitation extremes during a La Niña winter

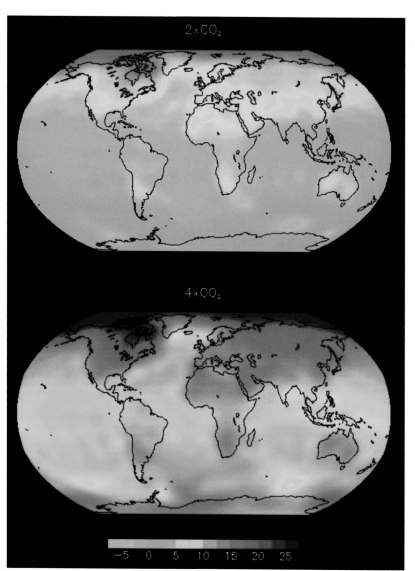

2×CO₂

4×CO₂

Plate 7.13   A general circulation model projection of global warming (in degrees Fahrenheit) if carbon dioxide is doubled or quadrupled; red indicates the most warming

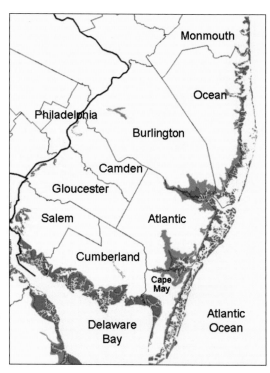

Plate 7.15   Coastal areas most vulnerable to rising sea level. Red areas are 5 feet or less in elevation; blue areas are between 5 and 11.5 feet

The Naturally Changing Coastline

Many of the biggest and most appealing shore communities sit on barrier islands, connected to the mainland by causeways and bridges. These islands have not been there forever. They have complicated histories and are continuously being reshaped and eroded. Wind plays an important but indirect role in these changes because winds are the driving force behind waves. With regard to waves, Benjamin Franklin had a keen eye in 1773 when he wrote:

> Air in Motion, which is Wind, in passing over the smooth surface of Water, may rub, as it were, upon that Surface, and raise it into Wrinkles, which if the Wind continues are the elements of future Waves.

Whether waves are whipped up by locally produced winds such as the sea breeze or by the gale-force winds of distant storms, they are constantly modifying the shoreline. The speed of the wind, how long it blows, and the distance of water it crosses all determine the waves' height. In general, stronger winds that blow for a long time over a great stretch of ocean produce the biggest waves. And the pounding on the shoreline is constant. Ten waves per minute breaking on the shore add up to more than 14,000 a day and about 100,000 per week. Given that a typical Atlantic wave during the winter exerts a pressure of nearly one ton per square foot, you can begin to imagine the long-term impact of waves on the shoreline.

Waves usually do not approach the shore head-on, but rather at an angle. This sideswipe creates a slow yet significant movement of water parallel to the shore called the **longshore current**. You have experienced the subtle forces of this current if you have been out in the water for awhile, looked back at the shore, and found yourself fifty yards down the beach from where you planted your umbrella. In addition to moving swimmers, the longshore current moves sand, particles, and other material, a transport known as **littoral drift** (*littoral* means "of the shore"). Debris can be moved hundreds of feet or more each day. The total amount of sand moved by this drift can be staggering, on the order of 200,000 cubic yards each year at Cape Henlopen, Delaware, for example. That is enough sand to build a one-foot-thick wall ten feet high and a hundred miles long. Nor'easters, with their strong northeast winds and pounding waves, are the most powerful influence on littoral drift along the Delaware and Jersey coasts. One big nor'easter can transport as much sand southward in a day or two as all other processes combined over the rest of the year.

The movement of sand up and down the beach is one way that the shoreline is being reshaped. Changes in sea

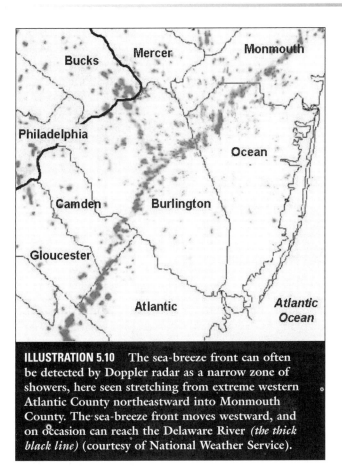

ILLUSTRATION 5.10 The sea-breeze front can often be detected by Doppler radar as a narrow zone of showers, here seen stretching from extreme western Atlantic County northeastward into Monmouth County. The sea-breeze front moves westward, and on occasion can reach the Delaware River *(the thick black line)* (courtesy of National Weather Service).

10–20°F cooler when the ocean air moves inland, even with full sunshine. That is the "negative" side of the sea breeze. But in the heart of summer, when temperatures regularly reach 90°F in Philadelphia and other inland areas, shore areas may warm up to the 80s by midday, but then see a drop to the 70s during the afternoon as the sea breeze develops. That refreshing breeze helps makes summer at the shore so inviting.

Because of the localized sea breeze, predicting air temperatures at the shore can be tricky if the wind forecast is not obvious and the ocean influence is not taken into account. For example, let's assume that the high temperature in Philadelphia on a late June day reaches 90°F in the warm air mass ahead of a cold front. Meanwhile, at the shore, the temperature reaches only 80°F that day before the sea breeze develops and cools things off. The next day behind the cold front, strong northwest winds blow across the entire area and temperatures are more uniform, with highs of 85°F in both Philadelphia and at the shore. So it actually got warmer at the shore behind the cold front, simply because the sea breeze could not overpower the northwest winds!

level are another. Sea levels rise and fall naturally over periods of hundreds and thousands of years and longer, exerting a subtle but significant influence on the location of the shoreline. Sea level has been rising very, very slowly for thousands of years since the end of the last ice age (see Chapter 7), and there is no sign that it will level off any time soon. According to Dr. Norbert Psuty of the Institute of Marine and Coastal Sciences at Rutgers University, sea level at Atlantic City rose about 12 inches between 1915 and 1995. Half of that is due to a true rise in the level of the ocean (with global warming likely a contributing factor), and the rest is the result of a slow subsidence of the land. On average, this rise in sea level amounts to less than one-sixth of an inch per year. But even that seemingly negligible amount takes a toll on the shoreline. Because of the gradual slope of most beaches, every increase in sea level of one foot means that between 100 and 1,000 feet of beach are lost to the sea. The rise in sea level also means that over time less and less powerful storms are able to cause significant damage to the beaches and anything built near them.

The key point in this discussion is that the beach is not a permanent structure, but rather a river of sand. The rearrangement of the shoreline is usually slow and subtle, except when a big storm hits. But regardless of how quickly change occurs, the sand is golden to those who have houses or hotels just behind the dunes. If the beach disappears, the value of expensive shore property collapses. The interests of property owners as well as the larger public that uses the shore for recreation have resulted in sometimes prodigious efforts to "protect" the beaches from nature and safeguard them for future use.

"Newjerseyization": Groins, Jetties, and Seawalls

Beach preservation efforts have been launched so often in New Jersey and to such a degree that the state has become famous for them. In many areas, it is not a pretty sight, and according to some, it has not necessarily saved the beaches. Professor Pilkey has a name for the process of trying to stem shoreline erosion. He calls it "Newjerseyization."

One way to protect the shore is broadly referred to as "hard stabilization." Jetties, groins, and seawalls fall in this category. A **jetty** is a long, permanent structure made of stone, concrete, or timber that starts at the shoreline and extends far out into the ocean for hundreds of feet. Jetties are built at inlets to halt the movement of sand up

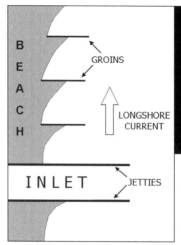

ILLUSTRATION 5.11 A schematic of jetties and groins on a shoreline. These man-made structures cause beaches to grow in some areas and shrink in others. Jetties fix the location of an inlet.

and down the shore to prevent the inlet from moving or changing its depth (see Illustration 5.11). Barnegat Inlet, which separates Island Beach State Park (to the north) from Long Beach Island (to the south), migrated to the south at rates varying from twenty to forty feet per year before jetties were constructed there in 1939. According to Richard Weggel, a Drexel University coastal engineer, jetties are "the only way you can maintain an inlet fixed at a location." While this may be so, Weggel contends that the inlet itself leads to problems: "Inlets are big sand traps . . . they are the real bane of shore erosion. That's where the sand ends up."

The first jetties at the Jersey shore were built in 1911 to stabilize Cape May Inlet. The jetties worked for the inlet and helped increase the beach in Wildwood (on the north side of the jetties). But the jetties also accelerated the beach loss in Cape May to the south by blocking the natural southward drift of sand that would replenish the beach. Decades of effort were required to overcome the effect of these jetties. Illustration 5.12 is a recent aerial view that shows the area around the Cape May jetties. It is clear that the presence of the jetties has built the beach to the north, while shrinking the beach to the south. There are now twenty-four jetties in New Jersey.

Groins are structurally similar to jetties, but are much shorter. They were built to keep the sand drift from shrinking small beach areas. Every time a groin was built, however, the beach on one side of the groin grew (because the natural movement of sand along the beach was blocked) while the beach on the other side suffered, especially if the groin was too high or too long. So a new groin had to be built to protect the stretch of beach that was shrinking. Then the beach on one side of that new

groin could not be naturally replenished with sand. So a new groin had to built. And so on. At this time there are 368 groins in New Jersey. Illustration 5.13 is an aerial shot of Beach Haven, New Jersey that shows several. Like the jetties at Cape May, the groins block the natural tendency for sand to drift southward along this stretch of coastline, so the beaches to the north of the groins continue to grow at the expense of the beaches to the south.

There is another possible way to help protect structures situated close to eroding beaches: build a seawall or bulkhead to stop the ocean. Properly constructed, seawalls will at least temporarily save the structures behind them. But seawalls are controversial and considered a last resort. Experts disagree about their long-term success. Professor Pilkey contends that the force of waves crashing into a seawall forces the water down with more force, ripping more and more sand from the beach and sending the sand back into the ocean. Over time, the beach shrinks. At some point, there is no beach left at high tide.

Professor Weggel views the situation differently. "People see seawalls on eroding beaches," he says. "That's because you put the seawall up on an eroding beach. They [people who object to seawalls] get the cause and effect screwed up." He continues: "I'm not an advocate of seawalls, just like I'm not an advocate of amputation. But if it's going to save my life, I'll have them amputate my leg."

ILLUSTRATION 5.13 An aerial view of Long Beach Island, in the vicinity of Beach Haven. Note the build-up of the beach to the northeast *(here, to the right)* of the groins and the loss of beach to the southwest *(left)* of the groins (courtesy of U.S. Army Corps of Engineers, Philadelphia district).

Along the Jersey and Delaware coasts, seawalls are most obvious in northern New Jersey, where entire towns have highly visible walls. Monmouth Beach is an example: at one time, there was no beach, plus a seawall so high with so much erosion that people in oceanfront houses could not even see the ocean, even from some two-story homes. (There is now a beach, thanks to replenishment—see the next section.) Weggel believes the seawalls in northern New Jersey have indeed saved entire towns from regular flooding. Much of South Jersey has seawalls too, but most of them are buried in sand: "They are there should they be needed in a big storm," he says.

Beach Replenishment

In terms of the economics of tourism and related industries, New Jersey's beaches are worth billions of dollars a year to the state. But the value of the beach goes far beyond money. The popularity of the shore makes it clear that it is a place worth saving. Groins, jetties, and seawalls are examples of armoring the beach against the power of erosion. Some believe these methods have caused more problems than they have solved. Beach replenishment, or adding sand to a beach, is a much less controversial alternative. But it is costly, and it is far from a permanent fix.

The loss of New Jersey beaches in the March 1962 nor'easter led to the first large-scale beach replenishment program in the United States. Since then, numerous projects have added incredible volumes of sand to many of the beaches in New Jersey and Delaware. A 1996 survey by Duke University estimated that over the years, almost sixty million cubic yards of beach has been replaced in New Jersey, at a cost of $312 million. That is enough sand to build a wall three feet wide and sixty feet high the entire length of the East Coast. (Of course, New Jersey beaches are worth billions of dollars to the state's economy each year, so this investment in preservation pays off.) In Delaware, the numbers are eight million cubic yards and $47 million. Currently, many individual projects are underway, with the Army Corps of Engineers responsible for the major ones. As an example of the magnitude of ongoing beach replenishment, one project in Ocean City, New Jersey initially placed approximately six million cubic yards of sand in 1992–93, with additions of about one million cubic yards every three years.

The replenishment projects are part of a seemingly never-ending cycle, since sand loss is an ongoing process. Pilkey believes that the loss comes quickly because the new sand cannot perfectly repeat nature. For example, the sand is typically thickened only on the upper part of the beach above the water level. This creates a more steeply sloped beach, so waves crash with more force. And more powerful waves make for more efficient erosion. For each beach replenishment project, the Corps of Engineers estimates how often additional sand will have to be added, and how long it will take for the "new" part of the beach to be lost—essentially, the life span of the replenished sand. Here is where the effectiveness of beach replenishment is often debated. Estimates given by the Corps are usually more optimistic than those of other researchers, including Pilkey. For example, the Corps gives the replenished beach at Ocean City about a twelve-year life span, while Pilkey estimates only about three years. That is a huge difference.

In reality, because much of the pounding on the shoreline occurs during storms, all it takes is one exceptionally powerful nor'easter or hurricane to undo much of a beach replenishment project. A major storm can erode the coast inland 100 feet or more in a day. In the powerful February 1998 nor'easter (see Chapter 4), about 80 percent of the new sand from a replenishment project at Sandy Hook, New Jersey was lost as several hundred feet of beach disappeared.

Nonetheless, Weggel, who is a veteran of the Corps of Engineers, points out: "What would your house look like if you didn't keep maintaining it? Anything that man constructs that has value is worth maintaining, whether it's a house or a beach." The important point is that beaches are not simply "natural" anymore. They are constructed spaces that require upkeep.

The Shore and the Future

The ongoing give-and-take between land and sea leaves the shoreline in a state of constant change. Only in recent decades have humans assumed a role in what was a completely natural process. Assuming that the slow sea level rise continues throughout this century, we will see more beach losses and more replenishment projects. But the potential shore problems go well beyond the aesthetic. The long-term future of shore life itself could be in jeopardy.

The development in the shore communities of New Jersey and Delaware since the March 1962 nor'easter has been astonishing. The value of shoreline property (adjusted for inflation) has risen from $8.7 billion in 1962 to $34.2 billion in 1997—an increase of 300 percent. Some communities have seen property values rise well above that average; these include a 527 percent increase in Atlantic City, 654 percent in Harvey Cedars, 697 percent in Sea Isle City, and 1,202 percent in Avalon. With the increased development comes heightened concern about even slight increases in sea level.

Records of tide heights show that it does not take as powerful a storm today to produce the same degree of coastal flooding as in the past. Norbert Psuty, the Rutgers University coastal scientist, estimates that if a storm as powerful as the March '62 storm were to occur today, it would lead to 10 percent higher water levels. The tremendous growth in property values means that damages would be many times higher as well. Although Professor Weggel believes beach replenishment projects are likely to keep up with a gradual sea-level rise, his concern increases if the rate of sea-level rise accelerates (as it might from global warming, for instance—see Chapter 7).

A comprehensive study of coastal erosion was released by the Federal Emergency Management Agency (FEMA) in 2000. One of the beaches studied closely was South Bethany, in Sussex County, Delaware, where the average erosion rate is a few feet per year. The study projected the extent of coastline erosion over the next sixty years (see Illustration 5.14) if erosion proceeds without human intervention. If the projections prove correct, the three rows of houses closest to the beach (marked with circles) are in danger of being lost to erosion over this period.

The battle between the advancing ocean and beachfront homes and hotels is a shoreline variation on the theme of the irresistible force meeting the immovable object. The casinos and hotels in Atlantic City cannot be relocated inland a few blocks. There is no place to go. While some retreat may have been possible after the '62 nor'easter, the population density makes such a "move" impossible now.

Would a repeat of a huge storm like the March 1962 nor'easter cause people to reevaluate the potential hazards of shore living? Recent evidence in coastal areas pounded by hurricanes suggests that it would not. Hurricane Opal caused great damage in the Florida panhandle in 1995, and instead of discouraging development, it was followed by a building boom and even higher property values. A similar boom occurred in the islands near Charleston, South Carolina after Hurricane Hugo. Unmistakably, the lure of the shore is strong. What happens to oceanfront homeowners who have saved for years to fulfill a dream of shore living? Does the government reimburse them for their inevitable loss? Or will they have to fend for themselves, spending more futile dollars armoring their beachfront? The issues are economic, environmental, psychological, and political.

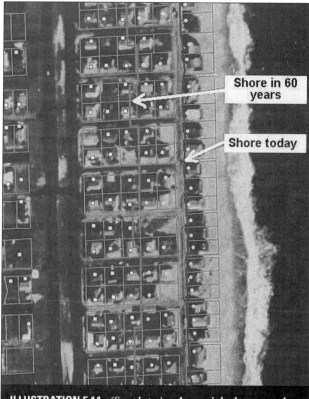

ILLUSTRATION 5.14 *(See plate.)* An aerial photograph of South Bethany, Delaware. According to a FEMA-commissioned study of future beach erosion, the shore is expected to erode inland sixty-five feet over the next sixty years (for this study, "shore" is defined by a vegetation line or the front edge of a dune or bluff). Dots mark three rows of houses projected to be lost to erosion (courtesy of Federal Emergency Management Agency).

Pilkey thinks that we need to choose: "You can have buildings or you can have beaches, but you cannot have both." Weggel does not see a crisis looming just yet: "Ultimately, we may lose the battle, but I don't see that happening for several hundred years."

Labor Day weekend marks the traditional close of the summer season at the shore and the beginning of the school year for most students. But the ocean water in mid-September is typically as warm as in early July, so September visits to the shore can still be very pleasant. To meteorologists, the relatively warm ocean water in September means a heightened awareness of the tropics, because warm ocean water and hurricanes are closely correlated. We will elaborate more on tropical storms and hurricanes and other features of meteorological autumn in the next chapter.

Hurricane Agnes, June 1972

In the latter half of June, the list of weather worries in the Philadelphia area is usually topped by excessive heat and humidity, along with the occasional severe thunderstorm. Tropical troubles are not high on the list yet, given that hurricane season does not typically heat up until later in the summer. But that did not stop a weakening tropical system named Agnes from producing the worst natural disaster in Pennsylvania history in late June of 1972.

The month of June 1972 had already been a wet one in Pennsylvania, with more days with rain than without in many parts of the state. By the time Tropical Storm Agnes developed off the Yucatan Peninsula of Mexico on June 16, Philadelphia was in the fourth day of what would become fourteen straight days with at least some rain. It turned out to be a period of weather that few Pennsylvanians would ever forget. (In a bit of trivial irony, the floods that resulted from Agnes forced Glenn, later known as "Hurricane," to miss his college graduation ceremony.)

Agnes became a hurricane on June 18 and made landfall on the Florida panhandle the next day (see Illustration 5.15). Even at its most intense, however, Agnes produced winds of only 85 mph, a minimal hurricane. But wind would not be the storm's legacy: from the time it came ashore, Agnes was primarily a big rainstorm. Once on land, Agnes moved to the northeast and weakened to a tropical storm and then a tropical depression. Despite that official demotion, the storm's flooding rains caused two deaths and more than $4 million in damage in North Carolina.

On June 21, Agnes moved off the coast into the Atlantic Ocean and restrengthened into a tropical storm, not uncommon for tropical systems taking a similar path (see Illustration 5.15). But then something out of the ordinary happened. Instead of moving out to sea, Agnes was captured by an upper-level cutoff low (see Chapter 2) approaching from the Midwest. This unusual early summer jet-stream pattern pulled Agnes back toward the northwest, eventually resulting in a second landfall in southeastern New York on June 22. The storm then moved in a slow, counterclockwise arc until finally dissipating over Pennsylvania.

The already-existing soggy conditions in Pennsylvania, combined with the long duration of the rain from Agnes, resulted in the worst natural disaster in state history. About three-quarters of the state received 6 inches or more of rain from the storm, most of it falling in the period June 20–23. Amounts averaged 4 to 8 inches in extreme southeastern Pennsylvania but 8 to 12 inches in the Susquehanna River Valley from Wilkes-Barre to Harrisburg. Record flood levels, most which still stand today, occurred along the Schuylkill River (see Illustration 5.16) and virtually every branch of the Susquehanna River. Even the Governor's Mansion in Harrisburg had to be evacuated by boat. The floods from Agnes caused forty-eight deaths and $2.1 billion in damage (1972 dollars) in Pennsylvania alone. By comparison, the tolls were light in New Jersey and Delaware, where the storm was blamed for one death in each state and a few million dollars in damage.

Agnes would have been a big news story in Philadelphia based solely on the rain and flooding in the local area. But as the flooding unfolded into the

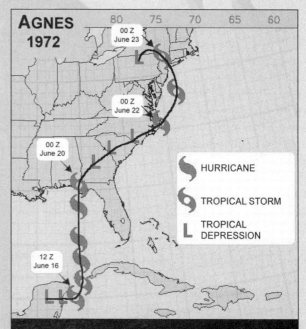

ILLUSTRATION 5.15 The track of Hurricane Agnes. The storm weakened to a tropical depression while inland over the Southeast, then restrengthened into a tropical storm over the Atlantic before looping back into Pennsylvania and dissipating. Agnes's rains produced the worst natural disaster in Pennsylvania history.

ILLUSTRATION 5.16 Agnes brought all-time record flooding on the Schuylkill River at Pottstown. The river crested at 29.97 feet, almost nine feet above the previous record flood level of 21 feet. Flood stage at Pottstown is 13 feet (Urban Archives, Temple University, Philadelphia, Pennsylvania).

biggest natural disaster in Pennsylvania history, Agnes became a huge news story. Herb Clarke had been weatherman at WCAU-TV for fourteen years when Agnes arrived. Herb, having a background in news (see "Story from the Trenches" in Chapter 1), wanted to cover it personally. Because some of the worst flooding was occurring in the Wilkes-Barre area (see Illustration 5.17), Herb traveled there every day to film a report and then returned to Philadelphia to do the weather on the 6 P.M. and 11 P.M. news programs.

By Friday, June 23, the rain had largely ended, but the rivers continued to rise. Parts of Wilkes-Barre were being evacuated because officials feared the river would top the levee that protected the city. Herb recalls: "People were mad. Some were still in their pajamas. They were sure the levee would hold. It [the water] had never come close to going over it. I went with the cameraman to a hotel and got onto the roof to film the flood. I was scared silly. We saw the river advancing, and by the time we started to leave, we had to wade through knee-high water just to get back to the car."

The water did top the levee, which had been built about three feet above the previous all-time record flood level. On Sunday, June 25, when Herb returned to Wilkes-Barre, the city was still flooded. Herb recalls: "Houses were underwater—hundreds of them just gone. People were wandering through the streets. Mud was hanging off the ceilings. Nothing was salvageable. People were even throwing out wedding pictures. The odor was horrible. Caskets were washing down the river." Overall, fourteen miles of streets were flooded and five bridges were destroyed. More than 25,000 homes and apartments were damaged or left uninhabitable.

The other big flood story was in Harrisburg. Another crew had been going there every day, usually by helicopter. Herb continues: "They needed to get film from us [in Wilkes-Barre]. I talked to the pilot and asked him how he was making out. He said, 'All I'm doing is flying and gassing up this thing, and it's 150 hours overdue for a checkup.'"

The next day (Monday, June 26) the station asked Herb if he would go to Harrisburg with the chopper

ILLUSTRATION 5.17   Downtown Wilkes-Barre under water and on fire in the aftermath of Hurricane Agnes. Thirty thousand residents had to evacuate. The water inundated more than 2,700 commercial establishments, including 150 factories and five shopping centers in the Wilkes-Barre area. One-third of Luzerne County's work force was left unemployed (from the collections of the Luzerne County Historical Society).

and relieve reporter Sid Brenner, who had been covering the story there. Herb agreed to go. But a little later the boss decided to do a special that night after the news, and Herb was needed to write that show. So Herb stayed put in Philadelphia, and Sid headed back to Harrisburg on the chopper.

While writing, Herb got a call from the police commissioner: "Herb? Are you Herb Clarke? I thought you were dead. My guys called me and said a helicopter crashed near Harrisburg." Herb recalls: "I went to the news director and said, 'I think you need to get in touch with the State Police in Harris-

burg. The chopper went down and everybody died.' Then I turned around, got a drink of water, walked back to my typewriter, and started typing, because I had to. And I wrote a little insert [on the crash] to put into the show."

Killed in the crash were Sid Brenner, political reporter for WCAU-TV; Lew Clark, senior cameraman for WCAU-TV; Del Vaughn, anchorman for the CBS-Radio network in New York; and Samuel Sedia, the pilot and a former Army helicopter instructor in the Korean War. ❖

CHAPTER 6

Autumn: September–October–November

The Hares, Rabbits, Foxes, and Partridges,
in September and the Beginning of October,
change their Colour to a snowy White.
—*Benjamin Franklin*

The heat of summer has eased; sunny dry days are common, as are foggy mornings; the trees change into beautiful colors. It is autumn, at first glance a likely candidate for having the most tranquil weather of all the seasons in the Delaware Valley. After all, winter has ice, snow, and bitter cold; spring brings gusty kite-flying winds and strong thunderstorms; while summer means heat, humidity, and gutter-gushing downpours. What is it about the period from September to November that favors relatively tranquil weather in the Philadelphia area? That is the first topic in this chapter.

But autumn has its share of stormy weather, with the threat of hurricanes looming above all else. The peak of hurricane season comes in September, and the Delaware Valley's position near the Atlantic means that forecasters always cast a wary eye to the south and southeast for tropical trouble. Since tropical storms and hurricanes are the most important type of hazardous weather to affect the Philadelphia area during autumn, they are the focus of much of this chapter.

How do tropical systems form? What is the greatest danger in a hurricane—is it the wind, the storm surge, or inland flooding? How do meteorologists forecast where a hurricane will go and whether it will strengthen or weaken? Is there a way to reliably predict months in advance how active a hurricane season will be? And just how vulnerable is the Philadelphia area (and the nearby shore) to hurricanes? We consider these questions in this chapter, as well as review the most significant tropical systems that have affected the area, from colonial days to Hurricane Floyd.

This chapter's "Story from the Trenches" focuses on Hurricane Floyd, which caused some of the worst flooding in Philadelphia-area history. How does a television station prepare its programming for a potentially huge storm and then carry it out when the weather forces continuous live coverage? And what do meteorologists tracking the storm look for in the observations and computer models in order to pin down the hardest hit areas?

To close the section on hurricanes, we ask "What if a major hurricane made a direct hit on Delaware Bay at one of the shore's busiest times of the year, Labor Day weekend?" It almost happened in September 1996, when powerful Hurricane Edouard was expected to come right up the Bay just as hundreds of thousands of shore-goers were saying farewell to summer.

As October turns to November, the threat from tropical systems lessens quickly, and meteorologists start to focus on the chill that lies ahead. Taking a similar strategy, we close the chapter on autumn—and the seasonal part of the book—by discussing several of the standard autumnal checkpoints along the road to winter. We will cover first freezes, early snowfalls, fall foliage, and "Indian Summer," several important hallmarks of the fall season in the Northeast.

AUTUMN: SEASON OF STABILITY

Autumn and spring are both transitional seasons between summer and winter. In terms of temperature, autumn and spring have a lot in common—roughly speaking, November's temperatures are comparable to March's, October's to April's, and September's to May's. But autumn weather in the Delaware Valley is much more tranquil than spring weather: spring is windier and wetter, with more thunderstorms. There is something fundamentally different about the atmosphere in autumn. It is all about "stability," a concept we first invoked back in the spring chapter.

In spring, sunshine gets more intense and the days get longer. In response, the ground and the lower atmosphere warm quickly. But higher up, far from the warming surface, the chill of winter lingers longer. This warm-under-cold layering is not very stable, leading to lots of overturning as warmer air rises and colder air sinks. Rising air makes showers and thunderstorms while all the overturning mixes faster-moving air from higher altitudes down to the surface, making spring windy as well.

In contrast, in autumn the ground and the lower atmosphere start to cool from the warmth of summer. But the air higher up retains the relative warmth of summer a little longer. The air near the ground is still warmer than the air at higher altitudes, but the difference is not as great as in spring. It is a more stable situation. So there is less rising and sinking air. That generally means fewer showers and thunderstorms, and less wind.

As evidence of autumn's relative tranquility, consider that in the period 1874 to 2001, there were only six periods of twenty or more consecutive days without any precipitation in Philadelphia. With one exception, each of those dry spells occurred between September 1 and November 30. The exception was a twenty-one day span that began on August 25 (see Table 6.1). And if we categorize precipitation by season, and then look at the driest seasons on record, the first six on the list are autumns.

There are references to autumn's dryness as far back as colonial times. Local weather historians Charles Peirce and Samuel Hazard wrote of a drought in the fall of 1793, when Philadelphia was in the midst of a devastating yellow fever epidemic:

TABLE 6.1 Longest Periods Without Precipitation in Philadelphia

Days	Period
29	Oct. 11, 1874 to Nov. 8, 1874
22	Sept. 1, 1884 to Sept. 22, 1884
22	Oct. 31, 1917 to Nov. 21, 1917
21	Aug. 25, 1874 to Sept. 14, 1874
21	Oct. 18, 1991 to Nov. 7, 1991
21	Oct. 19, 2000 to Nov. 8, 2000

September 1793: The earth was literally baked like powder and dust, except clay land which baked as hard as a pine board.

October 1793: Very dry weather and warm mostly thro' this month—very little rain for eight weeks past—the yellow fever raging in the city.

Sometimes, however, the relative tranquility of autumn weather is broken dramatically. Early autumn is prime hurricane season, and over the years the Delaware Valley and adjacent coastal areas of New Jersey and Delaware have been visited by many tropical systems, with just as many near misses. Although these tropical systems lose much of their punch by the time they get to our latitude, this area has the potential for a hurricane disaster. The combination of a large seasonal coastal population, low-lying areas, limited evacuation routes, and little hurricane experience puts the area at risk. We now take a detailed look at tropical storms and hurricanes.

HURRICANES: THE GREATEST STORMS ON EARTH

Hurricanes are sometimes called the "greatest storms on earth," and for good reason. Although some tornadic winds can be higher, the combined power of the wind, water, and waves of a major hurricane over large areas can make the damage caused by even a strong tornado look puny.

A **hurricane** is an intense area of low pressure that forms over warm tropical waters. Typically, hurricanes measure a few hundred miles across, but they vary widely in size as seen by a comparison of Hurricanes Andrew and Floyd, two of the most devastating storms in recent memory (see Illustration 6.1). Hurricanes occur in many parts of the world and go by different names—**typhoon**

ILLUSTRATION 6.1 Visible satellite images of Hurricane Floyd (September 14, 1999) and Hurricane Andrew (August 23, 1992) to the east of Florida. Note how much larger Floyd is than Andrew, yet both packed winds above 120 mph at the time (courtesy of NOAA).

in the northwest Pacific Ocean and tropical **cyclone** in the Indian Ocean and southwest Pacific—but they are really all the same beast. The only difference is that winds circulate counterclockwise around hurricanes in the Northern Hemisphere but clockwise in the Southern Hemisphere.

Though hurricanes are areas of low pressure, a hurricane's structure and its appearance on a satellite image differ significantly from the more common midlatitude low-pressure system. The most distinctive feature of a hurricane is its **eye**, the relatively calm and clear center of lowest pressure. The strongest thunderstorms, heaviest rain, and fastest winds surround the eye in a donut-shaped

ring called the **eye wall** (see Illustration 6.2a). In contrast, the heaviest rains and strongest winds in a typical midlatitude low, often called an extratropical storm, are found hundreds of miles from the low-pressure center, often near the storm's warm and cold fronts (see Illustration 6.2b). Hurricanes do not have fronts, because fronts separate warm and cold air masses—and where hurricanes form, there is nothing but warm air! However, hurricanes do have **spiral bands** of gusty showers and thunderstorms that pinwheel around the storm, sometimes out hundreds of miles from the eye (Illustration 6.2a).

Hurricanes that could potentially affect the Delaware Valley form in the Atlantic Basin, which includes the

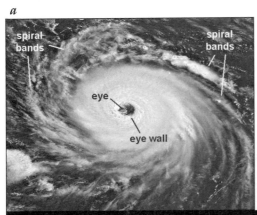

ILLUSTRATION 6.2 (a) Structure of a classic hurricane on visible satellite imagery. The eye of Hurricane Edouard in August 1996 is surrounded by the towering thunderstorms of the eye wall. Spiral bands of showers and thunderstorms pinwheel about the hurricane like arms of a spiral galaxy. (b) Structure of a typical extratropical storm on visible satellite imagery. Clouds and precipitation are concentrated near the cold and warm fronts, not wrapped around the center of lowest pressure as in a hurricane (courtesy of NOAA).

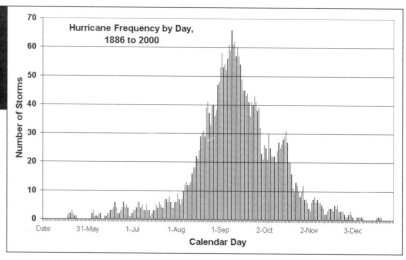

ILLUSTRATION 6.3 The total number of hurricanes reported in the Atlantic Basin for each day from May through December, for the period 1886 to 2000. Mid-August to mid-October is clearly the most active time, with the peak on September 9.

North Atlantic Ocean, Caribbean Sea, and the Gulf of Mexico. Officially, Atlantic hurricane season runs from June 1 to November 30, but August, September and October are by far the most active months (see Illustration 6.3). The peak time for tropical systems in the Atlantic is early to mid-September, coinciding with the time of warmest ocean waters, clear evidence of just how important warm seas are to hurricane formation.

Hurricanes begin humbly as disorganized clusters of tropical thunderstorms. These clusters often develop along weak troughs of low pressure, called **tropical waves**. Many tropical waves originate over Africa, moving off its northwest coast into the eastern Atlantic Ocean every few days during the summer and early fall. For a tropical wave to have any possibility of ever reaching hurricane status, several conditions must be met:

- Water temperatures must be at least 80°F, ideally to a depth of a few hundred feet. Warm water means lots of evaporation, and thus plenty of water vapor in the air to fuel the thunderstorms. Ocean waters off New Jersey and Delaware rarely get this warm for long stretches of time and over large areas. So for a tropical system to affect our region, it has to form farther south and move this way.
- The wind speed and direction must not change much from lower to higher levels in the atmosphere. Strong high-altitude winds or winds that change direction with height tend to blow the tops off thunderstorms, interrupting a budding hurricane's development.
- The tropical wave must be at least 350 miles (about 5° latitude) from the equator. This insures that the earth's rotation can give a twist to the system and

get its circulation going (through a complicated process called the Coriolis effect). No hurricanes have been observed to develop within 5° latitude of the equator, and few develop within 10° latitude (about 700 miles).

Once sustained winds increase to 20 knots (23 mph) and thunderstorms begin to wrap around a closed center of low pressure, the tropical wave is upgraded to a **tropical depression**. If the central pressure continues to lower, thunderstorms organize better around the center, and the winds increase to 34 knots (39 mph), then the tropical depression becomes a **tropical storm**. At this stage, the storm is given a name. If the tropical storm continues to strengthen and winds reach 64 knots (74 mph), the tropical storm is upgraded to a hurricane. Once a storm reaches hurricane strength, its intensity is measured on the **Saffir-Simpson scale** (see Table 6.2). A storm that reaches at least Category-3 intensity is considered a major, or intense, hurricane.

TABLE 6.2 Saffir-Simpson Hurricane Damage Potential Scale

Category	Wind (mph)	Storm Surge (ft.)
1: Minimal	74–95	4–5
2: Moderate	96–110	6–8
3: Extensive	111–130	9–12
4: Extreme	131–155	13–18
5: Catastrophic	>155	>18

ILLUSTRATION 6.4 Visible satellite image on September 26, 1998 shows a very rare occurrence and helps illustrates the logic behind naming tropical systems: there are four hurricanes in the Atlantic Basin at the same time—Georges, just west of Florida in the Gulf of Mexico; and Karl, Ivan, and Jeanne in the central Atlantic Ocean (courtesy of NOAA).

In an average year, about ten tropical storms form in the Atlantic Basin, with six of them becoming hurricanes. However, there is great variability from year to year. In just the 1990s, for example, the numbers ranged from nineteen tropical storms in 1995 (with eleven becoming hurricanes) to just six tropical storms in 1992 (four became hurricanes). But just looking at these numbers can be deceptive—one of the storms in 1992 was Andrew, the costliest hurricane in U.S. history.

The most often-asked question of meteorologists, when it comes to tropical systems, is how and why they get their names. Naming storms not only makes referencing them easier but also helps avoid confusion when several are active at the same time, as sometimes happens (see Illustration 6.4). Historically, many hurricanes in the West Indies were named for the particular saint's day on which the hurricane occurred. For example, two "San Felipe" hurricanes struck Puerto Rico in the 1800s, both on September 13. The use of women's names originated in World War II, when American pilots often named their airplanes after their wives or girlfriends, and meteorologists in the Pacific informally gave women's names to storms for ease of reference. The military connection continued after the war—from 1950 to 1952, hurricanes in the Atlantic were named using the phonetic alphabet (Able, Baker, Charlie, and so on). In 1953, the U.S. Weather Bureau switched to women's names, a tradition that continued until 1979 when they began to be alternated with men's names. Since then, six lists of multinational, easy-to-pronounce names have been used on a rotating basis (see Table 6.3). For example, the list used in 1979 was also used in 1985, 1991, and 1997, and will be used again in 2003. The only

exception to this rule occurs for hurricanes historic enough to have their names retired. Then a replacement name is chosen. A recent example of a retired name is Floyd in 1999, which will be replaced by Franklin in 2005.

The names used now for tropical storms and hurricanes in the Atlantic Basin are agreed upon by representatives of the World Meteorological Organization and the U.S. National Weather Service. Prior to that, there was less formality to the choosing of names. For example, one of us (Glenn) was working at the National Hurricane Center in the mid-1970s with a tropical specialist named Gilbert Clark. Clark had been assigned the task of coming up with the names for the entire decade. Gil used baby-naming books for ideas, and he also claimed to have chosen the names of his favorite movie stars, former girlfriends, and the secretaries at the Hurricane Center. In an incredible irony, a storm named Gilbert in 1988 became the most intense hurricane ever recorded in the Atlantic Basin. Gil, whose passion for hurricanes continued long after his retirement, probably could not have imagined a greater legacy.

HURRICANE DANGERS

Wind

Often, the wind is the focus of attention in hurricanes. The strongest winds are in the donut-shaped eye wall, but they are not evenly distributed in strength. In general, the winds are strongest ahead of and to the right of the eye, looking along the direction of the storm's movement (see Illustration 6.5). For example, this would be the north

TABLE 6.3   Tropical Storm and Hurricane Names in the Atlantic Basin

2002	2003	2004	2005	2006	2007
Arthur	Ana	Alex	Arlene	Alberto	Allison
Bertha	Bill	Bonnie	Bret	Beryl	Barry
Cristobal	Claudette	Charley	Cindy	Chris	Chantal
Dolly	Danny	Danielle	Dennis	Debby	Dean
Edouard	Erika	Earl	Emily	Ernesto	Erin
Fay	Fabian	Frances	Franklin	Florence	Felix
Gustav	Grace	Gaston	Gert	Gordon	Gabrielle
Hanna	Henri	Hermine	Harvey	Helene	Humberto
Isidore	Isabel	Ivan	Irene	Isaac	Iris
Josephine	Juan	Jeanne	Jose	Joyce	Jerry
Kyle	Kate	Karl	Katrina	Kirk	Karen
Lili	Larry	Lisa	Lee	Leslie	Lorenzo
Marco	Mindy	Matthew	Maria	Michael	Michelle
Nana	Nicholas	Nicole	Nate	Nadine	Noel
Omar	Odette	Otto	Ophelia	Oscar	Olga
Paloma	Peter	Paula	Philippe	Patty	Pablo
Rene	Rose	Richard	Rita	Rafael	Rebekah
Sally	Sam	Shary	Stan	Sandy	Sebastien
Teddy	Teresa	Tomas	Tammy	Tony	Tanya
Vicky	Victor	Virginie	Vince	Valerie	Van
Wilfred	Wanda	Walter	Wilma	William	Wendy

side of the eye for a hurricane moving up Delaware Bay. On this side of the eye, winds are blowing off the water, so they are not slowed by friction from trees and buildings on land. Also, on this side of the eye, the winds are in the same direction as the hurricane's movement. The two effectively add together in the same way as they would if you were on a train moving 60 mph and threw a baseball at 20 mph in the direction the train was moving. The ball would actually be moving 80 mph relative to the ground.

In contrast, winds are blowing offshore on the opposite side of the eye, opposing the direction of the storm's movement. These winds have come off land, where they have been slowed by friction from the rougher ground. The potential for significant wind damage is generally much lower on this side of the eye.

When a tropical system makes landfall, the strongest winds occur at the shore. Unless you have traveled inland from the coast in the middle of a hurricane, you cannot imagine just how dramatic the difference can be. Winds on the beach can easily be 50 to 100 percent stronger than those just a few blocks inland. The difference in damages can be just as dramatic. As a hurricane moves ashore, it loses access to its fuel, warm ocean water, so the storm weakens. Only the most intense hurricanes, such as Andrew in 1992, or those moving very fast, such as Hazel in 1954, can cause major wind damage many miles inland.

There is also a difference in wind speed vertically in a hurricane. Recent studies of hurricane damage show that winds 300 feet above the ground are stronger, on average, by about 20 mph (one Saffir-Simpson category) than those at ground level, where friction with the surface slows the wind a bit. So, for example, a hurricane packing 75-mph winds at the ground might have 95-mph winds at the 30th floor of an oceanfront high-rise. Thus, the upper floors of high-rise buildings are potentially much more vulnerable to wind damage than lower floors. But this does not mean that the safest place in a tall oceanfront building is the lowest floor. At that level there is danger from the storm surge, the rise in sea level that accompanies a hurricane when the eye reaches the coast (that is, when the hurricane makes **landfall**).

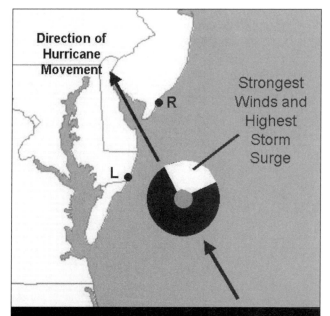

ILLUSTRATION 6.5 The strongest winds and highest storm surge occur in the right front quadrant of a hurricane. In this example, point *R*, to the right of the eye, would experience much higher winds and a much higher storm surge than point *L*. The track shown would be the "worst case" hurricane track for South Jersey and the immediate Philadelphia area, producing a long period of onshore winds and building waves.

Water: Storm Surge and Inland Flooding

Damage caused by hurricane winds can be tremendous, but water—in the form of the storm surge and inland flooding—presents the biggest threat to lives from hurricanes.

The **storm surge** is a rise in the water level of the ocean that occurs as hurricane winds blow water toward shore. As the eye wall crosses the coast, the water level can rise several feet in just a few minutes. The storm surge is greatest where winds are blowing strongest onshore—to the north of the eye for a hurricane making landfall on the east coast, for example (see Illustration 6.5). On the opposite side of the eye, water levels have actually been observed to drop as offshore winds push water out to sea.

The storm surge causes the greatest damage and potential loss of life right at the coast. On top of the elevated water level, the winds form huge waves. A cubic yard of water weighs about 1,600 pounds, and the rapidly moving, crashing ocean puts tremendous force on any structure in its path. Even well-constructed buildings

can be ripped apart (see Illustration 6.6). High-rises remain standing, but the ocean can plow right through the lower floors. The potential for destruction is greatest if the storm hits at high tide, when water levels are naturally higher anyway.

Away from the crashing waves at the beach, the destructive force lessens, but the storm surge can spread many blocks inland, flooding all low-lying areas. On barrier islands, the ocean can meet the bay, cutting off all escape routes. The storm surge of a strong hurricane can even create new channels, changing the island structure permanently. At the very least, the storm surge moves tremendous amounts of sand around, reshaping the beach.

a

b

ILLUSTRATION 6.6 The devastation caused by the storm surge of Category-5 Hurricane Camille in August 1969. The Richelieu Apartment complex on the Gulf Coast at Pass Christian, Mississippi: *(a)* Before Hurricane Camille struck; *(b)* After Hurricane Camille struck (courtesy of NOAA).

TABLE 6.4 Record Flood Levels on the Schuylkill and Delaware Rivers

Schuylkill River			
Gauge Location	Flood Stage (ft.)	Record (ft.)	Date
Reading	13	31.30	June 22, 1972
Pottstown	13	29.97	June 22, 1972
Philadelphia (Fairmount Dam)	11	14.67	June 22, 1972
		17*	Oct. 23, 1869

Delaware River			
Gauge Location	Flood Stage (ft.)	Record (ft.)	Date
Easton/Philipsburg	22	43.70	Aug. 19, 1955
Riegelsville	22	38.90	Aug. 20, 1955
New Hope/Lambertville	13	24.27	Aug. 20, 1955
Washington Crossing	20	27.77	Aug. 20, 1955
Trenton	20	30.30**	March 8, 1904
		28.60	Aug. 20, 1955
Burlington (tidal)		10.8	Aug. 20, 1955
Philadelphia (tidal)		10.8	Nov. 25, 1950
Reedy Island, Delaware (tidal)		9.5	Dec. 11, 1992

*Before official records were kept.
**Resulted from an ice jam.

In the past, most hurricane-related deaths occurred near the coast as a result of the storm surge. The Galveston hurricane of September 1900 remains the deadliest natural disaster in U.S. history. Between 6,000 and 10,000 people on this Texas island drowned in the storm surge of a poorly forecasted hurricane. In recent decades, storm surges have become less deadly in the United States because of better warning systems and a public that takes coastal evacuation notices more seriously. In the last thirty years, inland flooding caused by torrential rains has been responsible for more than half the deaths associated with tropical systems in the United States. A slow-moving tropical storm or hurricane—even a tropical depression—can dump half a foot or more of rain in a day or less, forcing streams and rivers out of their banks.

In the Philadelphia area, nearly every record flood level on the Schuylkill and Delaware Rivers resulted from the rains of a tropical system—either the Connie/Diane combination in 1955 or Agnes in 1972 (see Table 6.4). More recently, the remains of Hurricane Floyd brought widespread 6–12 inch rains to the Philadelphia area and surrounding counties, causing flash flooding on many streams, creeks, and rivers.

HURRICANE FORECASTING

The National Hurricane Center (NHC) in Miami, Florida, a branch of NOAA's Tropical Prediction Center, is responsible for tracking and forecasting tropical systems in the Atlantic Basin and the eastern Pacific Ocean. If tropical storm or hurricane conditions are possible for a specific coastal area within thirty-six hours, the NHC issues a **Tropical Storm Watch** or **Hurricane Watch**. If tropical storm or hurricane conditions are expected in twenty-four hours or sooner, the watch is upgraded to a **Tropical Storm Warning** or **Hurricane Warning**. Evacuations are recommended (sometimes required) for coastal residents when a hurricane warning is issued.

Imagine being Christopher Columbus and running into a tropical storm as he did on his second voyage in

1494, the first encounter of Europeans with a tropical system in the New World. Columbus's observations of that storm helped him greatly when he ran into a full-fledged hurricane on his fourth voyage in 1502. Over time, other keen observers began to notice phenomena that, taken together, foretold a hurricane's approach: the appearance of wispy high-altitude cirrus clouds fanning off the tops of thunderstorms, rising swells over the ocean surface, and a steadily decreasing air pressure. But even in the first half of the twentieth century, several hurricanes that were poorly predicted, such as the 1900 Galveston hurricane and the "Long Island Express" in 1938 (described later in this chapter), led to major losses of life in the United States.

Today, the network of weather satellites that covers the globe allows meteorologists to watch the formation of these storms and to track them over time. These satellites not only monitor clouds, but some can estimate ocean water temperatures, rainfall amounts, wave heights, and high-altitude winds. In addition, "Hurricane Hunter" aircraft from NOAA and from the Air Force Reserve's 53rd Weather Reconnaissance Squadron routinely fly into tropical systems, taking up-close-and-personal observations of pressure, wind, temperature, and humidity. In 1997, NOAA added a high-altitude jet to its research fleet, capable of flying high above and hundreds of miles away from the storm's center to get detailed information on the surrounding environment.

These observations are fed into sophisticated computer models that attempt to predict the storm's movement and intensity. These computer forecasts have steadily improved in recent decades, particularly in predicting the storm track. (Forecasts of storm intensity have only improved slightly.) Over the last three decades, official NHC forecasts of the tracks of tropical systems have improved about 1 percent each year for one-day forecasts and by almost 2 percent each year for three-day forecasts. The rate of improvement of track forecasts nearly doubles if you consider just the latter half of the 1990s. This is primarily the result of the new data gathered by the high-level jet and a new computer model developed at the Geophysical Fluid Dynamics Lab (a research arm of NOAA) in Princeton, New Jersey. To get a sense of current forecast capabilities, the average error in the official NHC forecast of the position of the storm's center is about 100 miles in a 24-hour forecast, 150 miles in a 36-hour forecast, and 230 miles in a 72-hour forecast.

Some of the biggest forecast errors (and greatest forecast challenges) occur near the East Coast, where storms tend to curve quickly. As tropical systems turn northward, they usually accelerate as they encounter stronger midlatitude upper-level winds. Because these winds are generally from the west, tropical systems tend to turn away from the coast. Exactly when and where that turn occurs makes the difference between a near miss such as Bonnie in August 1998 and a direct hit such as Hugo in September 1989. A storm moving northward and threatening the East Coast only has to shift east a small angle (perhaps 5 or 10 degrees) from the forecast track to completely miss an area under a hurricane warning. And if the eye stays offshore, the coast gets the weaker side of the storm.

Hurricane warnings often mean evacuations, and evacuations are expensive. There is the cost of boarding up homes and businesses in preparation for winds and waves, the cost of evacuation itself, and the dollars lost by businesses because of people leaving shore areas. The average preparation costs for evacuations are estimated at about $600,000 per mile of warned coastline, and the average hurricane warning covers about 350 miles of coast. The preparation costs per mile increased about eight-fold during the 1990s, primarily because of increased population and increases in property values. On average, evacuations are called for three times as often as prove necessary, leading to lots of frustration for the public (and forecasters). Floyd, in September 1999, led to the largest evacuation in U.S. history, with people leaving coastal areas from South Florida to New England. (The eye eventually came ashore in southern North Carolina.) As forecasts become more accurate, the size of the warning areas and the number of false alarms will be reduced. But one result of the increased development along coastlines is that the population at risk has grown by 4 to 5 percent a year, outpacing the 1 to 2 percent annual improvement in the forecasts of hurricane tracks.

A different type of hurricane forecasting occurs well before hurricane season even begins. In December of each year, Professor William Gray and a team of researchers at Colorado State University issue forecasts of how many tropical storms and hurricanes they believe will occur in the next hurricane season. They update their forecast the following April, June, and again in August to take advantage of new information. Their research has shown that tropical activity is correlated with several global and regional weather patterns. These include:

- El Niño and La Niña. Fewer Atlantic hurricanes tend to form when an El Niño is in progress, with more during a La Niña (see Chapter 7 for more on El Niño and La Niña). Historically, the probability of two or more hurricanes striking the United States in a particular year is 28 percent during an El Niño but 66 percent during La Niña.

- Rainfall in western Africa. Above average rainfall there in recent seasons tends to correlate with more Atlantic hurricanes.
- Air pressures over the eastern Atlantic. Lower-than-average pressure there in recent seasons tends to correlate with increased Atlantic hurricane activity, and vice versa.

Professor Gray has been issuing his long-range forecasts since the 1984 hurricane season. More important than the specific numbers is whether his forecasts can pinpoint active and non-active hurricane seasons—that is, years when the number of storms is above or below average. By this measure, his forecasts have been on target about two-thirds of the time. In April 1999, government forecasters from the National Oceanic and Atmospheric Administration also began issuing long-range hurricane forecasts, using many of the same predictors that Dr. Gray and his team developed.

ILLUSTRATION 6.7
Arrows mark the landfall sites of all major hurricanes (Categories 3, 4, and 5) that have come ashore in the eastern United States from 1899 to 2001. Note the lack of hurricane landfalls in areas where the coastline bows inward, such as from central Florida through Georgia and from northern North Carolina through New Jersey.

PHILADELPHIA AND COASTAL VULNERABILITY

Even though the ocean off New Jersey and Delaware might seem warm during the summer, the water is still too cold to support hurricane formation. At best, water temperatures might reach 80°F over small areas for short times in late August or early September. So for a tropical system to affect our area, it has to form elsewhere and move this way.

The great majority of tropical storms and hurricanes in the Atlantic Basin have no direct effect on the Delaware Valley and nearby coastal areas. Many curve northward before ever getting far enough west to affect the United States, eventually weakening over cooler waters. Others move westward through the Caribbean and into Central America. Florida is the primary target for U.S. hurricanes because it juts out into warm, tropical waters like a bad boxer who sticks his chin out too far. North Carolina is in a similar situation, its coast extending out into the Atlantic, so it too is very vulnerable to hurricanes. The Delmarva Peninsula and the Jersey shore are partially protected from tropical systems by the opposite type of geography, "tucked in" by comparison with Florida and North Carolina. The same is true of the Georgia coast and the east coast of central and northern Florida—notice the relative lack of landfalling major hurricanes in these areas in Illustration 6.7.

In order to make landfall on the Jersey or Delaware coasts, a tropical system needs to move toward the northwest. That direction of movement is rare at these latitudes where upper-level winds tend to be from the west, curving storms off to the north or steering them east. Only one hurricane in the last hundred years (and a weak one at that) has made landfall on the Jersey or Delaware coasts, coming ashore near Atlantic City in 1903. In September 1999, Floyd was downgraded to a tropical storm just before coming ashore near Cape May.

Hurricanes moving up the coast weaken once they move north of the ultra-warm Gulf Stream waters that shoot northeastward off of North Carolina (see Illustration 6.7). Slow-moving hurricanes tend to lose their strength because they spend an extended period of time over cool water. Only fast-movers, typically traveling at least 30 mph, can reach our area without significant weakening.

Another factor in a storm's potential impact is the timing of its arrival with respect to high and low tides. At the Jersey and Delaware coasts, the difference between high and low tide on a typical fall day is four or five feet, which can make a huge difference in the severity of coastal flooding during a storm. The flooding would be even worse during relatively high "spring" tides that occur about every two weeks when the moon is full or new.

The track of the storm determines what areas will feel the hurricane's most destructive punch. As an example, consider the effect of a hurricane approaching from the south on the New Jersey and Delaware coasts. Winds will be from the northeast on the hurricane's outer fringes. As the storm approaches, winds will increase, but not necessarily gradually. An outer spiral band of showers and thunderstorms may contain very strong gusts a hundred miles or more from the eye. If the storm stays offshore, winds along the coast will turn to the north and then northwest. The coast (and areas inland) will stay on the weaker side of the storm, at least in terms of wind. Coastal flooding can still occur, especially if the storm moves slowly, but it would be minimal because onshore winds probably would not last very long.

On the other hand, if the storm takes an inland track— for example, staying just west of Delaware Bay—the northeast winds would turn to the east and then the southeast. This persistent onshore flow would maximize coastal flooding because of the relentless push of water toward shore. (In the back bays, flooding typically occurs an hour or two after high tides at the beaches as the winds impede the return of water to the sea that naturally occurs at low tide.) This track also causes the worst flooding on Delaware Bay and the Schuylkill and Delaware Rivers because southeast winds drive water up the narrowing bay toward Wilmington, piling the water at the north end. A strong, long-lasting southeast wind could be devastating. Max Mayfield, director of the National Hurricane Center, says that a Category-3 storm moving on this track at 30 mph could cause a fifteen-foot storm surge up Delaware Bay to Wilmington and Philadelphia. "It's hard for people to believe," he adds.

Aside from the Delaware Bay influence, river flooding depends greatly on how much rain falls and how quickly. The rains from Floyd in September 1999 came fast and furious, first overwhelming streams and creeks and then smaller rivers, all in little more than twelve hours. The pre-storm ground and river conditions also play an important role. A ground saturated from recent rains plus rivers already at high levels make an area much more vulnerable to serious flooding. This was the situation with the record flooding from the combination of Connie and Diane in August 1955.

A hurricane's path is often plotted as a series of dots on a map, with each dot pinpointing the location of the eye at a particular time. But a tropical system's effects can extend out a hundred miles or more from the center. So it is possible to be "in" a hurricane but only experience fringe effects (gusty winds and heavy rain) instead of the full brunt of the storm (hurricane-force winds and a significant storm surge). Conditions can become dangerous in a short time and over short distances. Because the res-

idents of this area have only limited experience with tropical systems—fringe effects for the most part—it is easy to underestimate the danger.

Because of the risk of storm surge flooding at the shore, tens or hundreds of thousands of people might have to be evacuated inland if a hurricane appeared ready to make a direct strike. The roads on many of the area's barrier islands have such limited capacity that a large-scale evacuation anytime between Memorial Day and Labor Day would take a long time. In order to shorten the evacuation times, New Jersey plans to "reverse lane" the main roads off the islands—that is, use both sides of the road to move traffic out. Studies show that making Route 47 out of Wildwood one way would reduce the evacuation time from thirty-four hours to twenty. Reverse laning the Atlantic City Expressway would reduce the Atlantic County time from twenty-eight hours to twenty. And making Route 72 one way would reduce the Ocean County time from twenty-two hours to fifteen.

Even the reduced times make emergency managers nervous. To provide the twenty-hour warning necessary to make these evacuations possible, officials might have to make the decision when the hurricane is as far south as the Carolinas or even Florida (depending on how fast the storm is moving). But because storms that far away often curve out to sea, New Jersey would prefer to wait an additional six to twelve hours before making the evacuation decision. By that time, the storm track should be more clear-cut. Mike Augustyniak, hurricane program planner for the New Jersey Office of Emergency Management, realizes that such a delay could make a complete evacuation impossible. So the state has a "Plan B" that would stop the evacuation rather than strand people on the roads. For example, instead of allowing cars to continue on Route 72 into the Pine Barrens, where trees ripped down by hurricane-force winds could be disastrous, police would stop the cars before they reached the wilderness, directing people to take cover in nearby buildings. Max Mayfield, director of the National Hurricane Center, says that his "greatest fear" is of people being stranded on the road during a hurricane. The combination of reverse laning and "Plan B" are unique to New Jersey.

Despite these plans, the fact that evacuations resulting from hurricane warnings are called for three times as often as proves necessary would still lead to many evacuations that appear unfounded. There would undoubtedly be a lot of people angry over missing part of their shore vacations. But, as Mayfield says: "I'd rather people be mad at me than dead."

The casinos and other high-rises at the shore pose particular problems. The casinos bus thousands of people in daily. The buses then park miles away. If a quick evacuation were ordered, it would be difficult to get all those people back onto the buses. As a result, the plan has been to "vertically evacuate" guests, if necessary, to higher floors. But recent research showing that winds a few hundred feet above the ground are markedly faster than winds at the ground in many hurricanes puts that practice into question. A 90-mph wind at the ground floor of a casino might increase to 110 mph at the top floors of the tallest buildings. Mike Augustyniak says that "we're not sure" if the buildings (many with glass windows) can withstand that increase in wind. And since virtually no high-rises at the Jersey and Delaware shores have the storm shutters common in Florida, there is no added protection. As a result, people need to stay away from the higher floors, limiting the extent of the vertical evacuation. With the lowest floors susceptible to the storm surge, Augustyniak says: "They [the casinos] won't be able to house the 10,000 people they once thought they could." Similar problems could occur in any shore high-rise.

In summary, the Delaware Valley and the Jersey and Delaware coasts are somewhat protected from hurricanes by geography and a relatively cool ocean, but the area's vulnerability is still great. The large population has virtually no experience with the full force of a major hurricane. And the period of peak population at the coast partly overlaps with the time of greatest hurricane threat. The combination of a large population lacking hurricane experience and low-lying escape areas that can flood hours before the highest winds arrive, makes meteorologists and emergency management officials uneasy. A hurricane-caused disaster can happen here. Many experts feel it is only a matter of time. Statistical studies based on historical records give a sense of the vulnerability: the best estimates are that a Category-1 hurricane will come within seventy-five miles of Atlantic City an average of once every 22 years; a Category-2 once every 50 years; a Category-3 once every 87 years; a Category-4 once every 190 years; and a Category-5 once every 500 years or more.

Of course, the area does have plenty of experience with nor'easters, and their effects along the coast can be devastating as well. In fact, seven of the top ten highest water levels observed at Atlantic City, and nine of the top ten at Lewes, Delaware were caused by nor'easters, not hurricanes. But a major hurricane has the potential to generate more destruction in twelve or fewer hours than was experienced during even the most severe nor'easter

on record at the New Jersey and Delaware coasts, the March 1962 storm that lasted three days (see Chapter 4).

HISTORICAL DELAWARE VALLEY AND SHORE HURRICANES

Examples of this potential "big one" can be found in the historical record. Here is a look back at some of the tropical storms and hurricanes of the past that have visited our area. Much of the information was taken from *Early American Hurricanes: 1492–1870,* by American weather historian David Ludlum. The most severe hurricanes on record to affect this area were observed in the nineteenth century. A repeat of the tracks of some of these storms would probably produce the biggest natural disaster in Philadelphia-area history.

Eighteenth and Nineteenth Centuries

November 2, 1743

A strong storm that may have been tropical in origin affected the East Coast from Philadelphia to Boston. This was the famous storm that prevented Benjamin Franklin from observing a lunar eclipse in Philadelphia and enabled him to deduce that storms do not necessarily move with the surface wind (see Chapter 1). If this was a tropical system, it may have been the first one accurately measured by scientific instruments. Professor John Winthrop of Harvard College took pressure readings starting in 1742, and his written records from Boston contain references to this storm.

September 3, 1775

A storm of tropical origin produced the highest tide ever known till that time at Philadelphia as great southeast gales pushed water up Delaware Bay. Striking just as the War of Independence was beginning, this storm came ashore in North Carolina and moved north, probably through Chesapeake Bay and into eastern Pennsylvania. The meteorological records of Philadelphia weather observer Phineas Pemberton read: "Sept.3—Stormy and showery. A violent gale of wind from NE to SE the preceding night with much heavy rain, lightning, and thunder—a remarkably high tide in [the] Delaware this morning."

August 19, 1788

The winds from a very small but fast-moving tropical system caused tremendous damage to forests along a narrow path from Cape May to the Hudson River and northward into Maine. A similar storm today would likely be devastating in these now heavily populated areas. In Philadelphia, "a tremendous downpour of tropical proportions" was reported on August 18–19, with 7 inches of rain.

September 3, 1821

The eye of this hurricane passed right over Cape Henlopen, Delaware, where the wind went calm for 30 minutes. The storm moved northward, its center passing just west of Cape May and continued up the coast along approximately the route of the present Garden State Parkway. The description from Cape May resident Charles Ludlam likely fits the experiences of many on the coast that day: "it was blowing a hurricane with intermittent gusts that drove in doors and windows, blowing down outbuildings, trees, fences, and overflowing the marshes." Communities surrounding Chesapeake Bay and the Delmarva Peninsula suffered serious flooding. A newspaper report read: "The crops are laid low, and the country exhibits one scene of widespread desolation and ruin." Destructive northeast winds downed many trees and chimneys in Philadelphia. This hurricane started a serious discussion about the nature of storms, later involving Dr. James Espy, The Franklin Institute's first meteorologist.

October 13, 1846

The storm arrived on nearly the same date and moved along a similar track as Hazel would 108 years later. The winds in Philadelphia shifted to the southeast, indicating that the storm moved west of the city. These winds piled water in the north end of Delaware Bay, and the lowlands along the Delaware near New Castle experienced the greatest storm surge in seventy years. At Philadelphia, it was reported that "the Delaware was lashed into a perfect fury and its roar would have drowned the thunder of Niagara itself." Structural damage from strong winds was reported throughout the Philadelphia area, and many buildings were blown down in southern New Jersey.

October 23, 1878

This was another tropical system whose center passed well west of Philadelphia, putting the city on the most dangerous side of the storm. (The headline in the Philadelphia newspaper *The Press* the next day read "A Terrific Tornado," illustrating the confusion that still existed at the time between tornadoes and hurricanes.) Hurricane-force winds and considerable damage were reported throughout the area. *The Press* story began:

> Philadelphia was visited, yesterday morning, by one of the most destructive storms in its history. Indeed, in the memory of the oldest citizens no such widespread

havoc among property, and loss of life, has been experienced by an gale, however severe.

In Philadelphia, the peak wind was southeast at 72 mph, and seven people were killed. Damage in the city was $2 million. In New Jersey, 150 buildings were unroofed in Camden. Cape May reported east winds up to 84 mph and an "extraordinary high tide," and many resort hotels were greatly damaged. Many vessels of all shapes and sizes were driven ashore on Delaware Bay. The Delaware and Christina rivers and Brandywine Creek flooded in Delaware. Much of South Wilmington was under water. Lewes had a storm surge of seven feet, and beaches were severely eroded at Rehoboth. There were eighteen fatalities in Delaware.

Twentieth Century

Here we review about a dozen tropical systems from the twentieth century that have, in one way or another, affected Philadelphia or nearby coastal areas. The tracks of the most significant storms are shown in Illustration 6.8.

September 16, 1903

According to the official archives of the National Hurricane Center, this was the only tropical system to make landfall on the New Jersey coast as a hurricane in the twentieth century. The storm approached from the southeast as a Category-1, coming ashore near Atlantic City. Officially, the wind there only reached 47 mph, but extensive damage was reported.

August 23, 1933

This storm, known as the "Chesapeake-Potomac" hurricane, came ashore near the North Carolina/Virginia border

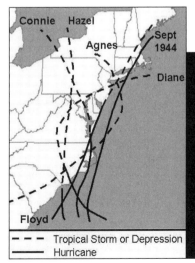

ILLUSTRATION 6.8 Tracks of some twentieth-century hurricanes that significantly affected the Philadelphia area. A solid line is used when the storm was a hurricane; a dashed line indicates tropical storm or tropical depression intensity.

ILLUSTRATION 6.9 The Fourth Street approach to the Ocean City, New Jersey boardwalk after the August 1933 "Chesapeake-Potomac" hurricane (Urban Archives, Temple University, Philadelphia, Pennsylvania).

and moved north to near Harrisburg. In Philadelphia, about 5.6 inches of rain fell over a three-day period. In Atlantic City, east winds reached 65 mph, and 11 inches of rain fell. Three deaths were reported in New Jersey and damage was $3 million in the state (see Illustration 6.9).

September 21, 1938

This storm, nicknamed the "Long Island Express," was a big surprise when it slammed into New England, resulting in one of the worst weather disasters in U.S. history. The hurricane caused more than 500 deaths and $300 million in damage (in 1938 dollars), much of it in Connecticut, Rhode Island, and eastern Massachusetts. The storm just grazed New Jersey, passing about one hundred miles east of Atlantic City where peak winds reached 61 mph.

September 14, 1944

Many who suffered through this hurricane at the Jersey coast were shocked to hear that it did not actually make landfall there. The eye stayed 50 to 75 miles offshore, but the damage at the shore was extreme. The hurricane followed a track similar to the 1938 "Long Island Express" and Hurricane Gloria in 1985 (see Illustration 6.8). It was a weakening Category-2 storm when it brushed New Jersey.

Two and a half miles of the Atlantic City boardwalk were destroyed (see Illustration 6.10), and many piers

were damaged. Winds reached 82 mph in Atlantic City and 49 mph in Trenton. In New Jersey, there were eight fatalities, 250 injuries, and $25 million in damage. On Long Beach Island alone, 300 homes were destroyed. Overall, the storm caused more than 400 deaths, mainly at sea. In Philadelphia, wind gusts reached 34 mph and about 5.5 inches of rain fell.

October 15, 1954—Hurricane Hazel

This storm produced the highest wind gusts ever officially recorded in Philadelphia (94 mph) and Wilmington (98 mph), and is still firmly in the memory of many area residents. Hazel moved rapidly from the Bahamas to make landfall as a Category-3 storm near the South Carolina/North Carolina border on the morning of October 15. Moving north at 50 to 60 mph, Hazel reached the Delaware Valley that evening, with the eye passing close to Harrisburg (see Illustration 6.8).

All across the Delaware Valley, Hazel's winds caused widespread destruction. Homes and buildings were unroofed, trees uprooted (see Illustration 6.11), and power lines felled. Signs were blown away, store windows

ILLUSTRATION 6.10 *(See plate.)* Margate City, New Jersey, just south of Atlantic City, in the aftermath of the September 1944 hurricane. The famous landmark "Lucy the Elephant" stands defiantly, but the boardwalk and many homes near the shore did not fare as well (Urban Archives, Temple University, Philadelphia, Pennsylvania).

ILLUSTRATION 6.11 Hurricane Hazel's record-setting winds in Philadelphia caused widespread damage, felling thousands of trees on October 15, 1954 (Urban Archives, Temple University, Philadelphia, Pennsylvania).

were smashed, and traffic was snarled because of fallen trees and other debris (see Illustration 6.12). More than 80 percent of the homes in Delaware lost power, some for more than three days. In Atlantic City, the peak wind was 66 mph. Damage in New Jersey was $7.5 million, with five deaths reported in the state.

ILLUSTRATION 6.12 *(See plate.)* The aftermath from Hurricane Hazel: two boats from the Wissinoming Yacht Club that were battered ashore at Delaware Avenue and Devereaux Street (Urban Archives, Temple University, Philadelphia, Pennsylvania).

Fast-moving hurricanes usually do not have time to produce lots of rain, and that was true of Hazel. Most places in the Philadelphia area got less than an inch, so Hazel will not be confused with Floyd. Hazel was a windstorm, the worst in the Delaware Valley in modern times.

August 13, 1955—Hurricane Connie

As with Hazel, Connie tracked well west of Philadelphia. After making landfall in North Carolina, the storm curved northwestward skirting the western shore of Chesapeake Bay (see Illustration 6.8). A schooner sunk on the bay, drowning fourteen people. Winds of 40 to 60 mph were reported in Philadelphia along with about 5.5 inches of rain and considerable flooding (see Illustration 6.13). Similar winds were recorded in Wilmington, which received 6.7 inches of rain.

August 18, 1955—Hurricane Diane

Diane arrived less than a week after Connie, but took a much different track. Diane made landfall in southern North Carolina, moved into Virginia, and then made a right turn so that its center passed almost directly over Philadelphia (see Illustration 6.8). The long trek over land reduced the winds to 30 mph, and Diane's rainfall was nothing spectacular locally, in the one to three inch range.

But farther north in the Delaware River basin, six to twelve inches fell. With the ground still saturated from Connie's rains, the Delaware River flooded to record

ILLUSTRATION 6.13 *(See plate.)* Flooding in Philadelphia from Hurricane Connie, August 13, 1955. A motorist, trapped by the rising waters of the Delaware River, climbs atop his car as the river floods the waterfront section of Delaware Avenue from Spring Garden Street to South Street (Urban Archives, Temple University, Philadelphia, Pennsylvania).

levels in many places north of Philadelphia (see Table 6.4)—22 feet above flood stage at Easton, Pennsylvania (see Illustration 6.14). There were ninety deaths along the Delaware and its tributaries in Pennsylvania. The Schuylkill River also topped its banks, putting some Philadelphia streets underwater. Overall, the flooding in Pennsylvania, New York, and New England contributed to nearly two hundred deaths and $4.2 billion in damages.

August 27–28, 1971—Tropical Storm Doria

The center of Doria moved north between Philadelphia and Atlantic City. As a weak tropical storm, Doria's winds were minimal, gusting only briefly to 38 mph in Philadelphia. But the rain was much more serious. Over a two-day period, 6.5 inches fell in Philadelphia and more than ten inches in Princeton, New Jersey. Flooding in parts of New Jersey was the worst since Connie and Diane in 1955.

June 22, 1972—Hurricane Agnes

Agnes still ranks as the most expensive natural disaster in Pennsylvania history, and one of the worst in U.S. history. Never more than a Category-1 hurricane, Agnes made landfall on the Florida panhandle on June 19. The storm

ILLUSTRATION 6.15   Flooding from Hurricane Agnes, June 1972. The Schuylkill River crested more than three feet above flood stage in Center City Philadelphia on June 23, 1972 (Urban Archives, Temple University, Philadelphia, Pennsylvania).

moved northeastward and back out into the Atlantic near the North Carolina/Virginia border. Agnes then turned northwestward toward New York, where it was caught up in the circulation of an upper-air cutoff low (see Illustration 5.15). This slowed the storm down, and helped it loop to the southwest into north-central Pennsylvania.

Persistent rains and the subsequent flooding caused the record damage. During the period June 20–26, about 75 percent of Pennsylvania received at least six inches of rain. Up to nineteen inches fell in parts of Schuylkill county, with a swath of ten to eighteen inches of rain all the way from York and Lancaster counties northeastward into the state of Connecticut. In Pennsylvania, the Susquehanna River basin experienced the most severe flooding, especially at Wilkes-Barre and Harrisburg. In Philadelphia, a relatively light four to five inches of rain fell, with less damage. However, because the Schuylkill River drains some of the areas that experienced the heaviest rains, many record flood levels that still stand were set along the river and its tributaries (see Table 6.4). The water was deep enough on Main Street in Manayunk to float a car. The Schuylkill's crest of 14.67 feet at Fairmount Park was the highest level it has reached there in the twentieth century (see Illustrations 6.15 and 6.16).

The entire state of Pennsylvania was declared a disaster area with 48 deaths and $2.1 billion in damage. Overall, Agnes's floods in the northeastern United States contributed to 122 deaths and $6.4 billion in damage. For personal recollections of Hurricane Agnes from Herb Clarke, and other photographs from the storm, see the "Story from the Trenches" in Chapter 5.

ILLUSTRATION 6.14   A week after Hurricane Connie, the rains from Hurricane Diane caused devastating flooding along the Delaware River, particularly north of Trenton. Here, the river surges over its banks at Easton, Pennsylvania, flooding sections of the city and inundating the bridge that connects Easton with Philipsburg, New Jersey (Urban Archives, Temple University, Philadelphia, Pennsylvania).

ILLUSTRATION 6.16 Boathouse Row is underwater as the Schuylkill River tops its banks in the aftermath of the rains from Hurricane Agnes (Urban Archives, Temple University, Philadelphia, Pennsylvania).

September 5, 1979—Hurricane David

David began its destructive path in the Caribbean a week earlier with 150-mph winds. By the time it reached the Delaware Valley, David was merely a fast-moving tropical storm with maximum winds around 50 mph and one to three inches of rain. But David's thunderstorms produced four tornadoes in the area. One was on the ground for three miles in New Castle County, Delaware, ripping several homes apart (see Illustration 6.17). Tornadoes also touched down in Chester and Berks counties in Pennsylvania, while a weak twister struck Cape May County in New Jersey. All told, the tornadoes caused one fatality, ten injuries, and several million dollars in property damage.

September 27, 1985—Hurricane Gloria

Gloria was at one time a Category-4 hurricane as it passed north of the Bahamas on September 25. It moved directly toward the Carolinas, leading to massive evacuations. But then the storm turned north, just missing North Carolina. On September 27, Gloria's eye passed just off the Jersey coast along a track similar to that of the September 1944 hurricane. But unlike that storm, Gloria had much less wind on its west side, so damage was confined to the immediate shore areas. In addition, Gloria was weakening as it moved up the coast, and the storm made landfall on Long Island at low tide, resulting in much less damage than had been expected.

September 16, 1999—Hurricane Floyd

Floyd was one of the largest, strongest, and wettest hurricanes in memory. On September 13, just north of the

Bahamas, its winds reached 155 mph, making it nearly a Category-5 hurricane (see Illustration 6.1). At that time, Floyd was heading toward the east coast of Florida. But the storm made a sharp turn to the north and accelerated, making landfall in North Carolina on September 15 as a weakened Category-2 hurricane. It then moved up the coast at around 30 mph, the center passing right over Cape May (see Illustration 6.8). This will not be recorded as a landfalling hurricane in New Jersey, however, because Floyd was downgraded to a tropical storm just before it reached the Garden State. Peak wind gusts reached only about 50 mph in Philadelphia and at the shore.

The rain with the storm produced all-time record flooding from North Carolina to New England. In Philadelphia, the 6.63 inches of rain on September 16 set an all-time one-day record, as did the 8.29 inches in Wilmington. In a matter of hours, rivers and creeks exceeded flood stage. The Schuylkill River at Fairmount Park rose from 6 feet just before sunrise to above the flood stage of 11 feet by 5 P.M., eventually cresting just over 14 feet around midnight. The heaviest rain in the Delaware Valley, in excess of ten inches in some spots, fell in a swath from northeastern Maryland through

ILLUSTRATION 6.17 The winds of an F2 tornado spawned by one of Hurricane David's thunderstorms ripped the roof off this house in Claymont, Delaware, just northeast of Wilmington. The tornado was two hundred yards wide and was on the ground for three miles (Urban Archives, Temple University, Philadelphia, Pennsylvania).

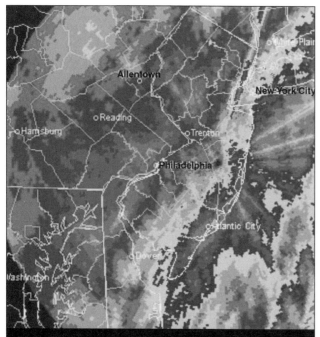

ILLUSTRATION 6.18 *(See plate.)* Doppler radar-estimated rainfall from Hurricane Floyd, September 16, 1999. Philadelphia is approximately in the center of the image. The areas stretching from northeastern Maryland through extreme southeastern Pennsylvania into northern New Jersey received at least eight inches of rain, with embedded pockets of more than a foot (courtesy of National Weather Service).

ILLUSTRATION 6.19 The Delaware River at Pennsville, New Jersey, just south of the Delaware Memorial Bridge, is whipped into a frenzy by Hurricane Floyd (courtesy of Kathleen Welch).

ILLUSTRATION 6.20 Flooding from Hurricane Floyd in Delaware County, Pennsylvania. Water from Darby Creek (in the background) inundates Macdade and Springfield Roads in Darby (courtesy of John A. Bobst).

Bucks and Montgomery counties in Pennsylvania into Somerset County, New Jersey (see Illustration 6.18). All-time flood levels occurred along Brandywine, Wissahickon, Perkiomen, Neshaminy, and Ridley Creeks and the Christina River, among others (see Illustration 6.19 and 6.20). Six deaths were reported in southeastern Pennsylvania, six in New Jersey, and two in Delaware, most from drowning (including people trying to drive cars through flooded roads). About 10,000 people had to be evacuated, mainly because of the stream and river flooding. Total damage in the United States from Floyd was estimated at $4.5 to $6 billion.

The television coverage of Hurricane Floyd is the subject of this chapter's "Story from the Trenches."

A Philadelphia-area Nightmare Hurricane Scenario

It has been about one hundred years since the eye of a hurricane made landfall on the Jersey or Delaware coasts.

And that storm, in 1903, was a minimal Category-1 storm. There have been some close calls since, including the September 1944 hurricane and Gloria in 1985, both of which passed about fifty miles off the Jersey coast.

For the New Jersey and Delaware shores to experience their worst-case hurricane disaster, the storm would have to move in an unusual way. A hurricane in the Atlantic must track northwestward to make a direct hit. In this case, the storm surge would be significantly higher than that coming from a glancing blow. And if the eye wall hit

Hurricane Floyd

When it comes to weather, a hurricane is the ultimate television story. Tornadoes come and go in minutes, snowstorms usually cannot be tracked for days before they hit, and floods are hard to get to and usually only affect a small percent of the audience. But a hurricane provides sustained drama over several days, potentially affects everyone in the audience, and is very "visual." In fact, it is hard to think of any television story that is as compelling and visual as a hurricane threat.

The "big news" nature of a hurricane depends on several factors. Most important for television is evidence of the storm's destruction before it reaches the East Coast. A weak hurricane that crosses Puerto Rico with high winds and flooding rains captured on video is likely to get more national media attention than a major hurricane in the open ocean that can be observed only through satellite images. Floyd started as one of those rather non-newsworthy stories.

Drought had dominated the Delaware Valley during the summer of 1999. From June 1 to mid-August, many locations across the area received only about two inches of rain, which is eight inches below average. Some local governments imposed water rationing. Many experts were saying that only the rains from a tropical system—or more than one—could break the drought. In early September, tropical moisture left over from Hurricane Dennis began to ease the water shortage. Although Dennis fizzled well to our south (in the Carolinas), its remnants came far enough north to contribute to sustained, significant rain in the Delaware Valley. Meanwhile, on September 8, Floyd formed about a thousand miles east of Puerto Rico.

When Floyd strengthened and became a major hurricane on Sunday, September 12, it was heading west toward the Bahamas. Newscasts and weather segments focused on the storm, showing the beautiful but ominous satellite pictures (see Illustration 6.1). But there was still no video from land showing strong winds or rain, since Floyd had stayed north of the Caribbean Islands.

Inside NBC-10 (and undoubtedly at other area stations), plans for special storm coverage had begun. Meetings on September 12 between the meteorologists and news managers focused on where and when the storm might hit. (Such speculation is never made on the air from responsible sources, since publicly predicting potential landfall that far in advance is risky and serves no real purpose.) Conversations centered on whether to send a reporter to "chase the storm," and if so, where and when. Most computer models were predicting that Floyd would come inland somewhere along the East Coast, and the storm was much stronger than Dennis. As a result, a meteorologist and a reporter were sent to the Carolinas to follow the storm. Other resources would be saved for the possible heavy coverage closer to home in midweek.

On Monday and early Tuesday, September 13 and 14, the east coast of Florida was part of the largest weather-related evacuation in U.S. history. Although virtually all computer models and the official National Hurricane Center forecast showed Floyd curving north and not making a direct hit on Florida, officials did not want to risk lives if the storm came closer than predicted. Floyd's winds had reached 150 mph, and video of the damage in the Bahamas appeared on television sets across the country.

For Philadelphia-area television stations, the main question now was when to increase coverage of the threat of Floyd coming this way. This is often a tough call. If newsroom staff starts working double shifts too soon, they may be "burned out" when the storm finally arrives. Or if the storm fails to materialize, viewers complain and forecasters' credibility suffers. So these are matters that concern the news managers as well as the weather staff. There were also more specific questions: What percentage of the news crews should go to the shore? When should hourly updates start? What are the plans for continuous coverage (that is, staying on air with constant news and weather, preempting the normal program schedule)? What expanded hours should the news and weather people work?

NBC-10's own computer model hinted that the most rain—more than six inches—would fall in the Philadelphia area, with less at the coast. So although the shore would be covered at multiple points, more staff and equipment were kept close to home than in a "typical" hurricane (where the effect would tend to be greater at the coast). Rain, not wind, would be the biggest threat, and the possibility of inland flooding was given high priority.

Continuous coverage on NBC-10 and other stations began early in the morning on Thursday,

September 16, as Floyd was moving quickly up the coast. Floyd was the only news story, and it stayed that way all day. Every reporter available was assigned to the storm in some way. In the weather center, three to four meteorologists were on duty at all times, focusing on different aspects of the storm and communicating with the news people. At least two meteorologists were used on air throughout the storm, with one providing the basic information (where the heaviest rain was, where the storm was heading) and the other concentrating on a more detailed analysis (for example, why the rain was heavier in Philadelphia than at the shore). Around 11 A.M., Jon Nese joined the group to provide a historical perspective on the storm.

When a television station is in continuous coverage, there are no scripts, and no strict format. Producers in the control room decide which live shots to take and when to give the meteorologists another segment. Managers in the newsroom oversee the coverage, and often give the producers suggestions ("Let's see more from the shore," for example). The meteorologists can request to be put on air as soon as possible when new information comes in. Aside from storm analysis and on-air updates, meteorologists consult with producers, reporters in the field, news writers, and the anchors, and even field a few calls from frightened viewers.

In covering a news story (weather-related or not), conflicting information is a concern to any local television station. This is especially true when a news event affects a large area (such as a hurricane), because many station personnel will be involved in the coverage. A coordinated effort is necessary so that a consistent message goes out on air. Sometimes, however, the conflicts come from beyond the local station. For example, if the story is of national importance the television networks will cover it—that means network anchors get involved. These anchors typically have little first-hand knowledge of what is going on in specific local areas. An example from Fort Myers, Florida during Hurricane Andrew in 1992 illustrates the dilemma that can face local television meteorologists when a network newsperson gives out inaccurate information. After Andrew had blasted southeastern Florida, one national anchorman went on air and said: "Next stop: Fort Myers." The problem was that the official National Hurricane Center forecast called for the storm to stay well south of that city. Local meteorologists working at the Fort Myers television station that was affiliated with that anchor's network had to immediately decide whether to let the comment pass or correct it on air (Glenn was one of those meteorologists). That decision became easy when frantic viewers started calling the station, wondering if they were next. In essence, a local retraction to that anchor's national reporting had to be made on air at the next available time: "Despite what you just heard, Andrew is not going to make a direct hit on the Fort Myers area."

A similar conflict occurred in the Delaware Valley during Floyd. In any historic storm, various public officials go on live television with their comments

the coast around the time of high tide, the rise in water level would be even worse.

Unlike the Florida coasts, which experience a drop in population during hurricane season, part of the peak vacation time at the Jersey and Delaware shores coincides with the peak of hurricane season. A hurricane threat in August or early September is much more of a problem at the shore than one in late September or October.

In late August 1996, Hurricane Edouard, a powerful Category-4 storm, was moving closer to the East Coast. For a time, National Hurricane Center computer models projected that Edouard would move up Delaware Bay on Labor Day weekend. In fact, Hurricane Watches and Tropical Storm Warnings were in effect for the Delaware and Jersey shores at various times from August 30 to September 1. Fortunately, Edouard curved out to sea, though the heavy surf and riptides it generated did contribute to two deaths at the Jersey shore. What follows is a combination of what did occur and informed speculation about what could have happened if the Delaware Bay forecast had been right (see Illustration 6.21).

August 30 (Friday)

Edouard, with 140-mph winds, continues to inch closer to the East Coast as Labor Day weekend approaches. The National Hurricane Center issues Hurricane Watches and Tropical Storm Warnings from Cape Lookout, North Carolina to Cape Henlopen, Delaware. Millions of vacationers are about to take their holiday trip to the shore, including hundreds of thousands to the New Jersey and Delaware beaches.

about what is happening. It is a way to reassure their constituents that everything is under control and that all precautions have been taken. At one point during a news conference on the afternoon of September 16, an official from the New Jersey Office of Emergency Management (with the governor at his side) said that "the storm is moving out to sea." This statement may have been intended to calm fears, or to indicate that the storm was weakening (which it was). But the implication that the storm was moving away and the danger had passed directly conflicted with what the meteorologists on NBC-10 (and undoubtedly other local stations) had been saying about the storm's movements. Immediately, the comment was clarified on NBC-10 by showing evidence that Floyd was actually coming right up the coast as predicted.

Floyd alone brought copious amounts of tropical moisture, but it dumped so much rain in the Philadelphia area in part because the storm interacted with a weak front that stretched southwest to northeast across the Delaware Valley. At that front, moist east winds blowing across New Jersey collided with cooler northwest winds that prevailed over much of Pennsylvania. The rising air that resulted increased the rain and the threat of flooding from northern Delaware into central New Jersey. This front was the "curve ball" that often gets thrown at meteorologists in a big storm—that is, some detail that makes the storm more (or less) severe in particular areas. If meteorologists failed to see the importance of that front, they might have underestimated the potential for serious flooding. After all, the storm was moving fast and weakening, and the shore (which was in the direct path of Floyd's center) was not getting the expected heavy rain or wind. More than six inches of rain fell on September 16 in Philadelphia, the highest one-day total in 125 years of record keeping, while shore areas had much less (see Illustration 6.18). As the eye of Floyd crossed Delaware Bay and headed toward southern New Jersey, the hurricane was downgraded to a tropical storm just before the center passed over Cape May. Thus, Floyd was not officially recorded as a direct hurricane strike on New Jersey.

As the rains ended in the Philadelphia area, the nearly fourteen hours of continuous coverage ended that evening with scores of news people exhausted, field crews drenched, and lots of commercials unaired. Despite the high cost of this type of coverage and the loss of commercial revenue, providing vital information to the public in such situations is probably local television's greatest responsibility. Such coverage requires long and pressure-filled hours from dozens of people, many out in the elements putting themselves at risk. Most agree that the adrenaline rush from the excitement and importance of the event helps get them through the long hours with little or no food or breaks. The "crash" comes after the storm, when exhaustion can overcome even healthy and fit people. Fortunately, most big storms are followed by a period of quiet weather and meteorologists can get some rest. But for the news people, the stories have just begun. ❖

August 31 (Saturday)

Edouard, still a powerful Category-4 hurricane, is now about six hundred miles south-southeast of Cape May. The official forecast track puts it over Delaware Bay in less than forty-eight hours. In late afternoon, Hurricane Watches are extended northward to Plymouth, Massachusetts. Forecasters and government officials express strong concern. Local newscasts lead with the hurricane and convey those feelings, but it is a beautiful day both inland and at the beach, and only a small percentage of vacationers cancel their plans.

September 1 (Sunday)

Edouard is now about three hundred miles away and is still a Category-4 hurricane over the warm Gulf Stream. Morning newscasts greatly increase their focus on the storm, but most people are already at the shore. It is another sunny day at the coast, although the surf is too rough for swimming and there is a stiff northeast wind.

The official National Hurricane Center forecast track still takes the storm into Delaware Bay, but several computer models now say it will curve out to sea without making landfall, as many storms do. A few past "false alarms" had prompted evacuations, only to see the storm curve away, putting the shore on the weak side of the storm. So most beachgoers take a wait-and-see approach to the situation.

By 5 P.M., it looks likely that the storm will continue to move north-northwestward and track up Delaware Bay. Hurricane warnings are issued from Wallops Island, Virginia to western Long Island. Evacuations of the barrier islands are ordered, but the weather is still nice.

ILLUSTRATION 6.21   Visible satellite image of Hurricane Edouard in late August 1996. The solid line shows Edouard's actual track, while the dashed line shows the predicted track around the time this image was taken (courtesy of NOAA).

A quick rush to leave occurs between 5 and 7 P.M., with massive traffic jams for those headed away from the coast. Reports of these tie-ups plus the good weather cause many to wait until daylight to act.

September 2 (Labor Day)

By daylight, it is too late for some. Edouard has weakened to a Category-3 hurricane over the cooler ocean waters, but winds near the center are still 125 mph and the eye is now less than two hundred miles away. Northeast winds at the coast have increased to 50 mph, and flooding has already cut off some evacuation routes at the high tide around 6 A.M. It is raining so hard that visibility is near zero, and the winds are too strong to allow boats to move to safer places.

The combination of flooded roads, increasing winds, high tide, and the late evacuation causes incredible traffic jams. As Edouard heads toward land, thousands of vehicles sit in traffic with no place to go. Cars are stuck on bridges, and boats are stuck in the bays. The Delaware Bay waters are piling up, causing the Schuylkill and Delaware Rivers to flood. Admiral Wilson Boulevard and other low-lying roads around Philadelphia are shut down by the flooding, so some of the fortunate few who left the shore in the morning are now stranded in Camden in 100-mph winds and torrential rains.

In the end, fatalities are in the hundreds, and the damage is in tens of billions of dollars in Delaware, New Jersey, and eastern Pennsylvania. Power is out for a week. Sand covers the main roads at the shore. The barrier islands are cut by the storm in several places, stranding thousands. And it could have been worse. The storm struck near first-quarter moon, which meant slightly reduced high tides.

The forecast was correct, but it still could not prevent one of the biggest weather-related disasters in U.S. history, and the biggest loss of life from a U.S. hurricane since the 1938 "Long Island Express," which killed about six hundred people.

LOOKING AHEAD:
SIGNS OF WHAT IS TO COME

Autumn is a bridge between the extreme seasons of summer and winter. And as such, it has features of both. The tropical threat is a sign of lingering warmth, as is Indian Summer. On the other hand, the arrival of colorful fall foliage and that first frosty night in autumn are checkpoints along the inevitable slide into winter. We close this chapter by considering these other meteorological highlights of autumn.

Early Chill

In Philadelphia, the average date of the earliest freezing temperature in the fall is around the first week of November, but it is earlier than that in outlying areas (see Illustration 6.22). The earliest official 32°F-or-below temperature in Philadelphia occurred on October 6, 1961, when the mercury dipped to 31°F (-1°C). However, keep in mind that official temperature readings are taken a few feet above the surface. At ground level, the temperature is usually lower at night, especially if the sky is

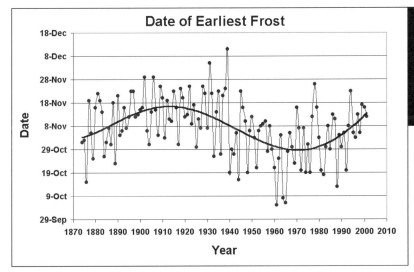

Date of Earliest Frost

ILLUSTRATION 6.22   Date of the earliest official freezing temperature in the autumn in Philadelphia, 1874 to 2001. The range is from October 5 in 1961 to December 11 in 1939. The thicker solid line is a mathematically smoothed version of the data.

clear and winds are light. So you can get frost on your grass or on your car even on nights when the "official" temperature does not drop to freezing.

The earliest-in-the-season measurable snow in modern times occurred on October 10, 1979, when 2.1 inches fell (see Illustration 6.23). Measurable snow has fallen in October in only two other years since 1884—on October 19–20, 1940 (2 inches) and October 30–31, 1925 (1.1 inches). Eleven other Octobers have seen a trace of snow—the flurries on October 9, 1895, make that the earliest official snowflake sighting in Philadelphia. Records from colonial times push that date a bit earlier. Entries from some Philadelphia-area diaries report that "snow this morning covered houses" on October 3, 1769, while on October 8, 1703, "we have had a fall of snow, and now the northwest wind blows very hard." Even considering these historical records, the first few days of October appear to be the early-season limit for snowflakes in the immediate Philadelphia area.

Of course, snow becomes more likely in November, though it is still not a sure thing. At least some snow was reported in 92 of the 118 Novembers from 1884 to 2001. But of those 92 Novembers, only 37 had enough to be

ILLUSTRATION 6.23   Snow coats pumpkins in Berwyn in eastern Chester County on October 10, 1979, the date of the earliest measurable snowfall on record in Philadelphia, 2.1 inches (Urban Archives, Temple University, Philadelphia, Pennsylvania).

measured. The 8.8-inch storm on November 6–7, 1953 stands out by two weeks as the earliest big snow on record in the month. The only larger November snowstorm was 9.2 inches on November 26–27, 1898. There have been a few "White Thanksgivings"—perhaps the most notable occurred in 1938, when temperatures stayed in the 30s the entire four-day weekend and two storms combined for more than 11 inches by Sunday morning. More recently, 4.6 inches of snow fell on the day before Thanksgiving and Thanksgiving Day in 1989, and the Turkey Day high temperature that year was only 32°F.

Fall Foliage

The dates of the first freeze or first snowflakes in autumn vary considerably from year to year, but there is one rite of passage of fall that can always be counted on to be reasonably timely. Seemingly on cue, leaves transform from the green of the warm season and put on a show of color.

The short-lived display of fall foliage is a natural response to the lowering sun angle and decreasing amounts of daylight that accompany the slide into winter. The appearance of fall colors signals that trees are beginning their winter "hibernation." In response to the decreasing amounts of sunshine, leaves stop producing food through photosynthesis. Any food left in the leaves moves slowly into the branches, trunk and roots for storage during the winter ahead. The colors seen in the leaves are from the pigments left behind.

Leaf colors are determined by the species of tree. Red maples, northern red oaks, and sumacs often feature striking reds and purples. The leaves of a few trees, including sugar maples and dogwoods, may turn red but can also be yellow. Leaves of trees such as birches, tulip poplars, and hickories are always yellow in the fall, never red.

Leaves that turn yellow or orange in the fall actually have just as much of those pigments in spring and summer (when the leaves are green) as they do in October when they are yellow or orange (the yellow pigment, called xanthophyll, is the same one that gives bananas their color). But the green pigment, chlorophyll, is so concentrated in the spring and summer that it masks the other colors. With the decreasing amounts of sunlight in the fall, the chlorophyll starts to break down, allowing the yellows and oranges to be unveiled. In contrast, the reds and purples result from pigments that are not made in the leaves until the fall.

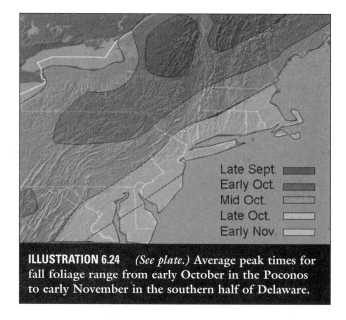

| Late Sept. |
| Early Oct. |
| Mid Oct. |
| Late Oct. |
| Early Nov. |

ILLUSTRATION 6.24 *(See plate.)* Average peak times for fall foliage range from early October in the Poconos to early November in the southern half of Delaware.

To a large extent, weather helps determine just how colorful the foliage will be. Warm sunny days and cool (but not freezing) nights are ideal for producing the most spectacular colors. A fall with cloudy days and warm nights leads to a more drab display. There is some variation from year to year because of specific weather conditions. For example, an early frost would end the colorful display prematurely, while a violent wind or rainstorm would not be good either because many of the leaves would be stripped from the trees.

Though the brilliance of the foliage does vary from autumn to autumn, there is not much year-to-year variation in the peak foliage time for a particular location. That is because the primary catalyst for the change in color is the decrease in the number of hours of daylight, and that is constant at a given location from year to year. However, there is a north-south variation in the peak foliage time, because daylight hours are lost a little more quickly farther north. Thus, the peak time ranges from mid-October in the Lehigh Valley to late October in the immediate Philadelphia area to early November in central and southern Delaware and southern New Jersey (see Illustration 6.24).

Indian Summer

The term **Indian Summer** is commonly used to describe a period of above-average temperatures in mid- or late-autumn, with mainly sunny and sometimes hazy days and cool nights. In the northeastern United States, it is generally agreed that at least one killing frost and usually a substantial period of cool weather must precede the warm spell for it to be considered a true Indian Summer.

Given the variability of daily weather, and the wide range of time over which Indian Summer conditions can be met, a period of Indian Summer weather is quite common in the Delaware Valley. And although Indian Summer does not occur every year, some years can have several such periods.

A typical weather map that brings Indian Summer conditions to the Delaware Valley shows a large area of high pressure along or just off the East Coast. This high might be the same one that delivered cold air and perhaps a frost a few nights before as it moved out of Canada. But as the high settles in along the coast, winds in the Delaware Valley come around to the south and southwest, bringing warmer air northward. This milder flow on the back side of the high may last for several days, and this same basic weather scenario can play out several times in the fall before the chill of winter sets in for good.

There is considerable uncertainty as to the origin of the term "Indian Summer." The earliest written uses in this country date back to the late 1700s. In one letter, dated "17 January 1778," a Frenchman named John de Crevecoeur writes: "Sometimes the rain is followed by an interval of calm and warmth which is called the Indian Summer; its characteristics are a tranquil atmosphere and general smokiness." The letter makes clear that the term "Indian Summer" was already in use during the Revolution.

The term may refer to the way Native Americans used the period of extra warmth to increase their stores for winter. Another possibility is that the early colonists invented terms such as Indian corn and Indian tea—and perhaps Indian Summer—to describe something that resembled the real thing as they had known it in England. European folklore does have Indian Summer equivalents, including "Old Wives' Summer" in central Europe and "St. Luke's Summer" and "St. Martin's Summer" in England.

Though there is uncertainty as to the origin of the term, and even some ambiguity as to its definition, one thing is for sure—Indian Summer never lasts!

CHAPTER 7

Philadelphia's Future Climate

> Everybody talks about the weather,
> but nobody does anything about it.
> —*Charles Dudley Warner*

Forecasting the specifics of the weather more than a week or so in advance has been a controversial subject for decades. One of the first serious attempts at making such a long-range prediction involved efforts to plan the D-Day invasion in 1944, probably the most important weather forecast in world history (see "Story from the Trenches—The D-Day Forecast" in Chapter 2). Even today, the best that meteorologists can do over time periods beyond about ten days is to describe (with varying degrees of certainty and success) some general features of temperature and precipitation averaged over periods of weeks, months, or seasons— for example, whether rainfall next month will be above average, near average, or below average.

In this chapter, we will explore monthly and seasonal forecasting—how the Climate Prediction Center does it, how detailed the forecasts actually get (it is not as specific as you might think), and when such forecasts tend to be most accurate. The key to monthly and seasonal forecasting these days is "teleconnections"—how an atmospheric or oceanic anomaly in one part of the world can have a ripple effect on the weather and climate in other areas, even thousands of miles away. In the past few years, two important teleconnections—El Niño and La Niña—have emerged into the public spotlight. We will explain what they are and whether they have an influence on the Philadelphia area. Among other important teleconnections that we will discuss is one that is not as well known—the NAO, or North Atlantic Oscillation—but which can have a dramatic impact on our weather, especially in winter.

In the remainder of the chapter, we turn to climate changes over longer time scales to see what (if anything) can be said about climate years and even decades in the future. We start thousands of years in the past, with the last Ice Age, and consider climate changes since then. In just the last few hundred years, beginning with the Industrial Revolution, humans have to be included as a possible cause of climate change. Global warming is the central issue in this discussion.

The subject of global warming has become entangled with political complications, but we will concentrate on the science, much of which is not at all controversial. We have tried to be clear about what is known and what is still in dispute. We describe how humans are altering the composition of the atmosphere and explain how those changes can result in rising global temperatures. We will also discuss the difficulties inherent in projecting future climate. And we explore the implications of global warming on a local scale, from the shore to the Poconos, and make an educated, plausible guess as to what the climate will be like in the Philadelphia area late in the twenty-first century. A few of the younger readers of this book will still be around at that time to observe how well the projection holds up.

FUTURE CLIMATE: MONTHS AND SEASONS

Over the years, various individuals and businesses have claimed they can make specific daily forecasts up to a year in advance. The truth is, as the science of meteorology stands at the beginning of the twenty-first century, forecasts of specific weather conditions beyond a week or two in the future are no better than coin flips most of the time (sorry, *Old Farmer's Almanac*). The official statement of the American Meteorological Society on the subject reads:

> Claims of skillful predictions of day-to-day weather changes beyond this limit [a week or two] have no scientific basis and are either misinformed or calculated misrepresentations of true capabilities.

The best that science can do over time periods beyond a week or two is to pin down with varying degrees of certainty some general features, such as whether temperature or precipitation will be above average, near average, or below average. That information will not help someone planning a wedding for a particular weekend next June, but there is great potential value in being able to foresee general features of the climate months or even seasons in advance. And the skill in such long-range forecasts has improved quite a bit in recent years.

The Climate Prediction Center (CPC) is the climate-forecasting cousin of the National Weather Service. Around the middle of every month, CPC forecasters issue outlooks for temperature and precipitation for the next calendar month and for all three-month periods during the coming year. The outlooks are simple, categorical, probability forecasts. The categories are "above average," "near average," and "below average".

As an example, suppose forecasters are preparing an outlook for temperature for next winter. They begin by assuming that each of the three categories has an equal likelihood of occurring—that is, warmer-than-average, near-average, and colder-than-average each have about a 33.3 percent probability of happening. Forecasters then adjust those percentages up or down for each region of the country to reflect their degree of confidence in a particular category. The adjustments are based on statistical techniques that relate current and future conditions to past weather patterns and also on teleconnections such

as El Niño (discussed later in this chapter). Usually, the adjustments are not very large; rarely would any of the probabilities be increased to 60 percent or greater. Often, no adjustments are made; that is, forecasters admit they cannot do much more than guess. The end product is a map of the entire country that shows areas where temperatures are expected to be above average or below average, with the remainder being regions where the data does not support a forecast one way or the other.

An example of a seasonal outlook issued by CPC is shown in Illustration 7.1a, the temperature forecast for the winter of 1997–98 (December through February). The shaded region labeled with "A" in the northern part of the country indicates where forecasters leaned toward above-average temperatures; the darker the shading, the greater the level of confidence in the warmer-than-average outlook. Even in the areas of greatest confidence, however, the forecast is not meant to be interpreted as a sure thing, as the media sometimes imply when reporting this information. Noticeably absent in the forecast map are any areas predicted to have below-average temperatures (there would have been a shaded region labeled with "B"). The Delaware Valley was right on the edge of the warmer-than-average zone, indicating a very small lean that way.

Illustration 7.1b shows actual temperatures for the winter of 1997–98, expressed as the departure from average (so positive means warmer-than-average). In agreement with CPC's forecast, colder-than-average temperatures were noticeably absent. The entire northern two-thirds of the country had a warmer-than-average winter, correlating reasonably well with the forecast. Though the match between the outlook and reality was far from perfect, this was one of CPC's best seasonal forecasts ever. We will discuss the reason behind that success shortly.

Tracking individual high- and low-pressure systems, the trademark of short-range weather forecasting, has little value in monthly and seasonal forecasting. Those highs and lows will not exist a month or season down the road. Instead of trying to pin down individual storms or fronts, forecasters focus on the larger-scale jet stream pattern and where the predominant troughs and ridges will be. If the positions of those large-scale dips and bends in the upper-air wind pattern can be determined with any accuracy, then there is a basis for adjusting the overall temperature and precipitation patterns.

One technique employed by monthly and seasonal forecasters looks to the past to predict the future, a procedure called **analog forecasting**. This is basically the same technique used by meteorologists who tried to forecast the weather for the D-Day invasion (see "Story from

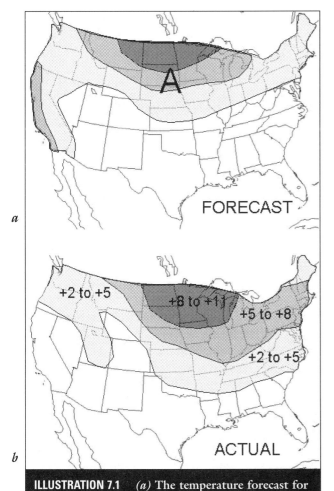

perature and precipitation patterns that developed, and incorporate those patterns into their seasonal outlooks.

Analog forecasting has been used for decades, but it has improved in recent years in monthly and seasonal forecasting. This is almost entirely because of the recognition that unusual changes, or anomalies, of ocean temperature or air pressure or wind in one part of the globe can affect the location and intensity of the jet stream in distant places. These far-away changes affect the overall weather pattern. A correlation between an atmospheric or oceanic variation in one part of the world and the general weather pattern at locations far away is known as a **teleconnection**. If large-scale changes in the ocean or atmosphere can be foreseen with any accuracy, and teleconnections are known for various parts of the globe, then there will be a higher degree of confidence in monthly and seasonal predictions.

Just in the last few decades, researchers have identified several teleconnections that can help explain, and thus often predict, temperature and precipitation anomalies in various parts of the globe over time scales of weeks, months, and seasons. Two phenomena in particular that have strong teleconnections to parts of the United States have emerged from research centers and university seminar rooms to become very familiar to the general public: El Niño and La Niña.

El Niño and La Niña

For hundreds of years, Peruvian fishermen had noticed that the Pacific Ocean waters off their shores occasionally turned unusually warm. Whenever that happened, their anchovy catch diminished greatly. This local connection was easy to explain. Anchovies feed on tiny animals that in turn feed in cooler, nutrient-rich waters that typically occupy that part of the Pacific. Warmer water is poorer in nutrients, so when warmer water takes over, the tiny animals do not survive. Then neither do the anchovies (or they move elsewhere to feed). This warming of the eastern Pacific off Peru happened every few years, but not regularly. The warming often began around Christmas, so the phenomenon was named "El Niño," which means "the Child."

The first time El Niño drew widespread public attention was during the winter of 1982–83, when storm after storm battered the California coast. Seeking explanations for this extreme weather, scientists turned to the large, unusually warm area of water in the central and eastern equatorial Pacific Ocean for a possible connection (see Illustration 7.2). Meteorologists and oceanographers began to realize that this ocean-temperature anomaly

the Trenches" in Chapter 2), and by Wally Kinnan in Philadelphia in the early 1960s when he issued some of the earliest five-day forecasts. In analog climate forecasting, meteorologists determine the average positions of the troughs and ridges in the jet stream in the weeks and months leading up to the present. Then they look for previous years during which a similar pattern occurred during the same season—that is, they look for analogs. Forecasters then follow the evolution of the upper-air flow during those analog years, note the general tem-

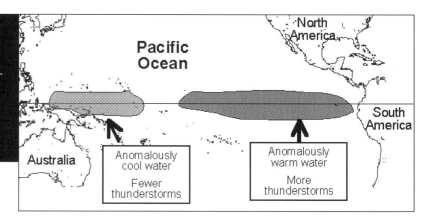

ILLUSTRATION 7.2   El Niño is a Pacific Ocean phenomenon characterized by unusually warm waters near the equator in the central and eastern Pacific and unusually cool waters in the western tropical Pacific. The warmer-than-average waters lead to more thunderstorms than usual, while less rain than average falls over the anomalously cool waters.

could have much more far-reaching effects, well beyond the tropics. It might be possible to explain (at least partly) warmer winters in the northern United States, rainier weather in Florida, and drought in Australia and the Philippines, for example.

After the 1982–83 El Niño, a network of buoys was installed across the tropical Pacific Ocean to closely monitor ocean temperature changes. These buoys measured water temperatures not only at the ocean surface, but hundreds of feet below. Over time, scientists realized that the unusually warm surface waters characteristic of an El Niño do not just suddenly appear, but rather they are preceded by changes in deeper waters. These observations, and a better understanding of the oceanic and atmospheric processes behind El Niño, led to the development of statistical techniques and computer models to predict it. Most of these models forecast that an El Niño would develop during 1997 and persist into 1998. This consistency among models gave forecasters increased confidence in their seasonal outlooks for areas where teleconnections to El Niño had been established. The U.S. outlook was for a stormy winter in California, a rainy one in Florida, and an unusually warm winter in the northern States . . . and it all happened.

El Niños come and go irregularly, every three to eight years or so, though by most measures their frequency has increased in the 1980s and 1990s. No two are exactly alike: the 1982–83 and 1997–98 events were considered "strong" because the temperature anomalies in the Pacific Ocean were unusually large, about 3–5°F, which is a big temperature increase for such a wide expanse of water (see Illustration 7.2). Other events are classified as "weak" or "moderate" when the tropical Pacific warms less. The most dramatic teleconnections occur with the strong events, but even weak El Niños can have a rippling effect elsewhere on the globe.

The El Niño teleconnection starts in the central and eastern tropical Pacific Ocean with the huge patch of unusually warm water, covering an area about twice the size of the United States. The air above that extensive stretch of ocean becomes anomalously warm and moist, generating more thunderstorms than usual (see Illustration 7.2). Sometimes satellite images show a plume of moisture stretching from the vicinity of the El Niño-warmed waters northeastward into the midlatitudes. This feed of tropical moisture has been dubbed the "Pineapple Express," and it tends to occur more often and with greater intensity during El Niños (see Illustration 7.3).

These examples of direct connections to El Niño are easily observed. But most of the time, outside of the tropics, the connection to El Niño is more subtle, occurring indirectly through changes in the jet stream pattern. When the jet stream is altered, certain storm tracks in the midlatitudes become more likely and others less likely. Cold air or warm air outbreaks become more frequent in some areas and less common in others. It is a bit like taking a die and weighting it so that a certain number (say two) comes up more often than it would randomly. You will still roll one, three, four, five, and six occasionally, but two will come up more frequently than the expected one-out-of-six times.

The jet stream is strongest in winter, and that is also the time of year when the jet lingers over the United States. So seasonal forecasts based on teleconnections to El Niño have shown the most promise for U.S. forecasters in winter. Figure 7.4 shows El Niño teleconnections for the United States during winter, for both temperature (Illustration 7.4a) and precipitation (Illustration 7.4b). What stands out the most is warmth in the northern United States (particularly the northern Plains and Rockies) and wetness in Florida, parts of the Midwest, the desert Southwest, and California. The Philadelphia area

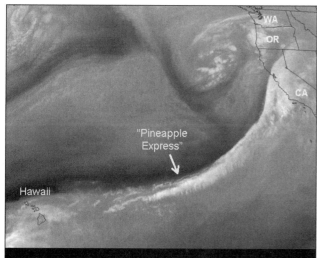

ILLUSTRATION 7.3 A water vapor image from November 1996 shows the "Pineapple Express," a channel of moisture that stretches from the mid-Pacific northeastward to the West Coast of the United States. This feed of moisture is typically more intense and happens more frequently during an El Niño (courtesy of NOAA).

(and the Northeast in general) is in an area of relatively weak teleconnections from El Niño. There is a very slight tendency toward above-average precipitation and above-average temperature in winter, but it is a much weaker signal than in other parts of the country. Any tendency toward warmer and wetter winters in the Philadelphia area is most likely to show up in a strong El Niño, such as the ones in 1982–83 and 1997–98.

One significant El Niño teleconnection is not obvious in Illustration 7.4, but it has a great impact on the Philadelphia area. In El Niño years, nor'easters tend to be more common, giving the potential for more snowstorms on the eastern seaboard. This teleconnection held true during the strong El Niño winter of 1997–98, when there were several big nor'easters. But the winter was so warm that very little of the precipitation fell as snow. The Philadelphia area and much of the eastern part of the country had a warm, fairly wet winter, but there was very little snow.

The duration of El Niños varies, typically from six months to a year. Then, for reasons still not completely understood, the waters of the central and eastern Pacific cool. The ocean may return to near to its average state, or ocean water temperatures may, like a giant seesaw, swing the other way to below-average temperatures. The same huge patch of ocean which, months earlier, was

3–5°F warmer than average, may now be 1–3°F cooler than average. It is an amazing process to watch. This opposite of El Niño is called La Niña.

As with El Niño's warmer waters, La Niña's cooler-than-average waters set in motion a rippling effect that can cause temperature and precipitation anomalies in other parts of the world, primarily through shifts in the typical jet stream pattern. Figure 7.5 shows La Niña teleconnections for the United States during winter, for both temperature (Illustration 7.5a) and precipitation (Illustration 7.5b). In general, they are opposite those associated with El Niño, but less extreme. Dryness in Texas and Florida, cold in the northern Plains and Pacific Northwest, and warmth in the southeastern states stand out the most. As with El Niño, teleconnections to the Philadelphia area are relatively weak with La Niña. Temperatures tend slightly toward warmer-than-average, but La Niña winters often feature wide swings in temperature, with times of bitter cold interspersed with milder periods. That kind of variability increases the likelihood of ice storms. But the tendency for fewer nor'easters during a La Niña winter leads to slightly below-average precipitation overall.

Despite all the publicity that El Niño and La Niña get with regard to stormy weather, neither is correlated with a snowy winter in the Philadelphia area. In fact, average snowfall is higher in "neutral" years, when neither El Niño nor La Niña is in control. There is nothing special about neutral winters—it is just that the slight tendency toward above-average temperatures during El Niño and drier-than-average conditions with fewer nor'easters during La Niña are not as favorable for snow. Going back about fifty years, we find the following statistics for average winter snowfall in Philadelphia:

All winters: 21.4 inches
Moderate to strong El Niño winters: 18.5 inches
Moderate to strong La Niña winters: 16.5 inches
"Neutral" winters: 24.5 inches.

A closer look at the individual winters that make up these averages reveals the link between nor'easters and El Niño and La Niña. The 18.5-inch average for moderate to strong El Niño winters is a bit misleading because the winters that make up that average either had little snow (less than 10 inches) or a lot of snow (more than 25 inches). There were plenty of nor'easters in those winters, as is typical for El Niños, but in some years it was simply too warm for much snow. The story is different during winters with strong to moderate La Niñas. With

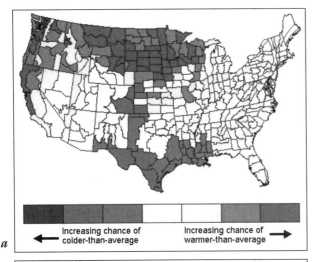

a

Increasing chance of
← colder-than-average

Increasing chance of
warmer-than-average →

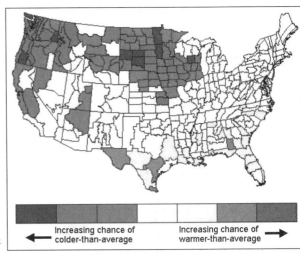

a

Increasing chance of
← colder-than-average

Increasing chance of
warmer-than-average →

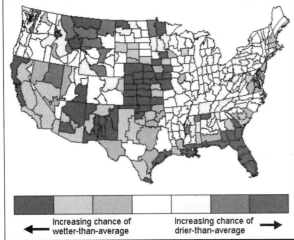

b

Increasing chance of
← wetter-than-average

Increasing chance of
drier-than-average →

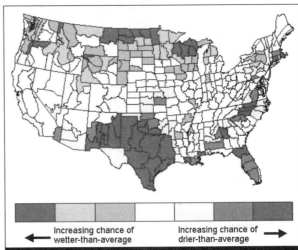

b

Increasing chance of
← wetter-than-average

Increasing chance of
drier-than-average →

ILLUSTRATION 7.4 *(See plate.)* *(a)* Based on historical records, the risk of warm or cold temperature extremes during an El Niño winter. The northern Plains are most likely to have warmer-than-average winters, while the western Gulf States are most likely to be cooler than average in an El Niño winter. No strong tendency in temperature is seen in the eastern states. *(b)* The risk of wet or dry precipitation extremes during an El Niño winter. Florida, the desert Southwest, California, and parts of the central Plains have much higher chances of wetter-than-average winters during an El Niño (courtesy of Climate Diagnostics Center).

ILLUSTRATION 7.5 *(See plate.)* *(a)* Based on historical records, the risk of warm or cold temperature extremes during a La Niña winter. The teleconnections are, in general, opposite of those during an El Niño. No strong tendency in temperature is seen in the eastern states. *(b)* The risk of wet or dry precipitation extremes during a La Niña winter. Dryness in Texas and Florida and wetness in the extreme northern Plains stand out (courtesy of Climate Diagnostics Center).

fewer nor'easters, none of the winters in this category had lots of snow. In fact, all had 23 inches or less.

Here is another way to interpret the relationship between winter snowfall in the Philadelphia area and the El Niño/La Niña teleconnection: only seven of the last fifty winters in Philadelphia have had fewer than ten

inches of snow. Of those seven winters, none were "neutral" winters. So there is little chance, at least historically, of having a relatively snow-free winter when neither an El Niño nor a La Niña is in progress.

Although El Niño and La Niña teleconnections tend to be strongest during winter, there is at least one link that is important during the summer and fall for the Philadelphia area. This teleconnection has been used successfully for more than a decade as a predictor, in a general sense, of how active (or inactive) the Atlantic hurricane season will be.

During El Niños, more thunderstorms than usual form over the central and eastern tropical Pacific Ocean. Upward motions in these thunderstorms add momentum to high-altitude winds, and those winds spread eastward into the Atlantic Ocean. Strong upper-air winds disrupt the organization of budding tropical systems, literally by blowing their tops off. As a result, fewer Atlantic hurricanes tend to form in El Niño years, while La Niña years tend to have more Atlantic storms than average. In fact, the historical probability of two or more hurricanes making landfall in the United States during an El Niño year is 28 percent, but that probability increases to 48 percent during neutral years and 66 percent during a La Niña year. Consequently, property damage and losses in the United States from tropical storms and hurricanes is significantly lower in El Niño years and considerably higher in La Niña years.

There are good examples of this teleconnection in recent years. The 1997 hurricane season overlapped with the early stages of the 1997–98 El Niño. There were eight tropical storms that year with three becoming hurricanes, below the averages of nine and six. In contrast, the active hurricane seasons of 1998 (fourteen storms, ten hurricanes) and 1999 (twelve storms, eight hurricanes) occurred during La Niña years.

This Atlantic hurricane teleconnection is one example of how El Niño's media-hyped reputation as a villain is not supported by the facts, at least not in regard to hurricanes in the United States. In addition, the tendency for milder winters in the north during an El Niño means less demand for heating fuel and fewer winter-related traffic accidents. Statistics show that El Niños save billions of dollars and hundreds of lives each year in the United States. Overall, it is La Niña with its more active Atlantic hurricane season that causes the most economic loss.

Here is a summary of teleconnections to moderate or strong El Niños and La Niñas that are relevant to the Philadelphia area:

Moderate to Strong El Niños

1. Milder than average winters
2. More nor'easters, but not necessarily more snow
3. Fewer Atlantic hurricanes and a reduced tropical threat

Moderate to Strong La Niñas

1. Tendency for alternating warm and cold periods, averaging out to slightly milder than average
2. Fewer nor'easters, and a slight tendency for below-average snow
3. More Atlantic hurricanes and an increased tropical threat

Other Teleconnections

The connections between El Niño and La Niña and climate anomalies in other parts of the globe are indisputable, especially during strong events. In the words of one senior meteorologist at NOAA, "The El Niño/La Niña events are no-brainers. Their effect on extreme weather events is well known. The real challenge is linking the neutral years with characteristic jet stream patterns." And while El Niño is the most well-known teleconnection for seasonal forecasting, it is not the only one.

One of the other cycles now coming under closer scrutiny is the **Pacific Decadal Oscillation** or PDO. This cycle is based on changes in water temperature in the Pacific Ocean over periods of several decades and is credited by some meteorologists as a major factor in long-term U.S. weather changes. In the "warm" phase of the PDO, warmer-than-average water temperatures dominate the equatorial Pacific and along the coasts of the Americas and up toward Alaska, while most of the rest of the North Pacific Ocean is relatively cool. The opposite temperature anomalies reign in the "cool" phase of the PDO.

While El Niños and La Niñas come every few years or so, the PDO tends to stay in the same phase for two or three decades. The PDO was in its "warm" phase in the 1980s and 1990s and is credited by some experts as a major reason for warming in the United States during that period and for the increase in El Niños. The PDO appears to have switched phases in the late 1990s, perhaps signaling an upcoming period of more La Niñas and cooler weather over the United States. Scientists are still trying to understand if the PDO is a distinct phenomenon or just another way of looking at the decade-to-decade variation in the frequency of El Niños and La Niñas.

If changes over the Pacific Ocean can influence climate in North America, it is no surprise that changes over the Atlantic can also. One cycle that has been identified over the Atlantic Ocean is called the **North Atlantic Oscillation** (NAO). It is not a cycle in water temperature but rather a variation in air pressure at high altitudes. Though the NAO has not received the publicity that El Niño and La Niña have, the NAO can help explain a substantial portion of the winter climate variability over the eastern United States, and thus the Philadelphia area.

The NAO is tracked by comparing the high-altitude air pressure in the far North Atlantic, near Iceland, with the high-altitude pressure measured at a location several thousand miles to the south near the Azores Islands. Sometimes, the pressure in the far North Atlantic is anomalously low, and the pressure near the Azores is abnormally high. This large pressure difference results in fast upper-air westerly winds blowing across the Atlantic (see Illustration 7.6a). This mode is known as the "positive phase" of the NAO. In this phase, a strong, persistent, zonal (predominantly west-to-east) jet stream dominates in the Atlantic. The fast flow means that cold air masses invading the eastern United States and storms forming near the East Coast tend to move quickly out to sea, while the zonal flow leads to weaker storms in general. As a result, the positive phase of the NAO typically leads to milder winters in the eastern U.S. with fewer storms and thus less snow.

At other times, the high-altitude pressure difference from north to south in the Atlantic is unusually small or even reversed. This is the "negative phase" of the NAO. This mode is characterized by anomalously high pressure in the far North Atlantic; sometimes a "blocking high" forms, slowing and diverting the typical west-to-east flow of the jet stream (see Illustration 7.6b). As a result, in the NAO's negative phase the westerly jet stream flow across the North Atlantic is weaker and often becomes very wavy. Cold air masses are not as quick to move off the North American continent, and the wavy jet stream favors more frequent and more powerful storms. Together, these factors raise the odds for snow along the East Coast.

The correlation between East Coast storminess and the negative phase of the NAO has been on display several times in recent winters. For example, during the winter of 1995–96, a weak La Niña was in progress. La Niña winters tend to be slightly milder than average in our area, with fewer nor'easters. But at that time, the NAO was strongly in its negative phase. That winter was the snowiest on record in Philadelphia, highlighted by the Blizzard of January 1996. More recently, the winter of 1999–2000 was not very cold, and not very snowy. But the NAO switched briefly to its negative phase in January 2000, coinciding with the only big snowstorm of the season (see Chapter 2). And the "end-of-the-millennium" storm on December 30, 2000, which brought 10 to 20 inches of snow to Philadelphia and many locations to the north and east, also occurred during a time when the NAO had gone strongly negative.

As with El Niño and La Niña, accurate forecasts of the NAO would allow for much better weekly, monthly, and

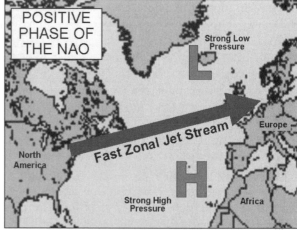

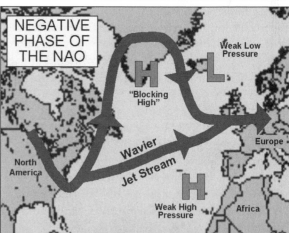

ILLUSTRATION 7.6   Average high-altitude flow pattern during the *(a)* positive phase of the North Atlantic Oscillation and *(b)* negative phase of the NAO. In the negative phase, abnormally high pressure near Greenland can form a block to the jet stream. This leads to a persistent upper-air trough near the East Coast, which favors storm formation there.

seasonal forecasts. But at present, the NAO is much less predictable than El Niño and La Niña, mainly because the NAO switches from one phase to another more frequently (sometimes from week to week), and the reasons for that variability are still poorly understood.

These cycles—El Niño/La Niña, the NAO, and others both known and yet to be identified—are evidence that the climate system is very complex, a mingling of variations on many different time scales. The success in measuring and predicting El Niño and La Niña has inspired new research into the ocean-atmosphere connection. As meteorologists and oceanographers more thoroughly understand these cycles and how to predict them, better monthly and seasonal forecasts will follow.

FUTURE CLIMATE:
YEARS AND DECADES (AND LONGER)

From month to month or season to season, the processes that control the variability of the climate are natural, though not necessarily regular or completely understood. However, climate changes over longer time scales—years, decades, even centuries in the future—now involve processes set in motion by humans as well as those caused by nature. People have become potential climate changers because our actions are changing the make-up of the atmosphere and the land. In many ways, the old saying that "everybody talks about the weather, but nobody does anything about it" is not completely true anymore.

Climate changes over centuries and longer time scales have a variety of natural causes, including changes in the output of the sun, variations in the earth's orbit, and deviations in ocean circulation. The most recent significant climate change to affect North America was the Ice Age that lasted from about 80,000 to 20,000 years ago. In the depths of this ice age, about 21,000 years ago, glaciers from the north had expanded as far south as today's northern New Jersey and northern Pennsylvania (see Illustration 7.7). Ice was a mile thick in what is now Central Park in New York City. Temperatures in the Philadelphia area, just to the south of the expanded glacier, were

ILLUSTRATION 7.8   Twenty-thousand years ago, the leading edge of an ice sheet straddled Hickory Run State Park in Carbon County, Pennsylvania, about seventy-five miles north of Philadelphia. Glacial processes at the edge of the ice sheet helped create this boulder field in the park (courtesy of Carol Sweeney).

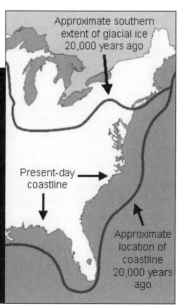

ILLUSTRATION 7.7   The U.S. coast has not always looked the way it does now. During ice ages, much of earth's water is locked up in glaciers, so sea level drops. Note that glacial ice penetrated well into northern Pennsylvania about 20,000 years ago at the height of the last Ice Age (United States Geological Survey).

Approximate southern extent of glacial ice 20,000 years ago

Present-day coastline

Approximate location of coastline 20,000 years ago

probably at least 10°F colder than at present. So much water was locked up in ice that sea level was 300 to 400 feet lower than it is today, and the Atlantic shoreline was 100 to 150 miles east of its present location.

About 20,000 years ago, global temperatures began to rise, and the ice sheet started to retreat to the north. The continual freeze-and-thaw activity at the leading edge of the glacier contributed to creating a boulder field in what is now Hickory Run State Park in Carbon County, about seventy-five miles north of Philadelphia (see Illustration 7.8). Dozens of glacial lakes in the Poconos are also reminders of the ice that once covered that land. By about 10,000 years ago, enough ice had melted (and the land had rebounded from the weight of the ice) so that sea level was close to where it is today.

Other natural climate swings have occurred since the end of the last Ice Age, but nothing comparable in magnitude. Some of these relatively minor climate variations have occurred over the last millennium. There is evidence that the period from around the year 1000 to 1300 was relatively warm, at least in the middle and high latitudes of the Northern Hemisphere. This was a time when the Vikings ventured over ice-free northern seas to settle Iceland, Greenland, and northern Newfoundland. By the middle of the 1500s and 1600s, however, the North Atlantic region had cooled, and mountain glaciers were advancing. This cooler period, known as the Little Ice Age, lasted until about the middle of the nineteenth century.

The most important point to realize about these climate changes, whether the relatively minor swings of the last thousand years or the significant chill of the last Ice Age, is that they were completely natural. But starting with the Industrial Revolution in the early 1800s, humans have slowly become part of the climate change process. The natural mechanisms remain, however, and separating the human influences on climate from the natural variability is at the heart of the global warming debate.

Global Climate Change

Any discussion of global climate change must begin with a few words of caution. In principle, science is an objective, nonpolitical discipline. Meteorology is often said to be an inexact science, but it is a science. There are laws of physics that govern how the atmosphere works. Political beliefs do not enter the debate on whether it will rain or snow tomorrow.

But when talk turns to projecting what the climate will be like many years or decades in the future, the situation changes dramatically. On these time scales, humans are one of the factors in climate change. As a result, large-scale responses to climate change might require countries as well as individuals to modify their behavior. Such changes then have wide-ranging political, social, and economic consequences, so the debate on climate change becomes much more than a scientific issue.

The controversy surrounding global warming demonstrates just how contentious and muddled the issue of climate change can be. Some scientists have been biased by politics, while some politicians, activists, and other nonspecialists lack a full understanding of the science or choose to ignore evidence that does not support their agenda. We call both "scienticians." Much of the public discussion of global warming has been dominated by scienticians on both sides, and by the media's tendency to focus on extremes.

In our brief attempt to separate fact from hype, we will rely on experts such as Dr. Jerry Mahlman, recently retired director of the Geophysical Fluid Dynamics Lab (GFDL) in Princeton, New Jersey. One of GFDL's missions is to objectively study and attempt to predict future climate. Mahlman has testified before Congress many times and is internationally respected as a nonpartisan expert on global warming. We will also depend on the conclusions of the Intergovernmental Panel on Climate Change (IPCC), a multinational group of hundreds of earth scientists organized by the United Nations. The IPCC is responsible for assessing the most current scientific, technical, and socioeconomic research in the field of climate change. Its most recent report was issued in 2001. Though the IPCC has its critics, its membership includes voices from both sides of the global warming debate.

Some aspects of global warming remain uncertain, but many are not controversial. We will start with those.

Science of Global Warming

The basic science supporting the case for global warming is sound. Global warming is best understood as an unnatural enhancement of a process that has been occurring naturally in the atmosphere for billions of years. This natural process, called the **greenhouse effect**, is indispensable to life on earth. It works as follows: our atmosphere is nearly transparent to sunlight—that is, most of the sun's energy that is not reflected back to space travels through the atmosphere and warms the ground. However, most of the energy that the earth emits, called **infrared radiation**, is absorbed by the atmosphere on the way out, primarily by two gases—water vapor and carbon dioxide (CO_2). The absorption by these **greenhouse gases** keeps the lower atmosphere about 60°F warmer than it would be without them, and in the process, keeps the earth habitable.

Human activities have been adding carbon dioxide and other greenhouse gases to the air, enhancing this natural greenhouse effect. This idea was first suggested in the late 1800s by Svante Arrhenius, a Nobel-prize winning chemist. He recognized that when wood and carbon-rich fossil fuels such as coal and oil were burned, carbon dioxide was produced and added to the atmosphere. "We are evaporating our coal mines into the air," Arrhenius wrote in 1896.

Carbon dioxide concentrations in the atmosphere have been increasing since the beginning of the Industrial Revolution. This is not in dispute—the conclusion is based on actual measurements, not estimates or theories. Two hundred years ago, there were about 280 molecules of carbon dioxide in every one million air molecules (commonly written as 280 parts per million, or 280 ppm). That level had been nearly constant for at least 10,000 years. By the turn of the twenty-first century, the concentration had risen to around 370 ppm (see Illustration 7.9), probably the highest level of atmospheric CO_2 in at least the last 400,000 years.

There is also no doubt that the burning of fossil fuels is the primary cause of this 30 percent increase in CO_2. A smaller contribution comes from deforestation—the clearing of trees. Wood is about half carbon, and when

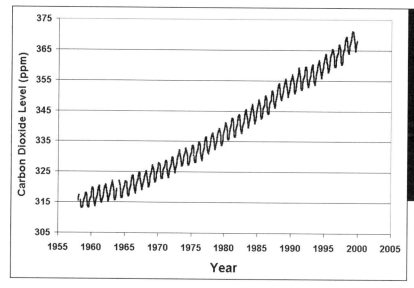

ILLUSTRATION 7.9 Atmospheric carbon dioxide levels measured at Mauna Loa, Hawaii from 1958 to 2000 show a steady increase as a result of human activity. The variations superimposed on the general increase result from the natural annual cycle of growth and decay of plant life (data from Trends Online, A Compendium of Data on Global Change, Carbon Dioxide Information Analysis Center, <http://cdiac.esd.ornl.gov/trends/co2/sio-mlo.html>).

forests are cleared, some of their carbon eventually returns to the air as CO_2. That return happens faster if the forests are burned, a common practice in the tropics. The current CO_2 level in the atmosphere would be even higher if natural processes, mainly involving the oceans, plants, and soils, had not removed about half of the CO_2 that humans have put into the air. In essence, nature has been taking in a lot of the carbon dioxide that humans produce and storing it elsewhere.

There is no sign that the increase in atmospheric CO_2 will stop anytime soon. Depending on what projection you believe, the level will reach 560 ppm—double the pre-Industrial Revolution value—sometime between 2050 and 2100. The rate of future increase depends on both CO_2 emissions and on how efficiently natural processes remove the excess CO_2 from the air—estimates of both rates vary widely.

There have been multinational attempts to reach consensus on how to reduce CO_2 emissions, with the most recent talks centered on the Kyoto Protocol, an agreement drawn up in Kyoto, Japan in 1997. The United States has resisted signing on to this agreement because of the economic consequences of reducing fossil-fuel emissions, and disagreements about whether implementing the measures would actually make a dent in future global warming.

The warming power of an enhanced greenhouse effect is not in question. It is on display now on our nearest planetary neighbor, Venus, where the average surface air temperature is a toasty 890°F. Although Venus is much closer to the sun than Earth, it is shrouded in dense clouds that reflect 80 percent of the incoming sunlight. If Venus's temperature were controlled only by the sunlight that reaches its surface, Venus would actually be colder than Earth. But Venus's atmosphere is 90 times thicker than Earth's, and 90 percent of that atmosphere is carbon dioxide. Venus is so hot because it has the solar system's most intense greenhouse effect.

Evidence from Earth's past also links carbon dioxide concentrations to average air temperature. Samples of ancient air trapped in ice cores drilled from the Antarctic and from Greenland show that CO_2 concentrations in the atmosphere have varied widely over time scales of thousands of years (see top graph of Illustration 7.10). These natural variations have many causes, including volcanic activity and changes in the rate of exchange of carbon dioxide with the oceans. In general, air temperatures have varied with the carbon dioxide changes (compare bottom and top graphs of Illustration 7.10)—when CO_2 concentrations are relatively high, so is temperature. Because centuries of information are crammed into thin slices of ice, it is tough to say whether the CO_2 variations caused the temperature changes, or vice versa. But regardless of which comes first, the two go hand in hand.

Though carbon dioxide gets most of the attention, it is not the only greenhouse gas that is increasing as a result of human activities. For example, levels of methane have more than doubled since preindustrial times and continue to increase. The chemicals known as chlorofluorocarbons (commonly called CFCs) that have been implicated in ozone depletion (see Chapter 5) are also greenhouse gases. Taken together, the greenhouse-enhancing potential of

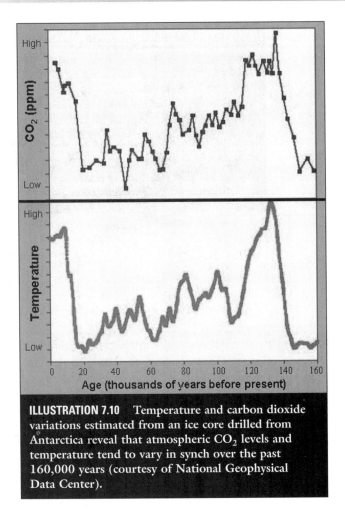

ILLUSTRATION 7.10 Temperature and carbon dioxide variations estimated from an ice core drilled from Antarctica reveal that atmospheric CO_2 levels and temperature tend to vary in synch over the past 160,000 years (courtesy of National Geophysical Data Center).

increases in methane, CFCs, and other gases is about one-half to two-thirds that of CO_2.

In short, greenhouse gases absorb some of the earth's outgoing infrared radiation, keeping the atmosphere much warmer than it would be otherwise. With more carbon dioxide and other greenhouse gases in the air compared to preindustrial times, the lower atmosphere should be warming up. But has it? And if so, are we responsible for the warming?

Temperature Trends

To assess global temperature variations, climatologists need data from thousands of locations across the globe, and over a long enough period of time so that any trends are meaningful. As in the Philadelphia area, precise measurements of air temperature have been made at a large number of locations for only the last hundred years or so,

and most of the observing sites are in the Northern Hemisphere. At some locations, urbanization has raised temperatures (the "urban heat island effect"—see Chapter 2), making detection of any warming caused by an enhancement of the greenhouse effect more difficult. Instrument changes and the relocation of some observation sites also potentially contaminate the data. And with 70 percent of the globe covered by water, relatively few observations of air temperature are available over the oceans.

Despite these difficulties, researchers have accumulated enough quality data to create time series of "global average surface air temperature" going back to the mid-1800s (Illustration 7.11). Though the details of the data are still debated, there is little disagreement about the trends: a slight warming from around 1900 to 1940, then a period of little change or even slight cooling into the 1970s, followed by a relatively rapid warming the last few decades. To emphasize the uncertainty in the amount of warming, the IPCC reports the trend as a range: 0.7°F to 1.4°F of warming over the last century, with more than half of that warming occurring since 1975. The six warmest years globally all occurred since 1990 and include, in order of decreasing temperature: 1998, 1997, 1995, 1990, and 1999. And this warming is not the urban heat island. Says Mahlman of GFDL: "We've taken the urban heat island completely out of the calculations. If anything, it's been overcorrected for."

On a regional level, temperatures during the twentieth century increased in most of the northeastern United States, while the number of three-day heat waves has approximately doubled in the eastern United States over the last fifty years. On average, however, the eastern U.S has not warmed up as much as other places. There is evidence that this may be partly caused by—of all things—air pollution. Burning coal or oil releases the sulfur in these fuels and produces the gas sulfur dioxide, which is converted in the air into tiny solid particles or liquid droplets called **aerosols**. More sulfate aerosols means more sunlight reflected back to space—a cooling effect. The same thing may be happening in other industrialized areas such as Europe and China, says Mahlman. So, in a remarkable irony, one human environmental impact—air pollution—may actually reduce another potential human impact—global warming. Thus, less air pollution could actually mean more warming, and this must be accounted for in any future climate projections.

Ocean temperature trends have been more difficult to diagnose, for several reasons. Fewer ocean temperature

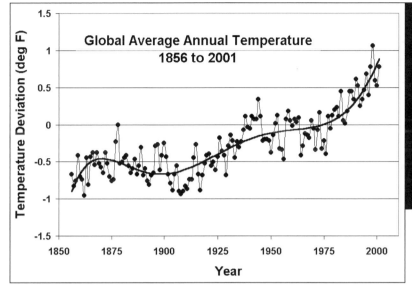

Global Average Annual Temperature 1856 to 2001

ILLUSTRATION 7.11   Global average annual temperatures are shown for the period 1856 to 2001. The thicker solid line is a mathematically smoothed version of the data. A value of zero indicates the long-term average. There is variation from year to year, but the general trend is up, with a more rapid warming seen since about 1980 (data from Trends Online, A Compendium of Data on Global Change, Carbon Dioxide Information Analysis Center, <http://cdiac.esd.ornl.gov/trends/temp/jonescru/jones.html>).

measurements are available than air temperature readings, and water changes its temperature much more slowly than air does. (For evidence on another scale, consider that the ocean off New Jersey is still below 70°F during June while air temperatures routinely surpass 80°F.) Researchers have detected a rise of about 0.1°F in the average temperature of the world's oceans since around 1950. A tenth of a degree may sound minuscule, but if just a fraction of the energy needed to warm the oceans by that small amount were added to the air, a significant warming would result. By taking in this extra energy, the oceans may have delayed warming that otherwise would have occurred in the atmosphere.

There are other temperature-sensitive indicators of recent warming worldwide. Measurements from tide gauges across the globe show that sea levels rose from four to eight inches, on average, during the twentieth century. An analysis of previously secret data gathered by U.S. Navy submarines suggests that the ice in the Arctic has thinned by about 40 percent over the last fifty years. There has been a widespread retreat of mountain glaciers outside of the polar regions over the last century. A study of the freeze and thaw dates of some lakes and rivers in the Northern Hemisphere found that the average time of freezing is about a week later than it was 150 years ago, while the average time of ice breakup is about a week earlier. More regionally, the length of time between the first and last days with snow on the ground in the northeastern U.S. has decreased by a week in the last fifty years, indicating a shorter snow season. Though some details of

these changes remain in dispute, few people argue with the trends. But do they result from an enhancement of the greenhouse effect caused by human activities? That is, are we seeing the first effects of global warming? Looked at another way: if not global warming, then what is it?

First, there is the inherent variability of climate to consider. Temperatures vary naturally from year to year, decade to decade, and over longer time scales, and they always have, long before our species roamed the planet. Is the recent trend in global average temperature more than what might occur naturally? A rise of 0.7°F to 1.4°F over a hundred years might not seem like much, but relative to the past few thousand years, the recent rate of temperature increase does stand out. Since the end of the last Ice Age about 10,000 years ago, average global surface air temperatures have only varied about 3°F from highest to lowest. Various statistical analyses suggest that the temperature change over the last century is quite unusual and difficult to explain by natural random processes, and that conclusion has been increasingly accepted over the last decade.

Other temperature observations strengthen the link between increases in greenhouse gases and the warming that has been detected in the lower atmosphere. That evidence comes from the stratosphere, the layer of air ten to thirty miles up. With CO_2 and other greenhouse gases accumulating in the lower atmosphere and keeping more of the earth's infrared radiation there, cooling should be observed higher up because less of earth's energy makes it out that far. In fact, this so-called "stratospheric cooling" has been observed, particularly since around 1980 when the increase in surface temperatures accelerated.

With evidence mounting that human activity is at least partly responsible for these climate changes, the IPCC in its 2001 report issued the strongest official pronouncement yet that human-induced warming is real. The report states that "most of the observed warming over the last fifty years is likely to have been due to the increase in greenhouse gas concentrations. Furthermore, it is very likely that the twentieth century warming has contributed significantly to the observed sea-level rise." These statements represent an important shift from the previous IPCC report in 1996, which concluded that "the balance of evidence suggests a discernible human influence on global climate." Mahlman of GFDL puts it this way: "It is my opinion that there is a better than 95 percent chance that more than half of the warming is due to human effects." Even the minority of respected scientists who are vocal critics of the idea that humans are contributing to global warming now largely accept that at least some of the temperature increase is related to increasing greenhouse gases.

Humans are conducting an unprecedented, large-scale experiment in the atmosphere. Most climate researchers believe that the debate on global warming has shifted from "Is it real?" to "How much, how quickly, and what are the consequences?" In Mahlman's words: "There is no significant uncertainty that the earth will warm up—the uncertainties are how much it will warm up, how fast it will warm up, and how the warming will be distributed geographically." And that is primarily where the real scientific debate lies as we begin the twenty-first century.

Future Climate Change: The Next Century

Climate variations of the past can provide clues to future climate change. But with humans in the mix now, there are no exact historical precedents. Greenhouse gases are accumulating at such a rapid rate that we are now moving into uncharted territory. For most projections of future climate change, researchers have turned to computer climate simulations produced by **General Circulation Models** (GCMs), cousins of the computer models used in short-range weather forecasting.

GCMs, like short-range computer models, are based on mathematical equations that represent how pressure, temperature, wind, and humidity change over time. In short-range weather models, the focus is the traveling high- and low-pressure systems and upper-air troughs and ridges that bring day-to-day changes in the weather. Other components such as the oceans and the ice-covered parts of the globe are represented simplistically. After all, the thickness of Arctic sea ice or the temperature of the deep Atlantic Ocean does not change much over a time span of a few days. But in the climate simulation problem, that simplification is not adequate. Over time periods of years, decades, and longer, changes in the underlying land, ocean, and ice-covered portions of the globe are critical. Each of these components itself requires another complicated computer model, adding to the complexity (and the uncertainty) that goes with modeling future climate.

Dozens of different GCMs have been developed in government and private research centers and universities worldwide. Though all are based on the same physical principles, they differ slightly in their formulation. For example, some have more complicated representations of the ocean and its circulation. Others might use a finer resolution to better capture the details of sea ice in the Arctic. No GCM, however, is capable of mimicking the current climate system perfectly, so one must be careful when assessing future projected changes from the models, particularly at local scales. Despite the model differences and the uncertainties inherent in computer modeling, projections from these GCMs do all basically agree that the average temperature of the lower atmosphere will increase during the next century. They also generally agree the warming would be greater in winter than in summer, greater at night than during the day, and greater in higher latitudes than in the tropics. They disagree on just how much warming there will be, how fast it will occur, and its detailed geographical distribution.

Of course, the timing of any future global warming will depend on how quickly carbon dioxide (and other greenhouse gases) builds up in the air. As a result, a standard global warming projection is tied to when the amount of CO_2 will increase to 560 ppm, which is twice its preindustrial value of 280 ppm (remember, it is 370 ppm now). Depending on who you talk to, this "doubled CO_2 earth" could happen as soon as 2050 or as late as 2100 (see top panel of Illustration 7.12). But a convenient way to express climate change predictions is to frame them in terms of a "doubled CO_2 earth."

Here is the IPCC's projection of future climate change, drawing on the results of many GCMs and the expertise of hundreds of climate researchers. In their 2001 report, the IPCC projects an increase in the average global surface air temperature of between 2.5°F and

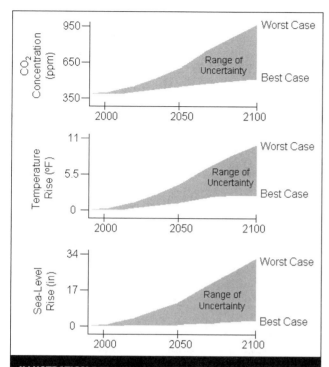

ILLUSTRATION 7.12 Projections of future atmospheric CO$_2$ concentrations *(top graph)*, global average temperature increase *(middle graph)*, and global average sea-level rise *(bottom graph)* through the year 2100, based on scenarios considered by the IPCC. The large envelope of uncertainly in future CO$_2$ levels contributes to the large range of uncertainty in temperature and sea-level projections.

10.4°F over the period 1990 to 2100 (see middle panel of Illustration 7.12), with a "most probable" increase being in the 3–5°F range. This range encompasses both uncertainties in the models and in future CO$_2$ emissions. Regional temperature changes will differ substantially from the global average, but land areas and polar regions will likely warm the most. Higher temperatures will produce, on average, more evaporation, leading to an increase in total precipitation across the globe. But again, the geographical distribution of the increase remains rather uncertain. The IPCC also projects that sea levels will rise, but the range of possibilities is huge, from just a few inches to nearly three feet by the year 2100 (see bottom panel of Illustration 7.12). This rise is caused by the melting of some ice on land and the expansion of water when it is heated (water expands slightly as it warms just as the liquid in a thermometer does).

It is very important to acknowledge the large ranges attached to these projections. Giving a range is a quantitative way to show uncertainty. Unfortunately, many reports about global warming focus on only the higher, more extreme number, or in some cases, just the lower end of the range (it often depends on the slant of the source), or fail to mention what the "most probable" increase is. Reports such as "warming of as much as 10.4°F" leave the public with a misleading representation of the state of climate change prediction. Both the higher and lower values should be mentioned in fair and balanced reporting, to emphasize the uncertainty.

Illustration 7.13 is a computer model projection made by a general circulation model run by Mahlman's group at the Geophysical Fluid Dynamics Lab. The top panel shows the predicted temperature changes (in °F) on a doubled-CO$_2$ earth, while the bottom panel extends the calculation to a quadrupled-CO$_2$ earth. As the color goes from yellow to orange to red, the amount of warming increases (see scale at the bottom). Though this is just one of many GCM projections, it illustrates two consistent features of all GCM projections: polar regions warm more than midlatitude and tropical regions, and the warming increases as the amount of CO$_2$ in the air increases.

Such GCM projections are not intended to give details for specific years in the future. They cannot be used to predict the average global temperature in the year 2035 or the total precipitation in the United States in 2070. Instead, they give insight into general trends and the meteorological logic that goes with them. The models supply a range of possible outcomes, and suggest those that are most probable. In a way, it is the climate-projection equivalent of predicting a snowstorm—instead of giving exact amounts of snowfall for specific towns, a range such as four to eight inches is given for a broad area. That is an expression of the limitations of the forecast.

In sorting out potential climate changes, it is useful to categorize them by how much confidence climate researchers have. Further cooling of the stratosphere is at the top of Mahlman's list as "virtually certain," while a global surface temperature increase, a rise in sea level, and an overall increase in evaporation (and thus global precipitation) are considered "very probable." An increase in precipitation extremes—both floods and droughts—falls in the "probable" category. More evaporation means more water vapor available for storms, so the probability increases that a given storm will produce a more extreme amount of rain or snow. On the other hand, increased

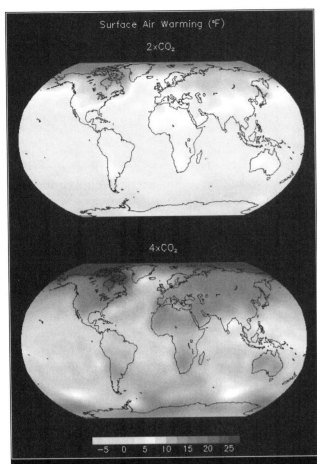

Surface Air Warming (°F)

2×CO₂

4×CO₂

−5 0 5 10 15 20 25

ILLUSTRATION 7.13 *(See plate.)* An example of a GCM projection showing the warming (in °F) caused by a doubling and quadrupling of atmospheric carbon dioxide. Note that the projected warming is greatest in North Polar regions, a consistent feature of GCMs. Based on this simulation, the eastern United states warms about 5°F from a doubling of CO_2 and about 15°F from a quadrupling (Geophysical Fluid Dynamics Lab).

evidence that the number and intensity of midlatitude low-pressure systems—the very storms that regularly affect us—will increase. The same can be said for severe events such as tornadoes. And climate researchers are not sure about the frequency and intensity of El Niños and La Niñas either.

Specifically, most climate modelers cite two fundamental sources of uncertainty in our understanding of the climate and thus in the reliability of GCMs: the oceans and clouds. The oceans cover 70 percent of the globe and are huge storehouses of carbon dioxide and heat, so their importance is obvious. The oceans have already taken in some of the excess CO_2 and excess heat from the atmosphere, but the extent to which that will continue in a globally warmed world is uncertain. The oceans circulate horizontally and vertically, mixing heat and CO_2 taken in at the top into deeper waters, but exactly how long this takes and how efficiently it happens are not well understood.

Clouds are the other major uncertainty, because they both cool and warm the planet. Clouds reflect sunlight back to space (cooling), but they also absorb the earth's outgoing infrared radiation (warming). The process that dominates depends on the type of cloud. Ice clouds, such as high, wispy cirrus clouds, are better absorbers than reflectors. But low, layered stratus or puffy cumulus clouds, which are mostly liquid water, are more reflective. At the present time, averaged worldwide, clouds do more cooling than warming. An increase in temperature is expected to lead to more evaporation, more water vapor in the air, and possibly more clouds, but we do not know what types of clouds. Thus the overall effect of clouds on temperature in a globally warmed world is uncertain. Moreover, water vapor is the most potent greenhouse gas. With more of it in the air on a warmer planet, the greenhouse effect should get even stronger, regardless of whether there are more clouds.

You can think of the oceans and clouds as the climate wildcards. Their behavior will go a long way in determining the pace of climate change: will it be a tortoise or a hare? And the speed at which climate change occurs is crucial to its consequences and society's vulnerability.

The question of the speed of climate change leads to another area of uncertainty that can be termed "surprises." Many climate changes and their impacts are likely to be extensions of current trends, so they may be at least partly foreseen. For example, a small increase in global temperature will probably lead to a proportionally small rise in sea level. But complicated systems do not neces-

evaporation dries the land more quickly, also making drought more likely, especially in summer in places away from the ocean. So as illogical as it may sound, the odds of both droughts and floods may increase on a warmed earth. "That's a hard thing to explain to people," says Mahlman. One of his GFDL studies projects that severe summer drought in the central U.S. will become six times more likely when CO_2 is doubled.

Many other potential climate changes—including some highly publicized severe weather events—still fall in the "uncertain" category. For example, there is no firm

sarily always work in a continuous way. Think about stretching a rubber band. At first, the more you stretch, the longer it gets. But at some point, the rubber band will break. Could the climate have such a breaking point too, a threshold beyond which change occurs suddenly? That is, might the climate at some point behave more like an on-off switch and go from one mode to another in relatively rapid fashion? This is not as far-fetched as it sounds. A good example is the appearance of the ozone hole (see chapter 5) in the 1980s, which was a complete surprise to scientists. Since we are moving into new territory with the climate, it is reasonable to expect things to happen that have never happened before or even been imagined.

With so many uncertainties about climate change on a global scale, GCM projections can only be relied upon to give a broad-brush picture of changes on a regional level. But we can mention some general points related to future climate change and its consequences in this area. Here, we depend on the "National Assessment of the Potential Consequences of Climate Variability and Change," a major ongoing research project aimed at understanding what climate change means for the United States on a region-by-region basis. This national assessment was timed to provide input to the 2001 IPCC report. Dr. Eric Barron coordinated the report for the mid-Atlantic region, which includes Pennsylvania, Maryland, Delaware, and New Jersey. Eric Barron is director of the Earth System Science Center at Penn State University and, like Jerry Mahlman, an internationally respected authority on the subject of global and regional climate change—past and future.

Based on various model projections, the Northeast has one of the lowest rates of projected future warming of any region in the United States. Nonetheless, that warming is projected to be at least 4°F, on average, by the year 2100, with the largest increases during winter. For precipitation, there is much more uncertainty: the model projections range from a 25 percent increase by the year 2100 to little change.

How would these climate changes reveal themselves locally? Over the next century, periods of extreme winter cold would likely become less frequent. Because the Philadelphia area is often on the rain-snow line during winter storms, warmer winters would favor more cold rains and less total snowfall, on average. Historically, there is a good overall correlation between higher winter temperatures and lower winter snowfall (see Illustration 7.14). But according to Mahlman, with more water vapor

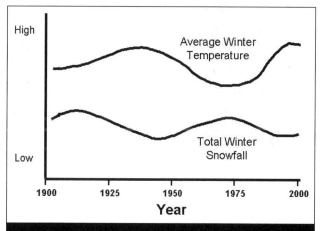

ILLUSTRATION 7.14   In general, winter snowfall and average winter temperature in Philadelphia are inversely correlated: higher temperatures tend to be related to lower snowfall, and vice versa.

in the air in a globally warmed world, infrequent but giant snowstorms would still be possible. (However, the huge January '96 snowstorm should not be considered evidence for warming—one storm, one season, or one year cannot be used to define a long-term change.) In general, winters would be less extreme and less stressful.

In summer, the opposite would be true. Extremes of summertime heat are thought to have a greater impact on human health than any other form of severe weather. Summers would be more of a threat to health and comfort, especially in large urban areas such as Philadelphia. Particularly at risk are the elderly, the young, and the chronically ill, people that are more physiologically vulnerable to heat stress. A recent study by Professor Kalkstein, the University of Delaware expert on the health effects of climate change, suggests that the number of heat-related fatalities in Philadelphia would at least double with a 2°F increase in average summer temperature. Higher summer temperatures coupled with more moisture in the air would lead to higher heat indices. For example, a rise in average temperature of 4°F would lead to an increase of about 6°F in the heat index because the moisture content of the air would go up too. Higher temperatures would also make ground-level ozone formation more likely, increasing smog pollution. Think of the hottest, most uncomfortable summer of the last twenty years in Philadelphia. There are some people living today who might experience that as a typical summer at some point in their lives.

ILLUSTRATION 7.15 *(See plate.)*   The areas in red, estimated by computer models to have an elevation of five feet or less, are most vulnerable to rises in sea level. A five-foot rise in sea level would not be necessary to inundate these areas: considering the affect of tides and other factors, these regions would be inundated during times of highest spring tides with just a 2.5-foot rise in sea level. The blue-shaded regions have an elevation between 5 and 11.5 feet (courtesy of Environmental Protection Agency).

Coastal areas are some of the most sensitive to climate change. Coastal population and property values continue to rise, with no end in sight to the growth. Illustration 7.15 shows the coastal regions that would be most affected by a rise in sea level. The red-shaded areas are estimated to have an elevation of five feet or lower above mean sea level. Approximately 400 square miles of New Jersey and 150 square miles of Delaware fall into this category. Though five feet is well above the IPCC's projected maximum rise in sea level by 2100 (which is around three feet), it is well within the expected range by the year 2200. The map does point to the areas along coastal New Jersey and Delaware that are most vulnerable, though it does not take into account any coastal protection efforts. As sea level rises gradually, less and less severe storms will do an equal amount of damage.

Coastal areas are also most sensitive to potential changes in tropical storms and hurricanes. On that subject, there are really two questions: Will there be more tropical storms and hurricanes in a globally warmed world? And will a given storm reach a greater intensity? Tropical systems feed on moisture evaporating from warm ocean water, and global warming will surely warm the oceans—so at this basic level, the number of hurricanes could increase. But there are numerous complicating fac-

tors. For example, Atlantic hurricane activity is also tied to El Niño and La Niña: historically, fewer hurricanes form in the Atlantic when an El Niño is in progress, and more during La Niña. So changes in the number of hurricanes also depend on changes in the frequency of El Niños and La Niñas, another major unknown.

The other question about intensity carries somewhat less uncertainty. With more water vapor in the air in a warmer atmosphere, any storm—be it a solitary thunderstorm or a tropical system—would have more moisture to work with. A recent GFDL study suggests that a hurricane of a given intensity today (say, a Category-3) would be one category stronger (that is, Category-4) later in the century if everything else is equal. Mahlman even suggests that "Floyds will become much more probable."

So in the early years of a new century, discussion of global warming is no longer about if, but rather when, how much, and where. The answers to many of the most important questions must still be framed in terms of ranges, probabilities, and levels of uncertainty. Often overlooked in the discussion of global warming is that not all of its consequences would necessarily be negative. Some climate scenarios suggested by GCMs may benefit some people, places, and sectors while harming others.

Philadelphia Climate in the Year 2075

Painting with the broad brush provided by GCMs and using the expertise of scientists such as Dr. Jerry Mahlman, we can make an educated guess about a typical year of weather in the Philadelphia area many decades in the future. We choose the year 2075, which is both a reasonable estimate for when a "doubled CO_2 earth" will be reality and a year that a few young readers of this book, and the children and grandchildren of many other readers, will be around to experience. We will assume that by then, average annual temperatures in the Philadelphia area will have increased by 3 to 4°F, a middle-of-the-road estimate taken from the latest IPCC report.

We present this summary of the weather of 2075 in Philadelphia in the form of an informal "Highlights of the Year" that might be written, for example, by a meteorologist in the local office of the National Weather Service. (Formal reviews are commonly prepared even today by local NWS officials.) For historical perspective, we include in the summary a few references to the climate early in the twenty-first century when there was still controversy about whether global warming was real or not.

A Summary of Philadelphia Weather of 2075

The year 2075 was another record-setting warm one in Philadelphia. The average annual temperature was 62.2°F, breaking the previous record of 62.0°F set just three years ago. The ten warmest years in Philadelphia history have all occurred since 2050. For comparison, the warmest year of the twentieth century, 1998, had an average temperature of 58.1°F—there has not been a year that "cold" since 2062.

The lowest temperature of the year, 20°F, occurred on January 15. A week later, the only big snowstorm of the winter dumped 7 inches of snow before the precipitation changed to sleet and then rain, leaving a slushy mess (such a changeover is common these days—all-snow storms are the exception). Total snow for the season was 15 inches, which is about average (at the turn of the century, the average was 20 inches). Ten years ago, the winter of 2064–65 had three big snowstorms and 58 total inches, but seasons like that are rare. Most of the ski industry in the Poconos has survived, but only be-cause of the new high-tech snowmaking machines that work at higher temperatures than the older ones.

The feel of spring arrived in the Philadelphia area in early March as temperatures reached 80°F on four consecutive days starting on March 7. The last freeze in the city occurred on March 20, and the suburbs were frost free by April 10. On average, the growing season is about three weeks longer now than it was at the turn of the century. The early arrival of spring also brought some humid days in April, and that humidity fed some drenching thunderstorms at the end of the month.

The average summer high temperature these days is 87°F, compared to 83°F back at the turn of the century. But this summer was exceptionally hot, with an average high of 90°F. June was the warmest on record, with fifteen days of 90°F-or-higher temperatures, and the temperature reached 90°F nearly every day in July. For the year, there were sixty days of 90°F or hotter weather spread out from April to September (at the turn of the century, the average was around thirty). The mercury topped 100°F on five days. There were also five nights when the temperature did not fall below 80°F, which coincided with the peak time for heat-related casualties at area hospitals. The highest temperature of the summer was 110°F, just shy of the all-time record for heat of 112°F that was set three summers ago (until 2020, the all-time record still stood at 106°F, a temperature record established in 1918).

On average, summers are more humid now—this summer the average dew point was 65°F, about 3°F higher than at the turn of the century. Excessive Heat Warnings for heat indices of 100°F or higher were issued on forty days, so frequently that they no longer always make the headlines. On five days the heat index reached 110°F, and it even approached 120°F on two days, which used to be unheard of. Ozone smog pollution reached unhealthy levels on twenty days. On a few of the smoggiest, only the top floors of the 100-story Rendell Building poked out of the murk. Given the heat, the ozone problem could be much worse, but most cars are electric, and more than half the commuters take public transportation. There is other good news about ozone: the stratospheric ozone layer has almost recovered to its natural state, so the heightened concern early in the century about increases in skin cancer related to ozone depletion has lessened.

With the heat and the related increase in evaporation, summer also brought a few dry spells. As usual, water restrictions were mandatory for a few weeks during the summer. But when it does rain, it really pours. Three times during the summer—on June 2, July 5, and August 14—torrential thunderstorms dropped between three and six inches of rain, leading to localized flash flooding. This "more active hydrologic cycle"—periods of drought interspersed with quick shots of heavy rain—has been common in the last few decades, and it is very disruptive. Overall, summer precipitation was about average— 13 inches for the June–July–August period—a slight increase from the average back at the turn of the century.

Hurricane season was also average in terms of numbers. Eight named storms formed in the Atlantic with four reaching hurricane strength (at the start of the century, the averages were around ten and six, respectively). Although the overall higher ocean temperatures favor more hurricanes, the increased number of El Niños in recent decades is primarily responsible for the decrease (changes in El Niño frequency were a big unknown at the turn of the century). One of this year's storms was Hurricane Carl, which threatened the East Coast in early September. With ocean temperatures over 90°F off the Florida coast, Carl strengthened rapidly from a Category-1 with 80-mph winds to a Category-5 with 160-mph winds in 36 hours. Carl curved and missed the coast, but it was a close call. Category-5 hurricanes that made landfall used to be extremely rare—only two in the United States in the twentieth century—but now the average is about one per decade. There are not more hurricanes overall, but they strengthen more than they used to owing to the warmer seas.

Even without hurricanes, there were problems at the shore once again this year. First, the gradual sea-level rise (on average, a tenth of an inch a year) caused by the warming ocean means that there is no beach at several Jersey and Delaware coastal points at high tide. With the higher sea level, less of a storm is needed to produce flooding, and the most vulnerable areas were inundated time and time again. One February nor'easter brought little snow but caused as much beach erosion as the infamous March 1962 nor'easter more than a century ago, setting record-high water levels that cut one of the barrier islands in three pieces. The higher temperatures that global warming has brought have resulted in a longer beach season and a rush to the shore to cool down, but the shore areas are, in general, becoming less habitable. The higher sea level has also caused more frequent flooding problems in Delaware Bay and increased saltwater intrusion farther up into the Delaware River.

As has become typical, summer-like temperatures lingered well into September, and even October was pleasantly warm. The first shot of chilly air arrived around Thanksgiving (though some years it has been possible to reach early December before breaking out the winter coat). Christmas was, as in most years, green and brown. But we did have some snow changing to rain for New Year's Eve, a fitting end to another record warm year. ❖

EPILOGUE

The Philadelphia area has been a center for advances in weather science and weather observation for hundreds of years, in no small part because of the wide variety of weather that this region experiences and the widespread public interest and discussion it generates. That variability can be traced to the Philadelphia area's location—about halfway between equator and pole, with thousands of miles of land to the west and thousands of miles of ocean to the east. With large seasonal variations, just about every type of weather has visited the Philadelphia area during its long and storied history, from blizzards, floods, hurricanes, droughts, and Arctic blasts to tornadoes, nor'easters, heat waves, and pollution episodes.

Yet the Philadelphia area has no single weather phenomenon for which it is renowned. We experience fewer blizzards than Buffalo and Chicago; fewer Arctic blasts than Minneapolis and Detroit; fewer tornadoes than Dallas and St. Louis; fewer heat waves than Atlanta and New Orleans; fewer thunderstorms than Tampa and Baltimore; and fewer pollution episodes than Los Angeles and Houston. Though we are near an ocean that is warm enough to revel in several months a year, we do not feel the regular danger of hurricanes. Though close to mountains that provide plenty of quality skiing, we do not endure months on end covered in a deep snowpack. Perhaps it is the rich variety of weather without imminent danger that helps makes the Philadelphia region a great place to live.

And what of the future climate of this area? If projections of global warming come to pass, future generations may become more adapted to summer heat, more used to less snow, and more farsighted about restricting development in floodplains and along vulnerable beaches. But the weather and the field of atmospheric sciences will surely provide some surprises in the decades to come. There will be new leaps in understanding on a par with linking El Niño to distant weather extremes, unexpected revelations comparable to the discovery of the ozone hole, and theoretical advances as revolutionary as weather prediction by computer. Whatever changes the science of meteorology brings in the future, there is one forecast that can be made with great confidence: the weather of the Philadelphia area will remain the subject of regular conversation and interest, and occasional controversy.

And, as meteorologists, we would not have it any other way.

APPENDIX A
Philadelphia Daily and Monthly Climate Data

Tables are given for each month. The first table for each month lists the official daily average ("normal") high and low temperatures and daily records for high and low temperature, liquid precipitation (rain plus melted snow and ice), and snow (which includes sleet). The second and third tables for each month give monthly averages and monthly extremes of temperature, liquid precipitation, and snow. Averages were computed using the period 1971–2000. All temperatures are in °F, while precipitation and snowfall are given in inches. The letter T stands for "trace," meaning that precipitation was observed, but measured less than 0.01 inches of liquid precipitation or 0.1 inches of snow.

Official record keeping for high and low temperatures and liquid precipitation began in 1872 (though limited data exist in 1872 and 1873), and for snowfall in 1884. An asterisk next to a year means that the record occurred in other years; only the most recent year is given. Statistics are valid through December 2001.

Data courtesy of the National Weather Service, Mount Holly, New Jersey.

January Daily Statistics

Day	Average High	Average Low	Record High	Year	Record Low	Year	Record Precip.	Year	Record Snow	Year
1	40	27	62	1973	4	1881	1.60	1948	5.3	1971
2	40	26	67	1876	7	1968*	1.68	1925	6.5	1925
3	40	26	63	2000	−3	1879	1.50	1999	1.5	1988
4	40	26	68	1950	2	1918	1.51	1982	4.5	1942
5	39	26	66	1997	−2	1904	2.24	1873	4.6	1887
6	39	26	71	1950	4	1896	2.00	1962	9.2	1893
7	39	26	65	1950*	7	1988	2.49	1874	27.6	1996
8	39	25	69	1998	2	1970	1.20	1923	3.8	1988
9	39	25	69	1930	1	1970	1.54	1936	3.7	1974
10	39	25	63	1950	−5	1875	1.28	1881	6.6	1965
11	39	25	66	1975	1	1982	1.29	1991	7.4	1954
12	39	25	72	1890	1	1981	2.19	1915	5.0	1893
13	39	25	70	1932	0	1981*	1.46	1978	7.1	1964
14	39	25	73	1932	0	1914	2.27	1968	7.2	1910
15	39	25	71	1932	2	1964	1.23	1884	5	1892
16	39	25	61	1990	0	1977*	1.09	1924	7.6	1945
17	39	25	63	1913	−7	1982	1.66	1978	4.4	1978
18	38	25	64	1990	−4	1982	1.06	1930	3.5	1984*
19	38	25	65	1951	−5	1994	1.40	1936	7.4	1961
20	38	25	65	1951	−3	1984	1.89	1978	9.3	1978
21	38	25	62	1959	−6	1985	1.32	1979	3.4	1917
22	39	25	70	1906	−7	1984	1.03	1987	8.8	1987
23	39	25	71	1906	−2	1936	2.32	1998	11.9	1935
24	39	25	69	1967*	0	1882	1.61	1953	5.8	1908
25	39	25	66	1967	−2	1963	1.83	1978	8.5	2000
26	39	25	74	1950	4	1961	1.28	1986	4.4	1963
27	39	25	69	1916	4	1927	1.64	1976	4.1	1941
28	39	25	67	1916	3	1961*	1.63	1943	8.8	1922
29	39	25	64	1975*	−5	1963	1.23	1925	5.0	1904
30	39	25	70	1947	8	1948	1.79	1939	5.6	1966
31	39	25	65	1947	3	1948	1.30	1878	3.0	1949
Month			74	1950	−7	1984*	2.49	1874	27.6	1996

January Monthly Statistics

Average High Temperature	39.0°F
Average Low Temperature	25.5°F
Average Temperature	32.3°F
Average Liquid Precipitation	3.52 inches
Average Snowfall	7.5 inches
Average Heating Degree Days	1020
Average Cooling Degree Days	0

10 Coldest		10 Warmest		10 Wettest		10 Driest		10 Snowiest		10 Least Snowy	
Year	Temp. (°F)	Year	Temp. (°F)	Year	Precip. (in.)	Year	Precip. (in.)	Year	Snow (in.)	Year	Snow (in.)
1977	20.0	1932	46.2	1978	8.86	1955	0.45	1996	33.8	1913	T
1893	24.1	1950	43.3	1979	8.74	1981	0.50	1918	28.9	1933	T
1918	24.1	1933	42.7	1915	6.74	1970	0.74	1978	23.4	1934	T
1970	24.5	1913	42.3	1936	6.44	1992	0.88	1935	21.5	1973	T
1982	24.7	1890	41.8	1949	6.06	1940	0.96	1893	20.0	1995	T
1945	24.8	1937	41.4	1937	5.71	1946	1.39	1961	19.7	1890	0.1
1961	25.0	1880	41.1	1884	5.46	1916	1.50	1912	16.1	1931	0.1
1875	25.2	1998	41.0	1953	5.01	1880	1.51	1925	16.0	1983	0.2
1940	25.3	1990	40.3	1999	4.89	1876	1.52	1966	16.0	1919	0.5
1981	25.3	1949	39.8	1923	4.78	1985	1.55	1922	15.9	1924	0.5
										1998	0.5

February Daily Statistics

Day	Average High	Average Low	Record High	Year	Record Low	Year	Record Precip.	Year	Record Snow	Year
1	39	25	66	1989	2	1920	2.92	1915	5.3	1934
2	40	26	61	1988	−4	1961	1.49	1973	3.6	1955
3	40	26	62	1991	4	1881	1.64	1939	8.0	1886
4	40	26	69	1991	2	1886	1.34	1920	7.6	1995
5	40	26	69	1991*	−2	1918*	1.62	2001	8.7	1907
6	40	26	65	1884	−3	1895	3.86	1896	12.4	1978
7	40	26	59	1990	5	1910*	1.35	1942	9.8	1967
8	40	26	63	1925	−2	1934	1.15	1890	6.0	1974
9	40	26	64	1990	−11	1934	1.17	1931	5.5	1906
10	41	26	65	1925	−6	1899	0.76	1926	8.7	1926
11	41	27	66	1887	−6	1899	1.85	1983	21.1	1983
12	41	27	70	1999	2	1979	1.35	1988	7.7	1899
13	41	27	66	1951	3	1983*	2.04	1930	10.0	1899
14	41	27	67	1990*	2	1979	1.26	1935	6.9	1914
15	42	27	74	1949	3	1943	1.09	1953	6.5	1958
16	42	28	73	1954	2	1888	1.84	1873	7.3	1996
17	42	28	68	1976	−2	1896	1.10	1902	11.0	1902
18	42	28	69	1891*	0	1979	1.35	1956	3.6	2000
19	43	28	68	1948	3	1903	1.65	1927	13.9	1979
20	43	28	70	1939*	1	1979	2.40	1921	8.0	1947
21	44	28	72	1930	6	1968	1.44	1902	7.5	1929
22	44	28	68	1997*	7	1963	1.20	1878	7.0	2001
23	44	29	75	1874	6	1963	1.65	1909	6.5	1987
24	44	29	74	1985	2	1889	1.84	1979	5.6	1966
25	45	29	79	1930	4	1894	0.90	1960	6.4	1885
26	45	29	69	1890	10	1990*	2.16	1912	5.1	1934
27	45	30	74	1997	6	1900	1.55	1958	3.1	1940*
28	46	30	68	1976	9	1934	1.40	1902	6.5	1941
29	46	30	69	1972	10	1884	1.03	1896	1.0	1968
Month			79	1930	−11	1934	3.86	1896	21.1	1983

February Monthly Statistics

Average High Temperature	42.1°F
Average Low Temperature	27.5°F
Average Temperature	34.8°F
Average Liquid Precipitation	2.74 inches
Average Snowfall	6.6 inches
Average Heating Degree Days	855
Average Cooling Degree Days	0

10 Coldest		10 Warmest		10 Wettest		10 Driest		10 Snowiest		10 Least Snowy	
Year	Temp. (°F)	Year	Temp. (°F)	Year	Precip. (in.)	Year	Precip. (in.)	Year	Snow (in.)	Year	Snow (in.)
1934	22.2	1925	42.2	1896	6.87	1991	0.75	1899	31.5	1892	T
1979	23.0	1998	41.8	1979	6.44	1877	0.84	1979	27.6	1952	T
1885	24.3	1890	41.4	1899	6.20	1901	0.86	1983	26.1	1959	T
1978	24.6	1909	41.3	1939	6.12	1980	0.96	1907	24.1	1973	T
1895	25.4	1954	41.2	1884	5.70	1892	0.98	1978	19.0	1981	T
1875	25.7	1990	41.2	1915	5.55	1987	1.17	1934	18.5	1984	T
1905	26.3	1976	40.9	1902	5.49	1879	1.19	1967	18.4	1998	T
1963	26.4	1927	40.6	1971	5.43	1992	1.31	1947	17.8	1951	0.2
1901	27.1	1884	40.1	1972	5.09	1977	1.33	1958	16.9	1954	0.2
1907	27.1	1930	40.1	1883	5.04	1978	1.35	1914	16.8	1977	0.2

March Daily Statistics

Day	Average High	Average Low	Record High	Year	Record Low	Year	Record Precip.	Year	Record Snow	Year
1	46	30	76	1972	9	1980	1.46	1925	8.6	1941
2	46	31	77	1972	8	1886	0.93	1994	5.0	1914
3	47	31	73	1991*	10	1886	2.20	1906	7.9	1960
4	47	31	74	1974	7	1943	1.61	1993	4.1	1944
5	47	32	79	1976	5	1872	1.40	1920	8.8	1981
6	48	32	71	1935	10	1978	2.06	1932	6.8	1962
7	48	32	74	1974	9	1960*	1.71	1967	4.0	1969
8	48	33	80	2000	12	1986	1.16	1995	8.0	1941
9	49	33	73	2000	12	1996*	0.94	1893	6.9	1976
10	49	33	73	1986*	7	1984	1.43	1994	6.9	1907
11	50	34	72	1977	12	1960	1.69	1952	3.5	1937
12	50	34	83	1990	12	1900	2.64	1912	10.5	1888
13	50	34	84	1990	8	1888	1.91	1984	11.7	1993
14	51	35	85	1990	12	1888	1.56	1956	3.0	1919
15	51	35	81	1990	11	1993	2.79	1912	4.0	1900
16	51	35	82	1945	16	1916	1.08	1923	4.1	1978
17	52	35	86	1945	10	1885	1.04	1993	3.5	1892
18	52	36	78	1989	10	1916	1.13	1968	5.4	1956
19	52	36	78	1918	9	1876	2.27	1975	3.7	1906
20	52	36	83	1945	8	1885	1.76	1958	9.6	1958
21	53	37	85	1948	6	1885	2.24	2000	4.7	1932
22	53	37	80	1948	11	1885	1.90	1977	3.0	1914
23	53	37	78	1938*	14	1885*	1.00	1991	2.8	1896
24	54	37	76	1988	15	1896	1.24	1989	2.2	1990
25	54	38	82	1910	16	1878	2.72	1876	1.7	1965
26	54	38	80	1921	21	1960	1.62	1978	0.5	1933*
27	55	38	83	1921	21	1894	1.78	1919	1.0	1891
28	56	38	83	1945	18	1923	1.98	1932	T	1984*
29	56	38	87	1945	14	1923	1.24	1912	1.9	1942
30	56	39	86	1998	18	1887	1.61	2001	0.6	1942
31	57	39	81	1998*	21	1923	1.57	1980	2.5	1890
Month			87	1945	5	1872	2.79	1912	11.7	1993

March Monthly Statistics

Average High Temperature	51.3°F
Average Low Temperature	35.1°F
Average Temperature	43.2°F
Average Liquid Precipitation	3.81 inches
Average Snowfall	3.3 inches
Average Heating Degree Days	676
Average Cooling Degree Days	2

10 Coldest		10 Warmest		10 Wettest		10 Driest		10 Snowiest		10 Least Snowy	
Year	Temp. (°F)	Year	Temp. (°F)	Year	Precip. (in.)	Year	Precip. (in.)	Year	Snow (in.)	Year	Snow (in.)
1885	30.8	1921	52.5	1912	9.10	1910	0.38	1941	17.7	1945	0
1960	32.6	1945	51.2	1932	7.43	1966	0.68	1914	15.2	2000	0
1875	34.5	1946	49.8	1980	7.01	1885	0.69	1958	13.4	Many	T
1888	35.2	1903	49.4	1876	6.71	1915	1.00	1993	12.4		
1916	35.5	1977	48.8	1983	6.70	1927	1.11	1960	12.2		
1984	35.5	1929	48.4	1993	6.61	1987	1.16	1888	12.1		
1892	35.8	1936	48.3	1899	6.47	1986	1.25	1978	12.1		
1896	35.9	1898	48.0	1994	6.44	1894	1.45	1956	10.9		
1941	36.1	1910	48.0	2000	6.41	1981	1.61	1934	10.3		
1887	36.3	2000	48.0	1953	6.27	1995	1.67	1984	10.3		

April Daily Statistics

Day	Average High	Average Low	Record High	Year	Record Low	Year	Record Precip.	Year	Record Snow	Year
1	57	39	81	1978*	14	1923	1.48	1948	6.6	1924
2	57	40	84	1967	25	1964*	1.71	1970	3.0	1965
3	58	40	87	1963	27	1894	2.28	1915	19.0	1915
4	58	40	80	1892	25	1943*	1.24	1987	1.0	1886
5	58	41	81	1942	22	1881*	1.23	1923	3.0	1898
6	59	41	87	1942*	24	1982	1.76	1939	3.5	1982
7	59	41	90	1929	19	1982	1.16	1962	2.4	1990
8	59	42	90	1929	23	1982	2.67	1940	4.1	1916
9	60	42	84	1991*	26	1977	2.07	1906	7.0	1917
10	60	42	87	1922	22	1985	1.78	1983	0.7	1894
11	60	42	84	1887	28	1966*	1.03	1919	1.0	1918
12	61	43	92	1977	18	1874	1.83	1933	1.0	1928*
13	61	43	89	1977	26	1940	1.88	1961	1.1	1940
14	61	43	91	1941	24	1950	2.28	1970	1.8	1923
15	62	44	88	1941	27	1943	1.00	1906	0.3	1892
16	62	44	88	1976*	27	1943	2.43	1986	0.5	1888
17	63	45	92	1976	27	1980*	1.16	1910	T	1949*
18	63	45	94	1976	25	1875	2.16	1924	T	1997*
19	63	45	91	1896	20	1875	1.44	1943	1.3	1983
20	63	45	92	1941	27	1897	1.71	1874	0.6	1983
21	64	46	89	1976	27	1875	2.11	1927	0	
22	64	47	90	1985	30	1875	0.57	1976*	T	1993
23	65	47	91	1960	33	1982	1.55	1928	T	1986
24	65	47	86	2001*	33	1930	1.80	1966	T	1956*
25	65	48	92	1960	33	1956*	2.11	1874	T	1919*
26	66	48	92	1990	35	1967*	2.42	1977	0	
27	66	48	93	1915	36	1967	1.59	1982	0.1	1967
28	66	49	90	1990	34	1946*	2.19	1923	T	1898
29	67	49	90	1974*	29	1874	0.85	1874	0	
30	67	49	91	1888	34	1874	3.29	1947	0	
Month			94	1976	14	1923	3.29	1947	19.0	1915

April Monthly Statistics

Average High Temperature	62.0°F
Average Low Temperature	44.2°F
Average Temperature	53.1°F
Average Liquid Precipitation	3.49 inches
Average Snowfall	0.6 inches
Average Heating Degree Days	363
Average Cooling Degree Days	10

10 Coldest		10 Warmest		10 Wettest		10 Driest		10 Snowiest		10 Least Snowy	
Year	Temp. (°F)	Year	Temp. (°F)	Year	Precip. (in.)	Year	Precip. (in.)	Year	Snow (in.)	Year	Snow (in.)
1874	42.3	1994	59.4	1874	9.76	1985	0.52	1915	19.4	Many	0
1875	45.5	1921	58.5	1983	8.12	1881	0.61	1917	7.0		
1943	46.8	1941	57.8	1913	6.87	1899	1.04	1924	6.8		
1907	47.4	1915	57.3	1973	6.68	1963	1.13	1916	5.1		
1966	47.8	1977	57.2	1947	6.58	1942	1.14	1971	4.3		
1940	48.3	1910	57.0	1929	6.44	1896	1.19	1982	4.0		
1876	48.7	1960	56.7	1939	6.40	1992	1.26	1898	3.1		
1975	48.7	1945	56.5	1895	6.14	1946	1.43	1965	3.0		
1881	48.8	1976	56.5	1970	6.12	1922	1.52	1990	2.4		
1879	48.9	1878	56.4	1940	6.06	1968	1.57	1996	2.4		
				1982	6.06						

May Daily Statistics

Day	Average High	Average Low	Record High	Year	Record Low	Year	Record Precip.	Year	Record Snow	Year
1	67	49	90	1942	34	1978	1.77	1976	T	1963
2	68	50	87	2001*	35	1960	2.49	1989	0	
3	68	50	90	2001*	35	1986*	1.36	1985	0	
4	68	51	91	2001	35	1957	1.88	1921	0	
5	69	51	90	1928	33	1966	1.81	1948	0	
6	69	51	90	1986	36	1891	2.00	1956	T	1891
7	69	52	94	1930	35	1970	2.69	1908	0	
8	70	52	93	1936	37	1947	1.42	1998	T	1898
9	70	52	93	1936	33	1947	1.50	1873	0	
10	70	52	93	1896	33	1966	1.35	1989	0	
11	71	53	93	1896	28	1966	1.82	1981	0	
12	71	54	91	1881	37	1962	1.58	1943	0	
13	71	54	92	1881	37	1967	1.40	1948	0	
14	72	54	92	1956	40	1996	1.48	1930	0	
15	72	54	93	1900	39	1966	1.40	1981	0	
16	72	55	90	1998*	40	1984*	2.03	1940	0	
17	72	55	92	1974	37	1956	1.33	1985	0	
18	73	55	94	1962	41	1973	0.98	1881	0	
19	73	55	96	1962	41	1882	2.36	1900	0	
20	73	56	94	1996*	45	1981*	1.64	1969	0	
21	74	56	95	1934	44	1994*	2.94	1894	0	
22	74	57	96	1941	43	1895	1.67	1947	0	
23	74	57	96	1925	41	1961	1.77	1951	0	
24	75	57	90	1964	40	1963	2.49	1999	0	
25	75	57	96	1991	40	1963	1.77	1952	0	
26	75	58	96	1880	43	1961	1.69	2001	0	
27	75	58	95	1991	42	1961	3.16	1918	0	
28	76	58	94	1941	41	1961	2.28	1982	0	
29	76	59	95	1991	42	1884	2.25	1990	0	
30	76	59	97	1991	41	1961	1.74	1908	0	
31	77	60	97	1991	42	1965	1.65	1993	0	
Month			97	1991	28	1966	3.16	1918	T	1963*

May Monthly Statistics

Average High Temperature	72.1°F
Average Low Temperature	54.8°F
Average Temperature	63.5°F
Average Liquid Precipitation	3.38 inches
Average Snowfall	0
Average Heating Degree Days	113
Average Cooling Degree Days	71

10 Coldest		10 Warmest		10 Wettest		10 Driest		10 Snowiest		10 Least Snowy	
Year	Temp. (°F)	Year	Temp. (°F)	Year	Precip. (in.)	Year	Precip. (in.)	Year	Snow (in.)	Year	Snow (in.)
1967	55.9	1991	70.8	1894	9.46	1964	0.47	1891	T	All	0
1882	57.1	1880	68.6	1948	7.41	1880	0.54	1898	T	other	
1917	57.1	1887	68.1	1947	7.22	1887	0.62	1963	T	years	
1907	57.6	1911	67.9	1983	7.03	1986	0.70				
1961	58.5	1944	67.4	1933	6.93	1977	0.70				
1924	59.2	1918	67.3	1984	6.87	1959	0.80				
1966	59.4	1896	67.2	1908	6.85	1914	0.90				
1997	59.5	1942	67.0	1918	6.81	1955	0.91				
1968	59.6	1914	66.8	1989	6.76	1903	0.93				
1874	59.9	1986	66.7	1946	6.09	1941	1.01				

June Daily Statistics

Day	Average High	Average Low	Record High	Year	Record Low	Year	Record Precip.	Year	Record Snow	Year
1	77	60	97	1895	44	1984*	1.58	1881	0	
2	77	60	98	1925	46	1907	2.96	1959	0	
3	77	61	97	1919	48	1986*	1.60	1879	0	
4	77	61	98	1925	47	1988*	1.34	1941	0	
5	78	61	100	1925	48	1990	2.31	1920	0	
6	78	62	100	1925	45	1945	1.95	1877	0	
7	78	62	98	1925	47	1958	1.79	1904	0	
8	78	62	95	1999	44	1977	2.00	1955	0	
9	78	62	98	1933	44	1957	2.03	1910	0	
10	79	63	97	1964	47	1988	2.08	1903	0	
11	79	63	95	1986	44	1972	2.04	1958	0	
12	80	63	95	1947	48	1979	3.05	1968	0	
13	80	63	95	1956	50	1979*	2.21	1982	0	
14	81	64	96	1994*	48	1978*	1.88	1947	0	
15	81	64	100	1994	47	1884	2.27	1908	0	
16	81	64	100	1991*	49	1963	3.30	1906	0	
17	81	64	98	1957	44	1964	2.41	2001	0	
18	81	65	96	1957	48	1950	1.93	1991	0	
19	82	65	100	1994	53	1946	1.64	1935	0	
20	82	65	98	1931*	51	1956*	1.64	1913	0	
21	82	66	99	1923	48	1968	1.62	1877	0	
22	82	66	100	1988	48	1963	2.47	1967	0	
23	82	66	97	1888	47	1963	2.12	1887	0	
24	83	66	99	1923	52	1947	2.61	1962	0	
25	83	66	98	1952	52	1979	2.55	1969	0	
26	83	67	100	1952	51	1960	2.77	1930	0	
27	83	67	97	1963	50	1965	3.27	1938	0	
28	83	67	97	1963	54	1961	1.58	1977	0	
29	83	68	102	1934	52	1888	4.62	1973	0	
30	84	68	100	1964	52	1988	1.56	1996	0	
Month			102	1934	44	1984*	4.62	1973	0	

June Monthly Statistics

Average High Temperature	80.6°F
Average Low Temperature	64.0°F
Average Temperature	72.3°F
Average Liquid Precipitation	3.29 inches
Average Snowfall	0
Average Heating Degree Days	12
Average Cooling Degree Days	235

10 Coldest		10 Warmest		10 Wettest		10 Driest		10 Snowiest		10 Least Snowy	
Year	Temp. (°F)	Year	Temp. (°F)	Year	Precip. (in.)	Year	Precip. (in.)	Year	Snow (in.)	Year	Snow (in.)
1903	65.8	1925	78.0	1938	10.06	1949	0.11	None		All	0
1907	66.8	1994	78.0	1906	8.04	1964	0.21				
1881	67.3	1943	76.7	1973	7.88	1966	0.41				
1878	67.4	1934	76.2	1975	7.57	1988	0.57				
1958	67.8	1923	75.8	1962	7.40	1995	0.62				
1916	67.9	1991	75.7	1969	7.31	1923	0.65				
1926	67.9	1930	75.2	1887	6.81	1960	0.71				
1927	68.6	1976	75.2	1879	6.77	1885	0.74				
1977	68.6	2001	75.2	1928	6.77	1971	1.01				
1972	68.7	1899	74.9	1920	6.76	1888	1.09				
		1957	74.9			1899	1.09				

July Daily Statistics

Day	Average High	Average Low	Record High	Year	Record Low	Year	Record Precip.	Year	Record Snow	Year
1	84	68	102	1901	52	1988*	1.04	1877	0	
2	84	68	103	1901	52	1965*	1.73	1987	0	
3	84	68	104	1966	54	1957*	3.66	1978	0	
4	84	69	103	1966	52	1986	2.08	1981	0	
5	85	69	100	1999*	55	1979*	4.38	1989	0	
6	85	69	98	1999*	52	1960	1.36	1930	0	
7	85	69	98	1994*	55	1979*	2.97	1941	0	
8	85	69	100	1993	53	1960	2.28	1964	0	
9	85	69	103	1936	53	1963	1.79	1952	0	
10	85	70	104	1936	53	1963	2.99	1931	0	
11	85	70	100	1988	55	1963	1.92	1967	0	
12	85	70	97	1936*	55	1999	1.40	1949	0	
13	86	70	98	1994	57	1998*	2.80	1991	0	
14	86	70	101	1954	52	1940	2.89	1994	0	
15	86	70	103	1995	56	1940	2.96	1926	0	
16	86	70	102	1988	55	1946	1.89	1979	0	
17	86	70	102	1988	56	1946	2.06	1911	0	
18	86	70	99	1953	57	1976	3.45	1994	0	
19	86	70	100	1930	58	1962	1.93	1877	0	
20	86	70	99	1930	57	1890	2.32	1938	0	
21	86	70	103	1930	51	1966	3.45	1988	0	
22	86	70	100	1957*	56	1944	2.98	1946	0	
23	86	70	99	1991*	58	1950	2.75	1887	0	
24	86	70	95	1987*	57	1985*	1.45	1958	0	
25	86	70	96	1999	56	1953	1.35	1902	0	
26	86	70	101	1892	56	1920	1.97	1988	0	
27	86	70	100	1940	53	1962	2.16	1873	0	
28	86	70	100	1941	53	1962	3.28	1969	0	
29	86	70	98	1892	59	1914	3.53	1980	0	
30	86	70	98	1988*	58	1981*	3.36	1971	0	
31	86	70	102	1954	57	1895	3.00	1889	0	
Month			104	1966*	51	1966	4.38	1989	0	

July Monthly Statistics

Average High Temperature	85.5°F
Average Low Temperature	69.7°F
Average Temperature	77.6°F
Average Liquid Precipitation	4.39 inches
Average Snowfall	0
Average Heating Degree Days	1
Average Cooling Degree Days	391

10 Coldest		10 Warmest		10 Wettest		10 Driest		10 Snowiest		10 Least Snowy	
Year	Temp. (°F)	Year	Temp. (°F)	Year	Precip. (in.)	Year	Precip. (in.)	Year	Snow (in.)	Year	Snow (in.)
1891	71.9	1994	82.1	1994	10.42	1957	0.64	None		All	0
1962	72.0	1995	81.5	1919	10.30	1983	0.68				
1884	72.5	1955	81.4	1989	9.44	1894	0.75				
1895	73.3	1993	81.4	1945	8.86	1881	0.96				
1960	73.3	1999	81.1	1926	8.40	1955	1.04				
1956	73.5	1887	80.7	1969	8.33	1944	1.05				
1984	73.9	1988	80.7	1889	8.29	1937	1.16				
1888	74.0	1952	80.1	1996	8.18	1999	1.22				
1914	74.0	1949	79.9	1988	8.07	1954	1.24				
2000	74.0	1931	79.8	1931	7.99	2001	1.30				

August Daily Statistics

Day	Average High	Average Low	Record High	Year	Record Low	Year	Record Precip.	Year	Record Snow	Year
1	86	70	97	1975*	56	1964*	2.01	1971	0	
2	86	70	99	1975*	58	1947*	2.86	1883	0	
3	86	70	98	1975	56	1959	5.63	1898	0	
4	86	70	98	1995	52	1965	2.74	1915	0	
5	86	70	98	1930	52	1951	1.60	1888	0	
6	85	70	103	1918	53	1957	2.74	1905	0	
7	85	70	106	1918	54	1964	3.70	1921	0	
8	85	70	100	2001	55	1989	4.40	1981	0	
9	85	69	101	2001	54	1964	2.36	1991	0	
10	85	69	98	2001*	54	1964	2.24	1984	0	
11	85	69	101	1900	55	1972	2.05	1899	0	
12	85	69	99	1900	53	1962	3.29	1955	0	
13	85	69	99	1881	51	1941	5.21	1873	0	
14	85	69	95	1988	50	1964	1.77	1977	0	
15	84	69	98	1988	50	1964*	1.54	1889	0	
16	84	69	98	1997	53	1979*	2.25	1917	0	
17	84	68	96	1995	50	1979	2.76	1928	0	
18	84	68	97	1995	52	1961	2.54	1901	0	
19	84	68	94	1937*	51	1963	3.73	1939	0	
20	84	68	99	1983	52	1949	1.87	1948	0	
21	83	68	96	1916	54	1949*	1.77	1888	0	
22	83	67	97	1916	52	1940	1.24	1887	0	
23	83	67	94	1916	53	1952	3.74	1933	0	
24	83	67	93	1998*	50	1957	1.70	1974	0	
25	83	67	97	1968	47	1963	3.03	1906	0	
26	82	67	100	1948	51	1963*	2.52	1972	0	
27	82	66	98	1948	51	1963*	4.77	1971	0	
28	82	66	99	1973*	49	1986	2.33	1978	0	
29	82	66	98	1991*	44	1986	0.83	1936	0	
30	81	66	100	1953	47	1986	2.32	1911	0	
31	81	66	101	1953	45	1965	3.41	1911	0	
Month			106	1918	44	1986	5.63	1898	0	

August Monthly Statistics

Average High Temperature	84.0°F
Average Low Temperature	68.5°F
Average Temperature	76.3°F
Average Liquid Precipitation	3.82 inches
Average Snowfall	0
Average Heating Degree Days	1
Average Cooling Degree Days	349

10 Coldest		10 Warmest		10 Wettest		10 Driest		10 Snowiest		10 Least Snowy	
Year	Temp. (°F)	Year	Temp. (°F)	Year	Precip. (in.)	Year	Precip. (in.)	Year	Snow (in.)	Year	Snow (in.)
1927	70.2	1980	79.9	1911	12.1	1896	0.46	None		All	0
1940	70.7	1995	79.8	1933	9.84	1964	0.49				
1946	70.7	2001	79.8	1955	9.70	1895	0.59				
1903	70.9	1900	79.2	1971	9.61	1943	0.65				
1963	71.2	1978	79.1	1905	9.57	1877	0.66				
1874	71.3	1991	78.9	1906	9.56	1980	0.80				
1962	72.0	1993	78.8	1901	9.42	1951	0.89				
1964	72.1	1973	78.7	1898	9.06	1916	0.94				
1883	72.4	1938	78.3	1942	8.93	2001	0.97				
1875	72.5	1988	78.2	1977	8.70	1876	0.98				

September Daily Statistics

Day	Average High	Average Low	Record High	Year	Record Low	Year	Record Precip.	Year	Record Snow	Year
1	81	65	97	1953*	49	1963	1.80	1952	0	
2	81	65	100	1953	47	1963	1.06	1926	0	
3	81	65	100	1953	51	1967*	3.37	1969	0	
4	80	65	93	1973	52	1997*	3.25	1935	0	
5	80	64	92	1985*	53	1994*	1.89	1935	0	
6	80	64	95	1983	48	1988*	2.00	1935	0	
7	79	64	102	1881	47	1984*	1.22	1999	0	
8	79	64	96	1939	45	1962	2.94	1929	0	
9	79	63	94	1884	49	1956*	1.59	1999	0	
10	79	63	97	1983	45	1956*	1.59	1909	0	
11	78	62	98	1983	43	1917	2.10	1971	0	
12	78	62	95	1931*	45	1940	4.60	1960	0	
13	77	61	95	1952	44	1958	2.80	1944	0	
14	77	61	92	1995	44	1975	4.69	1966	0	
15	77	61	95	1927	45	1895	2.74	1904	0	
16	77	61	94	1991	45	1984	6.63	1999	0	
17	76	61	94	1991	44	1986*	3.48	1876	0	
18	76	60	90	1891	42	1959	1.91	1918	0	
19	76	60	92	1983	43	1959*	1.95	1938	0	
20	75	59	90	1983*	45	1979*	3.04	1938	0	
21	75	59	97	1895	38	1956	2.44	1983	0	
22	75	59	97	1895	39	1962	4.65	1882	0	
23	74	58	97	1895	40	1983*	3.72	1882	0	
24	74	58	95	1970	40	1974*	2.08	1912	0	
25	73	57	92	1970	35	1963	1.56	1912	0	
26	73	57	92	1970*	38	1963*	2.79	1985	0	
27	73	57	91	1998	36	1947	1.85	1985	0	
28	72	57	91	1886	37	1957	1.82	1907	0	
29	72	56	89	1921	37	1942	1.87	1963	0	
30	71	55	88	1881	38	1942	2.38	1900	0	
Month			102	1881	35	1963	6.63	1999	0	

September Monthly Statistics

Average High Temperature	76.7°F
Average Low Temperature	60.9°F
Average Temperature	68.8°F
Average Liquid Precipitation	3.88 inches
Average Snowfall	0
Average Heating Degree Days	40
Average Cooling Degree Days	151

10 Coldest		10 Warmest		10 Wettest		10 Driest		10 Snowiest		10 Least Snowy	
Year	Temp. (°F)	Year	Temp. (°F)	Year	Precip. (in.)	Year	Precip. (in.)	Year	Snow (in.)	Year	Snow (in.)
1963	62.8	1881	75.4	1999	13.07	1884	0.20	None		All	0
1962	63.0	1930	74.1	1882	12.09	1906	0.36				
1875	63.8	1931	74.1	1960	8.78	1968	0.44				
1876	63.9	1921	72.8	1876	8.77	1915	0.46				
1917	64.4	1900	72.4	1966	8.70	1895	0.61				
1950	64.5	1895	72.3	1935	8.36	1970	0.82				
1984	64.7	1980	72.2	2000	7.95	1914	0.86				
1940	64.8	1970	72.0	1938	7.35	1941	0.88				
1887	64.9	1998	71.8	1904	7.21	1881	0.94				
1879	65.0	1891	71.6	1975	7.21	1878	0.96				
1924	65.0	1971	71.6								

October Daily Statistics

Day	Average High	Average Low	Record High	Year	Record Low	Year	Record Precip.	Year	Record Snow	Year
1	71	55	89	1927	34	1947	1.79	1902	0	
2	71	54	86	1927*	39	1966*	1.52	1929	0	
3	71	53	90	1919	33	1974	1.11	1985	0	
4	70	53	88	1941	34	1974	2.73	1877	0	
5	70	53	96	1941	31	1961	1.71	1906	0	
6	69	52	95	1941	32	1965	1.48	1873	0	
7	69	51	93	1941	37	1966*	1.60	1951	0	
8	69	51	84	1887	32	1964	1.89	1874	0	
9	68	51	86	1939*	35	1988*	2.17	1877	T	1895
10	68	50	90	1939	33	1979	2.10	1971	2.1	1979
11	68	50	86	1949	28	1964	1.06	1902	0	
12	67	49	88	1954	28	1964	1.62	1904	0	
13	67	49	87	1954	32	1988	0.90	1896	0	
14	66	49	88	1975	30	1988	1.66	1955	0	
15	66	48	87	1975	32	1876	1.04	1955	0	
16	66	48	88	1897	31	1876	1.60	1974	0	
17	65	47	89	1938	34	1961	1.87	1932	T	1970
18	65	47	85	1908	32	1982	1.11	1927	0	
19	65	47	80	1947*	29	1976	3.59	1966	0.5	1940
20	64	47	80	1938*	27	1974	2.72	1873	1.5	1940
21	64	46	86	1947	28	1952	2.21	1904	0	
22	63	46	83	1920	26	1940	2.71	1970	0	
23	63	46	87	1947	31	1969*	1.92	1883	0	
24	63	45	83	2001	25	1969	2.26	1917	0	
25	62	45	76	2001*	25	1962	3.82	1980	0	
26	62	45	78	1920	29	1960	2.11	1872	T	1962
27	61	45	80	1963	29	1936	1.45	1889	T	1957*
28	61	45	85	1919	28	1976	2.03	1953	T	1965*
29	61	44	79	1918	30	1965*	1.72	1953	T	1902
30	60	44	81	1946	27	1962	1.77	1935	0.7	1925
31	60	44	82	1946	26	1966	1.32	1895	0.4	1925
Month			96	1941	25	1969*	3.82	1980	2.1	1979

October Monthly Statistics

Average High Temperature	65.7°F
Average Low Temperature	48.7°F
Average Temperature	57.2°F
Average Liquid Precipitation	2.75 inches
Average Snowfall	0.1 inches
Average Heating Degree Days	268
Average Cooling Degree Days	19

10 Coldest		10 Warmest		10 Wettest		10 Driest		10 Snowiest		10 Least Snowy	
Year	Temp. (°F)	Year	Temp. (°F)	Year	Precip. (in.)	Year	Precip. (in.)	Year	Snow (in.)	Year	Snow (in.)
1876	50.9	1971	63.5	1902	6.66	1924	0.09	1979	2.1	All	0
1888	51.6	1947	62.7	1877	6.52	1963	0.09	1940	2.0	other	
1940	51.8	1931	62.6	1917	6.04	1892	0.30	1925	1.1	years	
1988	51.8	1879	62.5	1995	5.99	1921	0.40	1895	T		
1925	52.2	1920	62.4	1927	5.71	1879	0.41	1902	T		
1976	52.5	1990	61.9	1913	5.66	1952	0.49	1903	T		
1987	52.5	1949	61.8	1937	5.24	1920	0.64	1906	T		
1964	52.6	1900	61.6	1943	5.21	1918	0.69	1934	T		
1972	52.6	1995	61.3	1966	5.12	1922	0.71	1952	T		
1895	52.7	1919	61.2	1980	5.03	1928	0.73	1957	T		
		1954	61.2					1962	T		
		1975	61.2					1965	T		
		1984	61.2					1970	T		
								1972	T		

November Daily Statistics

Day	Average High	Average Low	Record High	Year	Record Low	Year	Record Precip.	Year	Record Snow	Year
1	60	44	84	1950	30	1968	1.81	1956	0	
2	59	43	81	1950	27	1976	1.65	1956	0	
3	59	43	80	1990	28	1951	1.58	1910	T	1951*
4	59	43	81	1974	26	1951	1.15	1910	T	1910*
5	58	42	80	1975*	26	1879	1.48	1877	T	1908*
6	58	42	79	1948	26	1962	1.41	1963	5.4	1953
7	58	42	76	1938	20	1962	3.99	1977	3.4	1953
8	58	42	78	1975	25	1960	3.07	1972	0.8	1892
9	57	41	78	1975	23	1976	1.57	1932	T	1991*
10	57	41	72	1985	23	1961	1.46	1875	T	1971*
11	56	41	74	1949	21	1961	1.03	1970	1.3	1987
12	56	40	75	1879	26	1976*	1.90	1975	0.4	1968
13	56	40	74	1902	24	1996	1.83	1904	0.2	1904
14	55	40	76	1955	19	1986	2.64	1972	0.7	1908
15	55	39	81	1993	19	1942	2.00	1892	1.6	1908
16	55	39	72	2001*	20	1933	1.46	1985	0.1	1906
17	54	39	72	1987*	21	1933	1.58	1935	0.3	1935
18	54	38	75	1921	20	1936	1.07	1911	T	1962*
19	54	38	75	1921	20	1936	2.13	1876	2.9	1955
20	53	38	76	1985	21	1984	2.59	1876	3.2	1961
21	53	38	74	1931*	19	1879	2.30	1952	0.3	1939
22	52	38	73	1883	14	1880	1.33	1920	1.8	1989
23	52	37	72	1931*	10	1880	1.80	1875	2.8	1989
24	51	37	71	1979	16	1880	2.22	1901	3.2	1898
25	51	37	73	1979*	21	1965	3.46	1950	4.4	1938
26	51	36	70	1946	18	1938	1.69	1900	6.0	1898
27	50	36	73	1896	16	1932	1.95	1944	6.9	1938
28	50	36	71	1990	17	1930	2.78	1945	2.0	1917
29	50	35	70	1927	15	1955	1.68	1971	1.9	1995
30	49	35	73	1933	8	1875	0.96	1972	4.9	1967
Month			84	1950	8	1875	3.99	1977	6.9	1938

November Monthly Statistics

Average High Temperature	54.8°F
Average Low Temperature	39.5°F
Average Temperature	47.1°F
Average Liquid Precipitation	3.16 inches
Average Snowfall	0.4 inches
Average Heating Degree Days	546
Average Cooling Degree Days	2

10 Coldest		10 Warmest		10 Wettest		10 Driest		10 Snowiest		10 Least Snowy	
Year	Temp. (°F)	Year	Temp. (°F)	Year	Precip. (in.)	Year	Precip. (in.)	Year	Snow (in.)	Year	Snow (in.)
1875	39.2	1931	54.0	1972	9.06	1976	0.32	1898	13.4	Many	0
1976	39.9	2001	52.9	1977	7.76	1922	0.43	1938	11.8		
1880	40.9	1975	52.7	1932	7.45	2001	0.56	1953	8.8		
1996	41.2	1994	51.7	1876	7.31	1973	0.64	1967	4.9		
1901	41.3	1985	51.2	1898	7.19	1908	0.67	1989	4.6		
1894	42.0	1902	50.9	1889	6.76	1936	0.69	1961	3.2		
1962	42.0	1927	50.9	1963	6.67	1890	0.80	1955	2.9		
1903	42.5	1948	50.9	1986	6.27	1974	0.81	1896	2.8		
1951	42.5	1999	50.9	1985	6.09	1917	0.86	1978	2.5		
1874	42.7	1934	50.6	1907	5.85	1981	0.95	1908	2.3		
1967	42.7										

December Daily Statistics

Day	Average High	Average Low	Record High	Year	Record Low	Year	Record Precip.	Year	Record Snow	Year
1	49	35	72	2001	11	1875	1.38	1920	0.8	1890
2	48	34	66	1970	11	1875	1.48	1986	2.0	1903
3	48	34	68	1950	9	1976	1.60	1967	2.0	1903
4	48	34	73	1998	12	1966*	1.69	1927	5.8	1921
5	47	33	71	2001	13	1886	1.52	1993	4.5	1886
6	47	33	72	2001	15	1901	1.16	1996	6.2	1910
7	47	33	73	1998	15	1926*	3.71	1914	2.8	1959
8	46	33	67	1980	14	1882	1.22	1972	3.5	1989
9	46	32	70	1966	6	1876	1.91	1978	2.9	1942
10	46	32	71	1966	4	1876	1.45	1969	4.3	1904
11	45	32	65	1971*	8	1880	2.25	1992	3.7	1960
12	45	31	65	1931	8	1988	1.45	1983	10.9	1960
13	45	31	65	1923	8	1960	2.90	1941	4.6	1915
14	45	31	69	1881	10	1960	2.02	1897	5.3	1951
15	44	30	66	1971	7	1960	1.13	1901	4.5	1916
16	44	30	68	1971	9	1876	1.54	1974	7.1	1896
17	44	30	64	2000	7	1919*	2.23	1888	7.3	1932
18	43	30	65	1990	4	1919	1.72	1977	6.0	1887
19	43	29	65	1929	4	1884	1.02	1934	9.2	1945
20	43	29	64	1957*	1	1942*	1.32	1951	3.2	1966
21	43	29	66	1895	1	1942	1.40	1902	2.7	1962
22	42	29	64	1998	6	1989	1.77	1983	2.8	1908
23	42	28	66	1990*	7	1989	1.62	1907	5.2	1963
24	42	28	64	1990	3	1983	1.73	1986	12.4	1966
25	42	28	68	1964	1	1983*	1.11	1945	5.5	1909
26	41	28	68	1964*	3	1983	1.90	1909	15.5	1909
27	41	28	64	1949	8	1989*	1.55	1964	3.4	1990
28	41	27	65	1982	9	1950	1.45	1942	3.0	1990
29	41	27	72	1984	−1	1917	2.80	1901	3.9	1935
30	41	27	65	1990	−5	1880	2.30	1948	9.0	2000
31	40	27	66	1884	0	1880	1.09	1911	3.8	1967
Month			73	1998	−5	1880	3.71	1914	15.5	1909

December Monthly Statistics

Average High Temperature	44.2°F
Average Low Temperature	30.6°F
Average Temperature	37.4°F
Average Liquid Precipitation	3.31 inches
Average Snowfall	2.1 inches
Average Heating Degree Days	863
Average Cooling Degree Days	0

10 Coldest		10 Warmest		10 Wettest		10 Driest		10 Snowiest		10 Least Snowy	
Year	Temp. (°F)	Year	Temp. (°F)	Year	Precip. (in.)	Year	Precip. (in.)	Year	Snow (in.)	Year	Snow (in.)
1876	25.4	1923	44.5	1996	8.48	1955	0.25	1909	22.4	1891	0
1989	25.5	1931	44.2	1983	7.37	1980	0.77	1904	18.9	1941	0
1960	27.6	2001	43.7	1914	7.35	1998	0.82	1966	18.8	1943	0
1963	27.9	1889	43.5	1969	7.23	1877	0.83	1960	17.5	1994	0
1917	28.0	1891	42.7	1901	6.72	1889	0.85	1945	13.9	2001	0
1880	28.4	1990	42.1	1902	6.63	1925	0.96	1910	12.5	Many	T
1958	29.4	1998	42.0	1973	6.34	1985	0.98	2000	10.5		
1904	29.8	1984	41.9	1986	5.89	1896	1.00	1887	9.6		
1945	29.8	1994	41.9	1967	5.88	1988	1.00	1932	9.5		
1910	30.3	1956	41.6	1909	5.86	1937	1.05	1962	9.5		
1976	30.3										

189

APPENDIX B
Climate Data for Wilmington, Delaware; Allentown, Pennsylvania; and Atlantic City, New Jersey

Monthly averages are computed from the 1971–2000 period, with data supplied by the National Climatic Data Center. HDD stands for Heating Degree Days, CDD for Cooling Degree Days. Daily maximum and minimum temperature records are courtesy of the National Weather Service, Mount Holly, New Jersey and are valid through December 2001. In all cases, temperatures are given in °F, while precipitation (rain plus melted snow and ice) and snowfall are given in inches.

Wilmington Monthly Averages

	Jan.	Feb.	March	April	May	June	July	Aug.	Sept.	Oct.	Nov.	Dec.	Year
Average Minimum	23.7	25.8	33.4	42.1	52.4	61.8	67.3	65.8	58.1	45.6	36.9	28.4	45.1
Average Maximum	39.3	42.5	51.9	62.6	72.5	81.1	86.0	84.1	77.2	65.9	55.0	44.4	63.5
Average Temperature	31.5	34.2	42.7	52.4	62.5	71.5	76.6	75.0	67.7	55.8	45.9	36.4	54.4
Average Precipitation	3.43	2.81	3.97	3.39	4.15	3.59	4.28	3.51	4.01	3.08	3.19	3.40	42.81
Average Snowfall	7.6	6.9	2.4	0.3	0	0	0	0	0	0.1	0.7	3.4	21.4
Average HDD	1029	862	682	377	132	15	1	2	51	296	566	878	4891
Average CDD	0	0	2	9	63	215	364	314	134	16	1	0	1118

Allentown Monthly Averages

	Jan.	Feb.	March	April	May	June	July	Aug.	Sept.	Oct.	Nov.	Dec.	Year
Average Minimum	19.1	21.0	28.9	37.8	48.3	57.7	62.6	60.7	52.7	41.1	32.7	24.0	40.6
Average Maximum	35.0	38.7	48.7	60.1	70.9	79.3	83.9	81.7	74.0	62.9	51.2	40.0	60.5
Average Temperature	27.1	29.9	38.8	49.0	59.6	68.5	73.3	71.2	63.4	52.0	42.0	32.0	50.6
Average Precipitation	3.50	2.75	3.56	3.49	4.47	3.99	4.27	4.35	4.37	3.33	3.70	3.39	45.17
Average Snowfall	8.4	8.9	5.8	0.7	0	0	0	0	0	0.1	1.3	6.1	31.3
Average HDD	1147	966	784	450	176	26	3	8	95	374	657	985	5671
Average CDD	0	0	1	6	45	164	292	235	83	7	0	0	833

Atlantic City Monthly Averages

	Jan.	Feb.	March	April	May	June	July	Aug.	Sept.	Oct.	Nov.	Dec.	Year
Average Minimum	22.8	24.5	31.7	39.8	49.8	59.3	65.4	63.7	56.0	43.9	35.7	27.1	43.3
Average Maximum	41.4	43.9	51.9	61.3	71.1	80.0	85.1	83.3	76.6	66.3	56.0	46.4	63.6
Average Temperature	32.1	34.2	41.8	50.6	60.5	69.7	75.3	73.5	66.3	55.1	45.9	36.8	53.5
Average Precipitation	3.60	2.85	4.06	3.45	3.38	2.66	3.86	4.32	3.14	2.86	3.26	3.15	40.59
Average Snowfall	5.2	5.4	2.6	0.3	0	0	0	0	0	0	0.4	2.2	16.1
Average HDD	1019	873	725	437	187	32	1	6	69	323	573	868	5113
Average CDD	0	0	1	5	44	168	322	269	110	15	1	0	935
Average Ocean Temp	37	35	42	48	56	63	70	73	70	61	53	44	54

Wilmington, Delaware: Record High Temperatures

Date	Jan.	Feb.	March	April	May	June	July	Aug.	Sept.	Oct.	Nov.	Dec.
1	65/1973	67/1989	75/1972	81/1978	88/1942	100/1895	103/1901	98/1917	97/1953	90/1927	85/1950	73/2001
2	64/2000	64/1967	77/1972	83/1967	90/1894	99/1895	106/1901	101/1955	100/1953	88/1927	81/1950	67/1970
3	64/2000	64/1991	71/1897	86/1963	90/1913	101/1895	102/1966	97/1957	100/1953	88/1919	77/1990	68/1998
4	69/1950	67/1991	76/1974	80/1950	88/1965	95/1925	102/1966	96/1995	94/1953	88/1941	79/1974	75/1998
5	66/1950	70/1991	79/1976	85/1942	88/1955	99/1925	100/1949	100/1896	93/1961	94/1941	81/1961	72/1982
6	70/1950	63/1938	71/1935	85/1929	89/1986	100/1899	98/1999	103/1896	98/1983	93/1941	79/1948	72/2001
7	65/1950	62/1938	72/1987	87/1929	91/2000	98/1899	98/1897	107/1918	95/1983	93/1941	75/1975	72/1998
8	66/1998	66/1965	81/2000	87/1929	90/2000	98/1914	98/1993	105/1918	95/1939	88/1941	77/1975	69/1980
9	67/1930	64/2001	74/2000	85/1959	92/1963	96/1984	100/1936	102/1896	94/1939	86/1939	76/1895	72/1966
10	64/1930	63/1990	76/1964	84/1922	98/1895	97/1964	103/1936	99/1913	98/1983	89/1939	71/1985	72/1966
11	66/1975	66/1960	73/1967	84/1922	96/1896	95/1973	101/1936	100/1896	100/1983	87/1954	74/1985	70/1899
12	63/1913	71/1999	85/1990	87/1977	91/1896	96/1914	97/1896	102/1896	94/1895	87/1954	74/1912	64/1983
13	70/1932	68/1951	85/1990	85/1977	89/1991	96/1956	98/1894	99/1896	95/1952	87/1954	73/1955	64/1923
14	70/1932	67/1990	77/1990	87/1941	93/1956	96/1956	100/1954	96/1918	92/1915	87/1975	76/1955	62/1991
15	71/1932	74/1949	80/1990	86/1896	97/1900	97/1994	99/1997	98/1988	92/1915	87/1975	80/1993	67/1971
16	69/1932	74/1954	82/1945	92/1896	90/1900	98/1957	100/1983	98/1895	94/1970	87/1897	73/2001	69/1971
17	63/1913	70/1976	83/1945	97/1896	94/1896	96/1952	100/1988	94/1999	94/1915	85/1938	71/1953	62/2000
18	65/1990	69/1981	78/1989	97/1896	94/1962	95/1957	100/1953	94/1995	90/1915	82/1908	74/1928	66/1990
19	66/1951	67/1997	78/1925	94/1896	95/1962	100/1994	100/1999	97/1914	94/1983	81/1963	75/1921	62/1929
20	66/1951	71/1930	81/1945	90/1985	96/1996	97/1895	100/1895	100/1983	93/1983	82/1969	75/1985	66/1895
21	62/1959	70/1953	86/1948	92/1985	95/1996	97/1923	102/1957	94/1968	97/1895	85/1947	75/1900	65/1895
22	67/1906	69/1974	82/1948	94/1985	94/1941	98/1988	101/1957	96/1914	98/1895	81/1979	71/1940	64/1998
23	69/1906	72/1985	78/1994	89/1960	94/1925	95/1988	99/1991	93/1968	98/1895	86/1947	73/1979	66/1990
24	71/1967	78/1985	79/1929	84/2001	94/1925	102/1894	96/1910	95/1914	91/1970	82/2001	73/1979	64/1990
25	67/1967	78/1930	80/1939	92/1915	92/1991	97/1952	96/1987	96/1968	93/1970	79/2001	74/1973	70/1982
26	75/1950	76/1930	77/1921	92/1990	91/1914	99/1952	99/1894	98/1948	92/1895	77/1973	71/1896	70/1964
27	70/1916	73/1997	82/1998	92/1915	98/1914	97/1952	100/1894	98/1948	91/1998	79/1963	73/1896	65/1973
28	70/1916	66/1976	83/1989	89/1990	93/1941	97/1898	101/1894	100/1973	85/1954	83/1919	72/1896	67/1982
29	68/1975	69/1976	85/1945	91/1974	93/1955	98/1959	99/1896	98/1953	88/1945	77/1971	70/1998	74/1984
30	68/1947		86/1998	86/1942	98/1895	98/1964	100/1901	101/1953	88/1953	79/1946	72/1991	66/1990
31	66/1916		82/1998		96/1895		102/1954	101/1953		82/1946		67/1992
Month	71/1967*	78/1985*	86/1948	97/1896	98/1914*	102/1894	106/1901	107/1918	100/1983*	94/1941	85/1950	74/1984

*Record tied in other years. Only the most recent year is given.

Wilmington, Delaware: Record Low Temperatures

Date	Jan.	Feb.	March	April	May	June	July	Aug.	Sept.	Oct.	Nov.	Dec.
1	0/1918	−1/1935	6/1980	11/1923	30/1978	40/1945	48/1988	56/1921	47/1934	31/1947	27/1917	10/1936
2	1/1968	−4/1961	10/1980	20/1924	34/1943	44/1945	50/1952	55/1998	49/1948	34/1997	26/1976	11/1967
3	4/1918	0/1934	9/1925	24/1924	33/1941	43/1915	49/1929	52/1920	46/1916	32/1974	24/1911	9/1976
4	−1/1918	4/1996	5/1943	23/1943	34/1957	43/1926	50/1933	53/1964	47/1997	30/1974	24/1951	10/1940
5	4/1904	−10/1918	7/1978	24/1995	35/1920	46/1997	52/1979	49/1951	48/1997	33/1965	26/1983	12/1926
6	7/1942	−3/1918	11/1926	24/1982	37/1917	42/1945	50/1979	51/1957	43/1924	30/1965	24/1967	15/1940
7	10/1988	2/1948	11/1960	18/1982	33/1970	44/1958	53/1980	55/1997	43/1984	35/1935	24/1962	7/1926
8	0/1970	0/1895	12/1986	24/1982	34/1997	41/1932	53/2000	53/1989	47/1984	33/1935	21/1953	15/1970
9	−2/1970	−15/1934	9/1984	25/1977	32/1947	45/1980	54/1918	54/1989	46/1986	33/1988	24/1976	13/1960
10	1/1982	−2/1979	2/1984	22/1985	33/1966	46/1988	54/1927	54/1927	42/1938	33/1895	26/1995	8/1910
11	−1/1982	−1/1979	9/1960	26/1921	32/1966	41/1972	55/1934	52/1972	39/1917	29/1964	22/1961	10/1968
12	−3/1981	−1/1979	9/1934	25/1967	32/1907	46/1980	51/1945	50/1930	44/1917	30/1964	23/1957	8/1988
13	−2/1981	−2/1979	17/1896	22/1940	35/1928	48/1979	53/1933	49/1930	46/1967	32/1934	18/1911	7/1960
14	−2/1912	−4/1979	14/1926	23/1950	36/1996	46/1933	54/1940	47/1941	40/1911	30/1988	20/1986	10/1982
15	1/1964	2/1943	13/1993	25/1943	35/1939	42/1896	53/1940	48/1964	43/1895	31/1937	19/1942	9/1962
16	1/1977	2/1943	10/1911	25/1928	38/1959	47/1946	49/1929	52/1979	40/1984	29/1944	15/1933	5/1917
17	−10/1982	−1/1979	14/1911	27/1980	35/1983	44/1926	54/1946	49/1979	43/1986	33/1937	18/1933	6/1951
18	−4/1957	−6/1979	11/1916	30/1926	40/1967	45/1950	55/1925	51/1941	40/1990	28/1982	19/1924	0/1919
19	−5/1994	2/1936	12/1967	29/1926	41/1976	50/1954	56/1924	51/1924	40/1929	28/1976	15/1936	8/1926
20	−8/1985	5/1978	16/1923	27/1926	42/1929	48/1956	53/1929	48/1949	40/1920	27/1974	17/1984	2/1942
21	−14/1985	6/1968	12/1965	28/1925	44/1950	45/1940	53/1929	50/1949	37/1956	25/1974	19/1984	−1/1942
22	−14/1984	7/1968	18/1930	30/1975	41/1895	44/1940	56/1950	48/1982	40/1962	26/1940	17/1964	5/1989
23	−4/1936	7/1963	14/1934	28/1933	40/1929	47/1992	54/1977	47/1923	37/1947	28/1969	17/1964	4/1989
24	2/1963	6/1923	15/1940	30/1930	36/1963	49/1947	56/1985	48/1952	36/1974	24/1969	16/1956	−5/1983
25	−1/1945	5/1914	17/1940	31/1936	39/1956	49/1979	56/1953	48/1942	37/1997	30/1962	18/1970	−7/1983
26	2/1961	8/1970	20/1983	32/1919	40/1925	51/1979	51/1920	49/1942	35/1940	26/1933	12/1938	−2/1983
27	1/1987	5/1934	22/2001	32/1933	39/1925	52/1979	51/1962	49/1944	32/1940	26/1936	13/1932	6/1950
28	−6/1935	−5/1934	20/1982	30/1934	41/1997	53/1970	53/1962	46/1986	38/1957	23/1936	14/1930	6/1933
29	0/1961	6/1980	13/1923	31/1922	40/1936	50/1937	57/1920	43/1982	34/1942	26/1941	11/1930	2/1933
30	5/1935		23/1941	33/1931	41/1961	43/1919	55/1956	45/1986	34/1942	26/1983	11/1929	−7/1917
31	2/1948		19/1923		41/1938		55/1928	47/1934		25/1925		−4/1917
Month	−14/1985*	−15/1934	2/1984	11/1923	30/1978	40/1945	48/1988	43/1982	32/1940	23/1936	11/1930*	−7/1983

*Record tied in other years. Only the most recent year is given.

194

Allentown, Pennsylvania: Record High Temperatures

Date	Jan.	Feb.	March	April	May	June	July	Aug.	Sept.	Oct.	Nov.	Dec.
1	61/1973	65/1989	67/1972	84/1978	87/1942	95/1937	97/1963	98/1955	97/1980	89/1927	81/1950	66/2001
2	59/1979	58/1983	72/1972	82/1967	90/2001	96/1925	102/1966	100/1955	99/1980	92/1927	80/1950	63/1998
3	63/1997	63/1991	71/1991	82/1963	91/2001	98/1925	105/1966	97/1930	99/1953	83/1950	77/1990	66/1998
4	66/1950	65/1991	73/1974	79/1928	92/2001	98/1925	102/1949	98/1944	94/1937	86/1926	79/1994	72/1998
5	62/1993	67/1991	72/1964	82/1928	92/1949	98/1925	98/1999	100/1955	92/1983	92/1941	75/1994	69/2001
6	60/1998	63/1938	76/1935	84/1929	92/1949	100/1925	100/1999	100/1924	96/1983	89/1941	78/1948	70/2001
7	61/1998	60/1938	70/1987	88/1929	94/1939	98/1925	97/1988	98/1924	91/1985	87/1946	74/1938	71/1998
8	67/1998	59/1965	79/2000	90/1929	93/1936	94/1984	97/1937	95/2001	97/1939	81/1943	74/1975	67/1980
9	65/1998	59/1990	70/2000	83/2001	92/1936	94/1984	103/1936	98/2001	93/1964	83/1939	75/1975	66/1980
10	52/1949	57/1990	70/1955	80/1955	91/1979	94/1984	103/1936	99/1949	96/1983	89/1949	70/1931	70/1946
11	66/1975	61/1981	71/1977	81/1955	90/1993	94/1984	100/1936	97/1949	96/1983	86/1949	72/1949	61/1979
12	60/1932	68/1999	81/1990	79/1996	89/1948	92/1961	98/1966	98/1944	93/1931	86/1928	71/1935	63/1931
13	67/1932	59/1937	83/1990	88/1941	90/1944	95/1956	99/1966	98/1944	94/1952	85/1954	71/1964	62/1923
14	71/1932	58/1990	81/1990	83/1960	89/1940	95/1988	100/1954	95/1944	91/1931	86/1975	73/1993	58/1929
15	68/1995	71/1949	79/1990	86/1994	89/1940	95/1994	98/1995	97/1988	95/1927	82/1947	77/1993	57/1929
16	62/1995	72/1954	80/1945	81/1976	89/1998	96/1957	101/1988	95/1944	95/1991	81/1963	71/2001	60/1971
17	61/1990	63/1976	84/1945	91/1976	92/1951	98/1957	99/1999	95/1944	93/1991	88/1938	71/1928	63/2000
18	62/1986	65/1981	77/1927	93/1976	96/1962	95/1957	98/1999	93/1995	90/1965	81/1963	76/1928	59/1984
19	60/1949	63/1997	70/1986	91/1976	97/1962	96/1994	99/1930	93/1937	93/1983	82/1963	70/1957	59/1957
20	56/1951	68/1930	81/1945	90/1941	92/1996	100/1923	101/1980	96/1983	92/1983	78/1936	75/1985	63/1957
21	63/1959	67/1953	75/1938	89/1985	92/1934	98/1923	101/1980	95/1937	91/1940	85/1947	73/1931	57/1957
22	63/1959	68/1974	82/1938	92/1985	96/1941	95/1941	101/1926	93/1937	93/1980	84/1979	73/1931	61/1998
23	62/1999	71/1985	80/1938	89/1960	95/1925	94/1923	99/1955	96/1936	93/1970	85/1947	72/1931	64/1990
24	66/1967	76/1985	78/1988	86/1960	90/1936	98/1923	95/1999	92/1947	92/1970	81/2001	69/1931	62/1990
25	64/1967	74/1930	77/1988	89/1960	90/1991	98/1923	95/1999	95/1948	92/1970	78/1963	72/1979	65/1964
26	72/1950	65/1976	75/1928	92/1990	91/1965	99/1952	98/1940	99/1948	90/1970	78/1963	67/1979	65/1964
27	66/1974	72/1997	83/1998	92/1990	91/1991	100/1966	98/1955	95/1953	88/1998	83/1963	62/1988	59/1949
28	58/1947	67/1976	85/1945	91/1990	93/1941	96/1948	97/1949	97/1973	85/1941	80/1984	70/1990	63/1982
29	57/1932	64/1976	86/1945	86/1974	95/1969	99/1934	96/1954	97/1953	86/1945	76/1989	66/1990	72/1984
30	66/1947		87/1998	92/1942	94/1987	97/1964	97/1940	97/1953	87/1986	79/1946	69/1933	64/1984
31	60/1974		86/1998		95/1939		101/1954	99/1953		81/1950		61/1972
Month	72/1950	76/1985	87/1998	93/1976	97/1962	100/1966*	105/1966	100/1955*	99/1980*	92/1941*	81/1950	72/1998*

*Record tied in other years. Only the most recent year is given.

Allentown, Pennsylvania: Record Low Temperatures

Date	Jan.	Feb.	March	April	May	June	July	Aug.	Sept.	Oct.	Nov.	Dec.
1	−3/1968	−5/1935	−5/1934	12/1923	30/1931	39/1945	48/1943	52/1998	42/1967	33/1993	24/1923	7/1967
2	−7/1968	−7/1961	8/1934	13/1923	30/1925	42/1993	49/1940	49/1947	44/1949	32/1997	24/1923	2/1967
3	6/1945	−2/1955	8/1925	20/1954	33/1966	43/1966	46/2001	48/1927	41/1967	33/1945	24/1951	8/1976
4	−4/1981	−1/1985	3/1943	20/1954	31/1957	41/1929	46/1933	50/1996	44/1946	29/1996	19/1951	8/1966
5	2/1970	−6/1996	8/1948	21/1995	31/1976	45/1950	50/1968	46/1951	44/1997	31/1996	22/1952	11/1926
6	−4/1968	−8/1996	7/1960	18/1982	36/1945	42/1945	50/1979	47/1957	42/1988	31/1996	19/1991	11/1940
7	−4/1942	−6/1935	1/1960	16/1982	32/1970	42/1945	49/2001	49/1964	43/1984	32/1958	23/1954	4/1926
8	−3/1942	−5/1935	7/1989	20/1982	30/1997	39/1932	51/2000	49/1997	42/1962	29/1954	20/1960	3/1926
9	−9/1942	−12/1934	7/1960	23/1972	31/1956	40/1957	48/1963	50/1989	42/1986	29/2000	19/1967	7/1989
10	−6/1942	−11/1934	4/1984	23/1950	28/1947	43/1988	51/1953	51/1999	39/1956	30/1929	21/1956	1/1958
11	−11/1942	−6/1962	6/1960	25/1938	31/1966	39/1972	50/1996	47/1972	39/1995	28/1943	19/1956	−2/1958
12	−6/1968	−6/1979	7/1960	24/1976	34/1963	44/1980	48/1945	47/1930	37/1967	25/1964	21/1926	2/1988
13	−3/1981	−6/1967	12/1926	24/1940	33/1996	45/1932	51/2001	45/1930	39/1967	31/1981	18/1996	5/1995
14	1/1974	−1/1979	14/1926	23/1950	29/1996	44/1933	51/1940	47/1941	40/1946	29/1963	17/1986	3/1958
15	−6/1948	−8/1943	8/1993	23/1943	35/1936	40/1933	51/1940	48/1964	38/1975	30/1987	18/1996	6/1960
16	0/1994	−7/1943	11/1993	23/1943	37/1939	44/1961	50/1946	49/1945	38/1923	26/1944	12/1933	1/1951
17	−8/1982	−3/1979	13/1967	28/1962	32/1956	47/1964	50/1946	46/1979	39/1986	25/1937	14/1933	−2/1951
18	−10/1954	−6/1979	5/1967	25/1948	35/1983	44/1958	52/1925	47/1981	36/1990	28/1974	16/1933	5/1945
19	−11/1994	0/1936	−1/1967	30/1990	35/1973	46/1999	52/1924	48/1981	37/1943	24/1974	11/1924	0/1951
20	−6/1994	−2/1936	6/1949	29/1961	40/1945	44/1926	48/1997	46/1944	37/1993	24/1972	12/1924	−2/1951
21	−15/1994	0/1936	10/1965	28/1956	38/1997	45/1968	49/1966	45/2000	34/1956	23/1972	17/1951	−3/1942
22	−12/1961	2/1968	15/1988	28/1981	40/1936	44/1940	53/1966	46/1988	36/1997	21/1940	17/1969	−2/1944
23	−9/1961	2/1972	13/1934	28/1945	38/1932	46/1992	51/1939	46/1924	35/1947	24/1997	16/1964	−3/1960
24	−6/1948	4/1948	14/1940	29/1930	33/1926	50/1988	51/1947	47/1971	31/1963	22/1969	16/2000	−5/1989
25	−4/1945	3/1964	15/1934	27/1966	33/1956	46/1979	51/1953	42/1927	33/1963	26/1962	16/1989	−2/1983
26	−3/1948	4/1970	17/1960	30/2001	39/1956	49/1965	48/1976	45/1944	35/1967	24/1952	16/1989	−3/1980
27	−5/1994	−2/1963	15/2001	30/1933	38/1972	51/1996	52/1962	43/1944	31/1947	25/1988	3/1938	−1/1950
28	−7/1935	−10/1934	11/1923	28/1947	38/1997	49/1970	52/1962	46/1986	30/1947	21/1936	12/1932	−8/1950
29	−9/1963	8/1980	13/1923	29/1947	38/1949	48/1970	53/1928	43/1982	32/2000	22/1940	12/1930	−1/1933
30	−5/1928		10/1970	29/1931	36/1949	48/1988	50/1997	41/1986	32/1942	25/1967	11/1955	−2/1933
31	−9/1948		11/1923		37/1996		51/1936	41/1934		21/1988	10/1929	−6/1963
Month	−15/1994	−12/1934	−5/1934	12/1923	28/1947	39/1972*	46/2001*	41/1986*	30/1947	21/1988*	3/1938	−8/1950

*Record tied in other years. Only the most recent year is given.

196

Atlantic City, New Jersey: Record High Temperatures

Date	Jan.	Feb.	March	April	May	June	July	Aug.	Sept.	Oct.	Nov.	Dec.
1	65/1966	69/1989	72/1972	80/1978	88/1985	94/1986	99/1968	97/1999	93/1995	85/1986	84/1950	72/2001
2	66/2000	68/1988	72/1997	89/1967	85/1992	93/1989	100/1966	98/1955	97/1953	84/1986	81/1950	64/1951
3	67/2000	62/1991	69/1967	86/1967	89/2001	88/1967	104/1966	95/1987	95/1984	84/1968	78/1990	69/1998
4	68/1950	68/1991	76/1974	77/1999	91/1965	90/1980	102/1966	98/1995	92/1985	87/1954	80/1974	73/1998
5	67/1950	72/1991	76/1946	78/1985	89/1955	90/1965	99/1999	94/1981	93/1985	90/1959	74/1975	73/1982
6	69/1950	63/1965	68/1959	83/1967	93/1949	93/1968	99/1999	94/1955	98/1983	87/1997	77/1961	73/2001
7	65/1965	58/1990	75/1974	84/1991	93/2000	96/1999	98/1986	97/2001	94/1983	83/1990	72/1965	77/1998
8	60/1998	69/1965	77/2000	87/1991	90/1964	98/1999	98/1993	100/2001	93/1985	83/1990	72/1965	71/1966
9	65/1965	67/1990	78/2000	85/1959	93/1963	96/1984	100/1993	103/2001	89/1959	83/1990	76/1986	72/1966
10	65/1950	60/2001	72/1986	79/1967	92/1965	99/1984	101/1993	97/2001	95/1983	84/1949	72/1985	68/1946
11	61/1950	63/1960	79/1967	80/1955	88/1993	97/1984	97/1966	97/1949	99/1983	83/1949	76/1985	67/1971
12	57/1995	62/1999	85/1990	85/1965	90/1959	93/1947	96/1966	94/1988	94/1961	83/1954	69/1992	64/1979
13	67/1995	71/1951	85/1990	83/1977	89/1991	96/1984	96/1966	95/1988	95/1952	85/1954	77/1966	69/1991
14	67/1995	72/1990	75/1990	81/1960	90/1991	95/1964	97/1992	97/1985	90/1972	83/1990	75/1989	64/1953
15	66/1995	72/1954	83/1990	90/1967	87/1957	97/1987	100/1995	100/1988	89/1957	86/1985	80/1993	61/1975
16	67/1953	70/1954	80/1945	84/1976	88/1998	97/1991	100/1983	96/1997	93/1998	84/1989	74/1965	66/1971
17	63/1990	70/1976	83/1945	88/1976	90/1986	93/1957	99/1955	95/1988	93/1991	81/1996	71/1955	65/1984
18	67/1990	69/1981	78/1989	86/1964	93/1987	93/1962	97/1953	93/1987	89/1965	78/1982	73/1953	70/1984
19	65/1951	73/1961	65/1945	92/1985	96/1962	96/1994	98/1963	94/1965	91/1983	80/1991	70/1953	57/1957
20	65/1951	66/1994	82/1945	85/1985	96/1996	94/1987	97/1991	98/1983	89/1983	85/1984	75/1985	60/1988
21	62/1974	67/1994	80/1948	89/1957	93/1996	97/1988	99/1957	93/1968	87/1965	83/1984	75/1991	58/1990
22	66/1967	68/1991	83/1948	90/1985	91/1992	100/1988	100/1957	93/1945	92/1970	83/1979	69/1953	65/1998
23	74/1967	72/1985	75/1994	85/1952	94/1964	98/1988	100/1991	95/1966	93/1970	78/1978	74/1992	65/1990
24	78/1967	75/1985	77/1994	85/2001	94/1964	96/1966	93/1987	93/1998	89/1970	81/2001	69/1979	65/1982
25	72/1967	68/1985	79/1963	89/1994	94/1991	96/1997	97/1965	97/1968	91/1970	79/2001	73/1973	70/1982
26	73/1950	64/1996	71/1986	87/1990	93/1965	98/1952	96/1963	102/1948	90/1970	80/1963	71/1946	65/1973
27	72/1974	73/1997	81/1998	94/1969	93/1991	97/1963	97/1963	98/1948	93/1998	80/1963	69/1981	67/1973
28	63/1947	68/1976	83/1945	86/1983	93/1991	106/1969	98/1999	97/1973	84/1954	81/1984	69/2001	68/1946
29	68/1975	70/1976	87/1945	92/1974	99/1969	98/1991	96/1949	96/1953	89/1945	76/1971	71/1998	75/1984
30	62/1974		87/1998	85/1985	96/1991	98/1991	96/1954	96/1973	88/1986	79/1950	72/1991	67/1990
31	63/1988		85/1979		96/1991		97/1983	97/1953		79/1950		65/1965
Month	78/1967	75/1985	87/1998*	94/1969	99/1969	106/1969	104/1966	103/2001	99/1983	90/1959	84/1950	77/1998

*Record tied in other years. Only the most recent year is given.

Atlantic City, New Jersey: Record Low Temperatures

Date	Jan.	Feb.	March	April	May	June	July	Aug.	Sept.	Oct.	Nov.	Dec.
1	5/1968	−1/1971	5/1980	12/1969	26/1978	37/1967	42/1988	53/2001	47/1991	36/1993	26/1968	6/1967
2	0/1968	−4/1961	8/1978	23/1964	32/1978	39/1966	46/1965	53/1998	44/1991	32/1997	25/1976	7/1967
3	6/1979	3/1955	10/1950	23/1995	35/1986	42/1994	48/2001	55/1998	39/1967	30/1974	19/1980	6/1976
4	3/1979	−3/1996	9/1950	24/1954	33/1976	40/1988	51/1986	51/1964	48/1976	27/1974	25/1951	6/1976
5	6/1969	−8/1996	10/1978	21/1995	32/1976	40/1976	49/1971	50/1992	44/1994	29/1965	23/1998	11/1966
6	4/1981	−3/1996	11/1978	24/1982	37/1972	44/1991	48/1979	50/1972	44/1994	27/1965	20/1967	14/1976
7	7/1988	4/1993	11/1978	20/1982	31/1980	44/1997	47/1980	50/1994	41/1976	32/1992	23/1991	10/1968
8	2/1981	3/1967	12/1960	25/1972	26/1980	42/1997	54/2000	50/1994	43/1988	33/1988	21/1993	11/1970
9	−2/1981	−2/1967	10/1996	19/1977	32/1947	41/1997	52/1969	50/1994	43/1986	29/2001	17/1967	6/1968
10	3/1982	−7/1979	5/1984	20/1985	29/1966	42/1969	53/1995	53/1972	38/1975	30/1989	22/1995	0/1968
11	−2/1982	−8/1979	9/1960	26/1997	25/1966	44/1987	55/1998	47/1972	41/1967	30/1964	21/1956	5/1968
12	−5/1981	−11/1979	11/1969	25/1976	34/1990	37/1980	51/1978	48/1968	38/1967	32/1996	24/1988	5/1988
13	−6/1981	−7/1967	16/1998	23/1990	32/1967	44/1990	49/1971	52/1987	39/1967	29/1988	22/1996	4/1982
14	1/1988	−6/1979	17/1992	22/1975	32/1994	40/1980	54/1980	50/1983	37/1975	28/1988	15/1986	2/1958
15	0/1988	10/1969	14/1993	27/1981	37/1966	44/1978	55/1999	50/1964	39/1975	30/1979	18/1986	8/1989
16	−4/1977	4/1979	15/1992	22/1981	33/1966	44/1997	54/1946	48/1972	42/1988	33/1994	16/1967	8/1968
17	−10/1977	−6/1979	17/1997	26/1980	34/1967	47/1997	53/1987	50/1980	40/1990	31/1978	14/1967	7/1951
18	−9/1977	−7/1979	12/1967	31/1990	39/1984	45/1992	53/1976	53/1954	36/1990	28/1978	21/1959	7/1980
19	−3/1994	8/1996	7/1967	28/2001	37/1992	42/1980	56/1989	50/1976	40/1990	25/1976	20/1997	8/1989
20	−4/1985	0/1966	16/1965	31/1989	31/1991	47/1956	54/1965	49/1976	34/1979	24/1992	16/1967	2/1980
21	−9/1985	3/1968	13/1965	25/1981	28/1992	44/1968	51/1966	49/1995	38/1956	26/1974	17/1967	2/1980
22	−8/1961	2/1968	11/1988	28/1988	39/1997	43/1992	53/2001	43/1969	38/1997	28/1974	16/1964	2/1980
23	4/1991	8/1963	17/1986	24/1989	40/1993	39/1992	51/1977	50/1988	36/1967	27/1997	10/1989	1/1989
24	1/1961	10/1948	19/1992	26/1989	38/1963	47/1988	54/1985	46/1971	32/1967	23/1969	11/1989	−1/1983
25	4/1963	10/1993	21/1986	31/1999	37/1956	47/1979	53/1969	47/1987	36/1967	29/1978	16/1989	−2/1983
26	3/1961	9/1990	20/1966	28/1967	40/1992	45/1979	49/1976	48/1995	36/1967	26/1987	20/1974	−2/1980
27	−1/1987	10/1990	16/2001	33/1964	36/1969	44/1965	54/1975	48/1969	35/1947	25/1988	16/1991	6/1983
28	−8/1987	8/1986	18/1990	30/1993	36/1994	47/1967	50/1977	44/1986	35/1965	26/1976	15/1996	−7/1950
29	0/1966	5/1980	22/1959	32/1992	37/1997	49/1995	53/1992	41/1986	36/1969	26/1985	16/1955	8/1977
30	−8/1952		24/1983	30/1992	39/1993	50/1988	53/1981	41/1986	32/1969	24/1967	13/1976	7/1962
31	5/1965		16/1969		36/1966		54/2001	40/1976		20/1988		4/1962
Month	−10/1977	−11/1979	5/1984*	12/1969	25/1966	37/1980*	42/1988	40/1976	32/1969*	20/1988	10/1989	−7/1950

*Record tied in other years. Only the most recent year is given.

198

NOTES

CHAPTER ONE

P. 1 Franklin quoted from *The Complete Poor Richard's Almanacks* (Franklin 1970).

Pp. 2–3 Franklin tornado figure: Bigelow 1904.

Pp. 3–4 Franklin meteorology: Abbe 1906.

P. 3 Campanius quote: Collin 1818.

P. 4 Douglass information: William Douglass to Cadwallader Colden, July 28, 1721 in Ludlum 1966a.

P. 4 Mr. De S. observations: Kirch 1737.

Pp. 4–5 Franklin's eclipse observation: Bache 1833.

Pp. 4–12 The following references were used extensively in constructing the story of the evolution of U.S. meteorology and weather observing systems: Fleming 1990, Miller 1931, Miller 1933, Bliss 1917.

P. 5 Weather observations taken by these individuals can be found in the manuscript collection of the American Philosophical Society under the heading Meteorology, Miscellaneous records, 1748–1822.

P. 5 Thomas Jefferson and weather: Randolph and Francis, 1895. Also Ludlum 1966b.

P. 5 The American Philosophical Society announcement is reproduced in Ludlum 1966a, p. 119.

P. 5 The College of Physicians charter is quoted in Fleming 1990, p. 6.

P. 5 Yellow fever epidemic: Foster, Jenkins, and Toogood 1998.

Pp. 5–6 Philadelphia medical journal: Barton 1807.

P. 6 Tilton and the Army network: Fleming 1990, pp. 13–16. Also Hagarty 1962.

P. 6 New York observing network: Fleming 1990, pp. 19–22.

Pp. 6–9 TFI involvement: Spitz 1944a.

Pp. 6–11 Espy: "Sketch of James Pollard Espy," 1888–1889.

P. 7 Espy quote: Espy 1831.

P. 7 Espy quote about simultaneous observations: Espy 1836.

Pp. 7–8 TFI and APS: American Philosophical Society 1871.

P. 8 Quote from APS/TFI circular: Espy 1834.

P. 8 First Report: Espy 1835.

Pp. 8–9 1837 funding: Spitz 1944a.

P. 9 Weather book: Peirce 1847.

Pp. 9–11 Smithsonian Network: Fleming 1990, pp. 75–86.

Pp. 10–11 Telegraphy: Abbe 1871.

P. 11 Brooks quote: J. Cecil Alter, "National Weather Service Origins," quoted in Fleming 1990, p. 143.

P. 11 Smithsonian observers around Philadelphia: *Smithsonian Institution Archives* 1849–75 (data from 1820). Also *Smithsonian Institution Annual Report* 1874.

Pp. 11–12 Early USWB history: Whitnah 1961, pp. 17–20. Also <www.nws.noaa.gov/pa/special/history/>

P. 12 Early Philadelphia USWB: Personal communication with Jason Franklin, meteorologist at National Weather Service, Boston (formerly at Mount Holly, N.J.).

Pp. 12–13 TFI involvement in the late nineteenth and early twentieth century: Spitz 1944b.

P. 13 ENIAC: <www.library.upenn.edu/special/gallery/mauchly/jwmintro.html>

P. 13 Ludlum: Robinson 1997.

P. 14 Early TV: Henson 1990, pp. 5–12.

P. 14 Ludlum article: Taylor 1984.

Pp. 14–16 Interview with Francis Davis, January 2001.

Pp. 15–16 Interview with Wally Kinnan, January 2001.

P. 16 Francis Davis, "Weather Is No Laughing Matter," quoted in Henson 1990, p. 9.

P. 16 AMS Seal of Approval: Jehn 1959.

P. 16 AMS seal numbers: American Meteorological Society 2001b, pp. 1711–17.

P. 17 Interview with Herb Clarke, January 2001.

P. 17 $2.50 per taxpayer: Personal communication with Gary Szatkowski, meteorologist-in-charge, National Weather Service, Mount Holly, N.J.

Pp. 20–24 Satellites and radar: Grenci and Nese 2001, pp. 45–53. Also <www.hq.nasa.gov/office/pao/History/weathsat.html>

CHAPTER TWO

P. 27   Franklin quoted from *The Complete Poor Richard's Almanacks* (Franklin 1970).

P. 27   Lorenz quote: Lorenz 1993, p. 77.

Pp. 27–28   Weather folklores: Fogel 1915.

P. 41   Fall line: Personal communication with Alan Cope, science and operations officer, National Weather Service, Mount Holly, N.J.

P. 44   Yogi Berra quote: <www.yogiberraclassic.org/quotes.htm>

P. 44   ENIAC: <www.library.upenn.edu/special/gallery/mauchly/jwmintro.html>

P. 44   History of numerical weather prediction: Thompson 1983.

P. 47   NWS forecast accuracy: Personal communication with Bill Lerner, National Weather Service headquarters, Silver Spring, Md.

P. 50   Ensemble Forecasting: Grenci and Nese 2001, pp. 423–24.

P. 52   Interview with Francis Davis, January 2001.

P. 52   Interview with Kenneth Spengler, February 2001.

CHAPTER THREE

P. 53   Franklin quoted from *The Complete Poor Richard's Almanacks* (Franklin 1970).

P. 54   Doesken and Judson 1997, p. 7.

P. 54   Snow removal costs: Personal communication with Cynthia Hite, Philadelphia Street Department.

P. 56   USWB quote from 1958 storm: U.S. Department of Commerce 1958.

P. 56   1958 storm map: Kocin and Uccellini 1990, pp. 104–6.

P. 60   USWB coldest day in 1934: U.S. Department of Agriculture, Weather Bureau 1934.

P. 60   Godfrey, Wilks, and Schultz 2002, available at <www.nssl.noaa.gov/~schultz/thaw/thaw.html>

P. 61   Updated wind chill: <www.noaanews.noaa.gov/stories/s720.htm>

P. 65   USWB quote from 1909 storm: U.S. Department of Agriculture, Weather Bureau 1909.

P. 66   Snowfall measurement for Blizzard of '96: Personal communication with Gary Szatkowski, meteorologist-in-charge, National Weather Service, Mount Holly, N.J.

Pp. 66, 73–74   Big snowstorms: Kocin and Uccellini 1990: pp. 8–9, 238–41.

Pp. 67–69   Interview with John Bolaris, January 2001.

Pp. 73–74   Snowstorms of 1888 and 1899: Ludlum 1983, pp. 120–29.

P. 75   Kalm observation: Peter Kalm, *Travels in North America* (London, 1770), quoted in Ludlum 1966a, p. 39.

P. 75   Isaac Norris quote: Logan manuscript in John F. Watson, *Annals of Philadelphia and Pennsylvania in the Olden Time* (Philadelphia, 1856), quoted in Ludlum 1966a, p. 41.

P. 75   Winter of 1740–41: Hazard 1828, pp. 24–25.

P. 75   Winter of 1764–65: *Pennsylvania Gazette,* March 28, 1765, quoted in Ludlum 1966a, p. 62.

P. 75   Observations of Mason and Dixon: Charles Mason's Daily Journal, manuscript in National Archives, quoted in Ludlum 1966a, p. 62.

Pp. 75–76   Winter of 1779–80, New York Harbor: Ludlum 1966a, p. 111.

P. 76   January 1780 temperatures: David Rittenhouse manuscript (Pennsylvania Historical Society), quoted in Ludlum 1966a, pp. 114–15.

P. 76   Muhlenberg quote 1779–80: The Journals of Henry Muhlenberg (Philadelphia, 1945), quoted in Ludlum 1966a, p. 113.

P. 76   Christmas Day 1776: The Writings of George Washington from the original manuscript sources 1745–1799, quoted in Ludlum 1966a, p. 98.

P. 76   Phineas Pemberton observations: Pemberton 1748–1822.

P. 76   Winter 1776–77 at Valley Forge: The Journals of Henry Muhlenberg, quoted in Ludlum 1966a, p. 102.

P. 76   Winter 1776 at Valley Forge: Christopher Ward, *The War of the Revolution* (1952), quoted in Ludlum 1966a, p. 102.

Pp. 76–77   Revolutionary War winters: Ludlum 1998.

P. 77   Winters of 1800s: Peirce 1847.

P. 77   Winter 1831: Ludlum 1968, pp. 11–13.

P. 77   Winter 1831: *Westchester Republican,* January 18, 1831, quoted in Ludlum 1968, p. 13.

P. 77   Winter 1836: Peirce 1847, pp. 26, 45.

P. 77   Winter 1836: Ludlum 1968, p. 28.

P. 77   January 1857: Ludlum 1968, p. 53.

Pp. 77–78   January 1866: American Philosophical Society 1869.

P. 78   Freezing of the Delaware River: Hazard 1828: pp. 24–26, 379–86.

P. 78   Freezing of the Delaware River: Personal communication with Walt Nickelsberg, senior service hydrologist, National Weather Service, Mount Holly, N.J.

CHAPTER FOUR

P. 79 Franklin quoted from *The Complete Poor Richard's Almanacks* (Franklin 1970).

P. 80 May 1816: Peirce 1847, p. 94.

P. 81 Late April snow: Peirce 1847, p. 82.

P. 81 May 1774: Diary of Jacob Hiltzheimer and *Germantown Zeitung,* May 19, 1774, quoted in Ludlum 1966a, p. 82.

P. 81 May 1803: Rittenhouse manuscripts of meteorological observations, quoted in Ludlum 1966a, p. 167.

Pp. 83–84 Nor'easters: Davis and Dolan 1993.

P. 84 '62 nor'easter: Cooperman and Rosendal 1962. Also the University of Delaware, Department of Geography, Delaware Coastal Management Program 1977.

Pp. 84–85 1920 storm and March 1962 storm: <www.nealcomm.com/rlw/northeas.htm>

Pp. 85–86 The "Perfect Storm": Davis and Dolan 1992, pp. 978–83. Also <www.noaanews.noaa.gov/stories/s444.htm> and <www.ncdc.noaa.gov/ol/satellite/satelliteeye/cyclones/pfctstorm91/pfctstorm.html>

P. 86 Bob Case story: Personal communication with Walter Drag, meteorologist, National Weather Service, Boston.

P. 87 Early tornado forecasting: <www.spc.noaa.gov/history/early.html> and <www.nssl.noaa.gov/~trapp/symposium-BAMS.html>

P. 89 Squall lines: Grenci and Nese 2001, p. 366.

P. 90 Lightning and flash floods: <www.esig.ucar.edu/socasp/weather1/golden.html>

Pp. 90–91 Lightning: Grenci and Nese 2001, pp. 201–2.

P. 92 Lightning, September 2000: Personal communication with Gary Szatkowski, meteorologist-in-charge, National Weather Service, Mount Holly, N.J.

P. 93 Windstorm in May 1933: U.S. Department of Agriculture, Weather Bureau 1933.

Pp. 93–94 Colonial hailstorm: *New American Magazine,* May 25, 1758, quoted in Ludlum 1983, pp. 170–71.

Pp. 96–97 Pennsylvania tornadoes: Forbes and Nese 1998.

P. 97 Early tornado definition: Ludlum 1970, Foreword.

P. 98 First tornado: *American Weekly Mercury* (Philadelphia), August 13, 1724, quoted in Ludlum 1970, p. 37.

P. 98 Franklin dust devil: *The Papers of Benjamin Franklin,* quoted in Ludlum 1970, pp. 38–39.

P. 98 1835 tornado: Bache 1837.

Pp. 98–99 1885 tornado: "Whirling Winds, a Tornado Visits Philadelphia and Camden" 1885.

P. 100 Delaware watershed: <www.state.nj.us/drbc/thedrb.htm>

P. 101 Mace Michaels story: "Viewers Irate after WFLA's Major Gaffe" 2000.

P. 102 Historic Delaware River floods: Ludlum 1983, pp. 181–83.

Pp. 102–3 January 1996 flood: <marfchp1.met.psu.edu/Flood/jan96.html>

CHAPTER FIVE

P. 105 Winchell quote: <members.aol.com/rogertiii/quotes.htm>

P. 108 Heat warning system: Kalkstein, Jamason, Greene, Libby, and Robinson 1996, available in part at <www.udel.edu/SynClim/phl.html>

P. 110 Hottest day 1918: U.S. Department of Agriculture, Weather Bureau 1918.

Pp. 110–11 Year without a summer: Peirce 1847: pp. 116, 134, 157.

Pp. 111–12 1843 flood: Y. S. Walter, "Report of a Committee of the Delaware Institute of Science on the Great Rain Storm and Flood of August 1843," quoted in Ludlum 1970, pp. 61–63.

P. 112 Drought quote: Peirce 1847, pp. 171, 192.

P. 112 Historical drought: Ludlum 1983, p. 192.

Pp. 112–13 Salt line: <www.state.nj.us/drbc/salt.htm>

P. 114 DRBC drought publication: Hogarty 1970.

P. 115 Particulates: Kaiser 2000.

P. 115 Tombstone weathering: Feddema and Meierding 1987.

P. 115 Donora disaster: <www.dep.state.pa.us/dep/Rachel_Carson/donora.htm>

P. 116 Philadelphia pollution: City of Philadelphia, Department of Public Health 1989.

P. 117 Ozone and health: <www.epa.gov/oar/urbanair/ozone/index.html>

Pp. 118–19 Good ozone: Grenci and Nese 2001, 507–12.

P. 120 Cape May 1802: Pilkey and Dixon 1998, p. 1.

P. 121 Franklin waves quote: Letter from Franklin to William Brownrigg, November 7, 1773, Willcox 1976, p. 468.

P. 121 Littoral drift: Personal communication with Tony Pratt, Delaware Department of Natural Resources and Environmental Control (DNREC).

P. 122 Psuty and sea–level rise: <marine.rutgers.edu/geomorph/slr.htm>

P. 122 Barnegat: <bigfoot.wes.army.mil/c803.html>

P. 122 Cape May jetties: Kaufman and Pilkey 1979, p. 166.

Pp. 122–23 Newjerseyization: Pilkey and Dixon 1998, pp. 1, 42–43.

Pp. 122–25 Interview with Richard Weggel, November 2000.

Pp. 123–24 Seawalls: Pilkey and Dixon 1998, pp. 38–45.

P. 124 Beach replenishment projects: <www.nap.usace.army.mil/cenap–dp/projects/nj_projects.htm> and <www.nap.usace.army.mil/cenap–dp/projects/de_projects.htm>

P. 124 Duke study on replenishment: <www.geo.duke.edu/Research/psds/cost.html> and <www.geo.duke.edu/Research/psds/psds_skidaway.htm>

P. 124 Pilkey and replenishment: Pilkey and Dixon 1998, pp. 8, 89.

Pp. 124–25 *Inquirer* article: "Land Rush at the Beach" 2000, available at <www.philly.com/specials/2000/coast/>

P. 125 FEMA study: <www.heinzcenter.org/publications/erosion/>

P. 125 Pilkey quote: University of North Carolina Center for Public Television 1990.

Pp. 126–27 Agnes: Romanelli and Griffith (1972). Also <www.publicaffairs.noaa.gov/pr97/jun97/noaa97-r228.html> and <marfchp1.met.psu.edu/Flood/agnes.html>

Pp. 126–28 Interview with Herb Clarke, January 2001.

P. 128 Helicopter crash: "Copter Pilot, 3 Newsmen Die in Crash," 1972.

CHAPTER SIX

P. 129 Franklin quoted from *The Complete Poor Richard's Almanacks* (Franklin 1970).

P. 130 Drought quote: Peirce 1847, pp. 171, 192.

P. 130 Drought quote: Hazard 1828, p. 381.

P. 133 Hurricane naming: <www.aoml.noaa.gov/general/lib/reason.html>

P. 134 New wind study: <www.nhc.noaa.gov/aboutwindprofile.html>

P. 136 Inland flooding: <www.nhc.noaa.gov/HAW/day3/inland_flooding.htm>

Pp. 136–37 Columbus: Ludlum 1963, pp. 1–8.

P. 137 Hurricane forecast accuracy: Max Mayfield, director of the National Hurricane Center, in a talk at the National Hurricane Center conference, Washington, D.C., April 2001

P. 137 Hurricane forecast models: <www.nhc.noaa.gov/aboutmodels.html>

P. 137 Costs of evacuations: Office of the Federal Coordinator for Meteorological Services and Supporting Research, quoted in Marks, Shay, and PDT-5 1998.

Pp. 137–38 Gray forecasting techniques: <typhoon.atmos.colostate.edu/forecasts/index.html>

P. 139 Delaware Bay storm surge: Interview with Max Mayfield, director of the National Hurricane Center, May 2001.

P. 139 Reverse laning and "plan B": Mike Augustyniak, Hurricane Program coordinator for New Jersey, in a talk at the National Hurricane Center conference, Washington, D.C., April 2001.

P. 140 New Jersey risk: Jack Beven, National Hurricane Center, in a talk at the National Hurricane Conference, Washington, D.C., April 2001.

P. 140 Nor'easter high tides: <www.udel.edu/dgs/breakwatertides.html>

P. 140 Hurricane 1743: Ludlum 1963, pp. 22–23.

P. 140 Hurricane 1775: Phineas Pemberton, quoted in Ludlum 1963, pp. 26–27.

Pp. 140–41 Hurricane 1788: *Pennsylvania Journal*, August 27, 1788, quoted in Ludlum 1963, pp. 31–34.

P. 141 Hurricane 1821: Charles Tomlin, *Cape May Spray* (Philadelphia, 1913); and *American Beacon*, September 13, 1821, quoted in Ludlum 1963, pp. 83–84. Also Ludlum 1983, pp. 92–96.

P. 141 Hurricane 1846: *Daily Advertiser*, October 14, 1846, quoted in Ludlum 1963, p. 94.

P. 141 1878 "tornado": "A Terrific Tornado," 1878. Also, Ludlum 1983, p. 99.

Pp. 141–42 1903, 1933, 1944 Hurricanes: Ludlum 1983, pp. 96–101. Also <www.nhc.noaa.gov/HAW/basics/historic_storms.htm#great>

Pp. 142–43 Hurricane Hazel: Anderson 1954. Also <www.nhc.noaa.gov/HAW/basics/historic_storms.htm#hazel>

Pp. 143–44 Connie and Diane: Anderson 1955. Also <www.nhc.noaa.gov/HAW/basics/historic_storms.htm#connie>

P. 144 1971 Doria: Ludlum 1983, p. 101.

P. 144 Agnes: "Agnes Floods State with Debris, 13 Dead" 1972. Also <www.nhc.noaa.gov/HAW/basics/historic_storms.htm#agnes>

Pp. 145–46 Floyd: <www.nhc.noaa.gov/HAW/basics/historic_storms.htm#floyd>. Also <www.nhc.noaa.gov/1999floyd.html> and <ftp://ftp.nws.noaa.gov/om/floyd.pdf>

P. 151 Early snow 1703: Watson, *Annals of Philadelphia and Pennsylvania in the Olden Time*, quoted in Ludlum 1966a, p. 80.

P. 151 Early snow 1769: Diary of Benjamin Horner, *American Daily Advertiser*, quoted in Ludlum 1966a, p. 81.

Pp. 152–53 Indian summer: <www.crh.noaa.gov/dtx/i-summer.htm>

CHAPTER SEVEN

P. 155   Warner quote: From an editorial in the *Hartford Courant,* August 24, 1897, quoted in Bartlett 1980, p. 603.

P. 156   AMS Policy: American Meteorological Society 2001a, pp. 701–3, also available on the Web at <www.ametsoc.org/AMS/policy>

Pp. 157–58   El Niño/La Niña: Grenci and Nese 2001, pp. 245–49. Also <www.elnino.noaa.gov/ and www.ogp.noaa.gov/enso/>

P. 161   El Niño and Hurricanes: Bove, Elsner, Landsea, Niu, O'Brien 1998, also available at <www.coaps.fsu.edu/~bove/hurr.shtml>

P. 161   NOAA scientist: Personal communication with Wayne Higgins, senior research meteorologist, Climate Prediction Center.

Pp. 161–62   North Atlantic Oscillation: <www.ldeo.columbia.edu/NAO/> and <www.cpc.ncep.noaa.gov/data/teledoc/nao.html>

P. 163   Coastline changes: <pubs.usgs.gov/circular/c1075/change.html#fig7>

P. 163   Boulder field: <www.dcnr.state.pa.us/stateparks/parks/hickory.htm>

P. 163   Past climates: Bradley 2000.

P. 164   Mahlman Senate testimony: <www.legislative.noaa.gov/mahlmantst0517.htm>

Pp. 164–65   Current CO_2 levels: Intergovernmental Panel on Climate Change 2001, also available at <www.usgcrp.gov/ipcc/wg1spm.pdf>

P. 166   Interview with Jerry Mahlman, January 2001.

Pp. 166–67   Global temperature graph: Jones, Parker, Osborn, and Briffa 2001 also available at <cdiac.esd.ornl.gov/trends/temp/jonescru/jones.html>

Pp. 166–67   Ocean temperature increase: Kerr 2000.

P. 167   Glacier loss: World Glacier Monitoring Service <www.geo.unizh.ch/wgms/>

P. 167   Other trends: Magnuson, Robertson, and Benson 2000.

P. 168   Mahlman quote, 95 percent sure: Interview with Jerry Mahlman.

Pp. 168–69   IPCC projections: Intergovernmental Panel on Climate Change 2001.

P. 169   GFDL model: <www.gfdl.noaa.gov/~tk/climate_dynamics/climate_impact_webpage.html>

Pp. 169–70   Mahlman categorization: Interview with Jerry Mahlman.

P. 171   Mahlman huge snowstorms: Interview with Jerry Mahlman.

P. 171   Kalkstein quote: <www.grida.no/climate/ipcc/regional/230.htm>

Pp. 171–72   U.S. National Assessment and Mid–Atlantic Regional Assessment: <www.nacc.usgcrp.gov/> and <www.essc.psu.edu/mara/results/overview_report/index.html>

P. 172   Sea–level rise maps: Titus and Richman 2001, available at <www.epa.gov/globalwarming/publications/impacts/sealevel/maps/maps.html>

P. 172   Global warming and hurricanes: <www.gfdl.gov/~rt/glob_warm_hurr.html>

BIBLIOGRAPHY
Print and Video Sources

Abbe, Cleveland. 1871. "Historical Notes on the Systems of Weather Telegraphy, and Especially Their Development in the United States." *American Journal of Sciences and Arts,* 3d ser., 2: 81–88.

———. 1906. "Benjamin Franklin as Meteorologist." *Proceedings of the American Philosophical Society* 45: 117–28.

"Agnes Floods State with Debris, 13 Dead." 1972. *Philadelphia Daily News,* June 23, p. 3.

American Meteorological Society. 2001a. "Statement on Seasonal to Interannual Climate Prediction." *Bulletin of the American Meteorological Society* 82: 701–3.

———. 2001b. "Television Seal of Approval." *Bulletin of the American Meteorological Society* 82: 1711–17.

American Philosophical Society. 1869. *Proceedings of the American Philosophical Society* 10: 329.

———. 1871. "The Part Taken by the American Philosophical Society and The Franklin Institute in Establishing Stations for Meteorological Observations." *Proceedings of the American Philosophical Society* 11: 516–20.

Anderson, G. C. 1954. "Highlights on Hazel." Memorandum to the chief of the U.S. Weather Bureau. Wilmington, Del., October 26.

———. 1955. "Report on Hurricane Connie." Memorandum to the chief of the U.S. Weather Bureau. Wilmington, Del., August 15.

"A Terrific Tornado." 1878. *The Press* (Philadelphia), October 24, p. 1.

Bache, Alexander D. 1833. "Attempt to Fix the Date of Dr. Franklin's Observations." *Journal of The Franklin Institute,* n.s., 12: 300–303.

———. 1837. "Notes and Diagrams, Illustrative of the Directions of the Forces Acting at or Near the Surface of the Earth, in Different Parts of the Brunswick Tornado of June 19, 1835." *Transactions of the American Philosophical Society,* n.s., 5: 407–20.

Barton, B. S. 1807. "Discourse on Natural History." *Journal of the Philadelphia Medical Society* 20: 61.

Bigelow, John. 1904. *The Works of Benjamin Franklin.* Volumes I and II. New York and London: G. P. Putnam's Sons, The Knickerbocker Press.

Bliss, George. 1911. *Climatology of Philadelphia, Pa.,* Philadelphia, Pa.: U.S. Department of Agriculture, Weather Bureau.

———. 1917. "The Weather Business: A History of Weather Records, and the Work of the U.S. Weather Bureau." *Scientific American Supplement,* no. 2172 (August 18): 110–11.

Bove, Mark C., James B. Elsner, Chris W. Landsea, Xufeng Niu, and James J. O'Brien. 1998. "Effect of El Niño on U.S. Landfalling Hurricanes, Revisited." *Bulletin of the American Meteorological Society* 79: 2477–82.

Bradley, Ray. 2000. "1,000 Years of Climate Change." *Science* 288 (May 26): 1353–54.

City of Philadelphia, Department of Public Health, Environmental Protection Division, Air Management Services. 1989. "A Brief History of Philadelphia Air Management Services" and "Analysis of Philadelphia's Air Pollution Emissions/Air Quality Trends."

Collin, Nicholas. 1818. "Observations Made at an Early Period on the Climate of the Country Along the River Delaware, Collected from the Records of the Swedish Colony." *Transactions of the American Philosophical Society,* n.s., 1: 340–52.

Cooperman, A. I., and H. E. Rosendal. 1962. "Great Atlantic Coast Storm, 1962." *Mariner's Weather Log* 6, no. 3: 79–85.

"Copter Pilot, 3 Newsmen Die in Crash." 1972. *Philadelphia Inquirer,* June 27, p. 1.

Davis, Robert E., and Robert Dolan. 1992. "The All Hallows' Eve Coastal Storm—October 1991." *Journal of Coastal Research* 8, no. 4: 978–83.

———. 1993. "Nor'easters." *American Scientist* 81: 428–39.

Doesken, Nolan J., and Arthur Judson. 1997. *The Snow Booklet: A Guide to the Science, Climatology, and Measurement of Snow in the United States.* Fort Collins, Colo.: Colorado State University, Department of Atmospheric Science.

Dolan, Robert, and Robert E. Davis. 1992. "An Intensity Scale for Atlantic Coast Northeast Storms." *Journal of Coastal Research* 8, no.4: 840–53.

"Editorial." 1897. *Hartford Courant.* August 24. Quoted in *Bartlett's Familiar Quotations,* 15th edition, edited by Emily Morison Bech (Boston: Little, Brown, and Co., 1980).

Espy, James P. 1831. "Observations on the Importance of Meteorological Observations." *Journal of The Franklin Institute,* n.s., 8: 389–406.

———. 1834. "Circular in Relation to Meteorological Observations, by the Joint Committee of the American Philosophical Society and The Franklin Institute." *Journal of The Franklin Institute,* n.s., 14: 382–85.

———. 1835. "First Report of the Joint Committee on Meteorology of the American Philosophical Society and The Franklin Institute." *Journal of The Franklin Institute,* n.s., 16: 4–6.

———. 1836. "Second Report of the Joint Committee on Meteorology of the American Philosophical Society and The Franklin Institute." *Journal of The Franklin Institute,* n.s., 17: 386–91.

———. 1841. *The Philosophy of Storms.* Boston: C. C. Little and J. Brown.

Feddema, J. J., and T. C. Meierding. 1987. "Marble Weathering and Air Pollution in Philadelphia." *Atmospheric Environment* 21, no.1: 143–57.

Fleming, James R. 1990. *Meteorology in America, 1800–1870.* Baltimore, Md.: Johns Hopkins University Press.

Fogel, Edwin M. 1915. *Beliefs and Superstitions of the Pennsylvania Germans.* Philadelphia: American Germanica Press.

Forbes, Greg, and Jon Nese. 1998. "An Updated Tornado Climatology of Pennsylvania: Methodology and Uncertainties." *Journal of the Pennsylvania Academy of Science* 71, no. 3: 113–24.

Foster, Kenneth R., Mary F. Jenkins, and Anna Coxe Toogood. 1998. "The Philadelphia Yellow Fever Epidemic of 1793." *Scientific American* 279, no.2: 88–93.

Franklin, Benjamin. 1970. *The Complete Poor Richard Almanacks.* Reproduced in facsimile with an introduction by Whitfield J. Bell, Jr. Volumes I and II. Barre, Mass. Imprint Society.

Godfrey, Christopher M., Daniel S. Wilks, and David M. Schultz. 2002. "Is the January Thaw a Statistical Phantom?" *Bulletin of the American Meteorological Society* 83, no. 2.

Grenci, Lee M., and Jon M. Nese. 2001. *A World of Weather: Fundamentals of Meteorology.* 3d ed. Dubuque, Iowa: Kendall-Hunt Publishing.

Hagarty, J. H. 1962. "Dr. James Tilton, 1745–1822." *Weatherwise* 15: 124–25.

Hazard, Samuel. 1828. "Effects of Climate on Navigation." *Hazard's Register of Pennsylvania,* pp. 23–26, 379–86.

Henson, Robert. 1990. *Television Weathercasting: A History.* Jefferson, N.C.: McFarland & Company.

Hogarty, Richard A. 1970. *The Delaware River Drought Emergency, Inter-University Case Program #107.* Indianapolis and New York: The Bobbs-Merril Co.

Intergovernmental Panel on Climate Change. 2001. "Summary for Policymakers, Climate Change 2001: The Scientific Basis—A Report of Working Group I." Geneva, Switzerland: IPCC.

Jehn, K. H. 1959. "Recognition of Competence in Weathercasting—The AMS Seal of Approval Program." *Bulletin of the American Meteorological Society* 40: 85–88.

Jones, P. D., D. E. Parker, T. J. Osborn, and K. R. Briffa. 2001. "Global and Hemispheric Temperature Anomalies—Land and Marine Instrumental Records." In *Trends: A Compendium of Data on Global Change.* Oak Ridge, Tenn.: Carbon Dioxide Information Analysis Center, Oak Ridge National Laboratory, U.S. Department of Energy.

Kaiser, Jocelyn. 2000. "Evidence Mounts that Tiny Particles Can Kill." *Science* 289 (July 7): 22–23.

Kalkstein, L. S., P. F. Jamason, J. S. Greene, J. Libby, and L. Robinson. 1996. "The Philadelphia Hot Weather Health Watch/Warning System." *Bulletin of the American Meteorological Society* 77: 1519–28.

Kaufman, Wallace, and Orrin Pilkey. 1979. *The Beaches Are Moving: The Drowning of America's Shoreline.* Garden City, N.Y.: Anchor Press/Doubleday.

Kerr, Richard. 2000. "Globe's 'Missing Warming' Found in the Ocean." *Science* 287 (March 24): 2126–27.

Kirch, C. 1737. "Obfervationes Quaedam Meteorologicae in Pennsylvania Habitae," *Miscellanea Berolinensia* 5: 123–31.

Kocin, Paul J., and Louis W. Uccellini. 1990. *Snowstorms Along the Northeastern Coast of the United States, 1955 to 1985.* Boston: American Meteorological Society.

"Land Rush at the Beach." 2000. *Philadelphia Inquirer,* March 5, p. A25.

Lorenz, Edward N. 1993. *The Essence of Chaos.* Seattle: University of Washington Press.

Ludlum, David M. 1963. *Early American Hurricanes, 1492–1870.* Boston: American Meteorological Society.

———. 1966a. *Early American Winters, 1604–1820.* Vol 1. Boston: American Meteorological Society.

———. 1966b. "Thomas Jefferson and the American Climate." *Bulletin of the American Meteorological Society* 47: 974–75.

———. 1968. *Early American Winters II, 1821–1870.* Boston: American Meteorological Society.

———. 1970. *Early American Tornadoes, 1586–1870.* Boston: American Meteorological Society.

———. 1982. *The American Weather Book.* Boston: Houghton Mifflin.

———. 1983. *The New Jersey Weather Book.* New Brunswick, N.J.: Rutgers University Press.

———. 1998. "The Weather of Independence." *Weatherwise* 51: 38–44.

Magnuson, John J., Dale M. Robertson, and Barbara J. Benson. 2000. "Historical Trends in Lake and River Ice Cover in the Northern Hemisphere." *Science* 289 (September 8): 1743–45.

Marks, Frank, Lynn K. Shay, and PDT-5. 1998. "Landfalling Tropical Cyclones: Forecast Problems and Associated Research Opportunities." *Bulletin of the American Meteorological Society* 79: 305–23.

Miller, Eric R. 1931. "The Evolution of Meteorological Institutions in the United States." *Monthly Weather Review* 59, no.1: 1–6.

———. 1933. "American Pioneers in Meteorology." *Monthly Weather Review* 61, no.7: 189–93.

Peirce, Charles. 1847. *A Meteorological Account of the Weather in Philadelphia, From January 1, 1790, to January 1, 1847, Including Fifty-seven Years.* Philadelphia: Lindsay & Blakiston.

Pemberton, Phineas. 1748–1822. "Observations." Manuscript collection of the American Philosophical Society, Meteorology, Miscellaneous Records.

Pilkey, Orrin H., and Katherine L. Dixon. 1998. *The Corps and the Shore.* Washington, D.C.: Island Press.

Randolph, Frederick, and Frederick Francis. 1895. "Thomas Jefferson as Meteorologist." *Monthly Weather Review* 23: 456–58.

Robinson, David A. 1997. "Necrology of David M. Ludlum." *Bulletin of the American Meteorological Society* 78: 2260–61.

Romanelli, Carl J., and William M. Griffith. 1972. *The Wrath of Agnes: A Complete Pictorial and Written History of the June 1972 Flood in Wyoming Valley.* Wilkes-Barre, Pa.: Media Affiliates.

Savadove, Larry, and Margaret Thomas Buchholz. 1993. *Great Storms of the Jersey Shore.* Surf City, N.J.: Down the Shore Publications.

"Sketch of James Pollard Espy." 1888–1889. *Popular Science Monthly* 34: 834–40.

Smithsonian Institution Annual Report. 1874.

Smithsonian Institution Archives. 1849–1875. Record Unit 60, Meteorological Project, Boxes 13, 14, 16 (data from 1820).

Spitz, Armand N. 1944a. "Meteorology in The Franklin Institute, Part I." *Journal of The Franklin Institute* 237, no. 4: 271–87.

———. 1944b. "Meteorology in The Franklin Institute, Part II." *Journal of The Franklin Institute* 237, no. 5: 331–57.

Taylor, Edward F. 1984. "Behind the Byline, David Ludlum." *Weatherwise* 37: 309–13.

Thompson, P. D. 1983. "A History of Numerical Weather Prediction in the United States." *Bulletin of the American Meteorological Society* 64: 755–69.

Titus, J. G., and C. Richman. 2001. "Maps of Lands Vulnerable to Sea Level Rise: Modeled Elevations along the U.S. Atlantic and Gulf Coasts." *Climate Research* 18, no. 3: 205–28.

University of Delaware, Department of Geography, Delaware Coastal Management Program. 1977. "Coastal Storm Damage, 1923–1974." Technical Report, no. 4, Newark, Del. (September).

University of North Carolina Center for Public Television. 1990. *The Beaches Are Moving.* Videocassette, 59 minutes. Port Royal, S.C.: Environmental Media. .

U.S. Department of Agriculture, Weather Bureau. 1909. Daily Local Record for Philadelphia, Pa., December 25, Form no. 1014-Met'l.

———. 1918. Daily Local Record for Philadelphia, Pa., August 7, Form no. 1014-Met'l.

———. 1933. Daily Local Record for Philadelphia, Pa., May 24, Form no. 1014-Met'l.

———. 1934. Daily Local Record for Philadelphia, Pa., February 9, Form no. 1014-Met'l.

U.S. Department of Commerce, Weather Bureau. 1958. Meteorological Record, Surface Weather Observations, Philadelphia, Pa., Form WBAN 10–B.

"Viewers Irate after WFLA's Major Gaffe." 2000. *Tampa Tribune,* June 19, p. 15.

"Whirling Winds, a Tornado Visits Philadelphia and Camden." 1885. *Philadelphia Inquirer,* August 4, p. 1.

Whitnah, Donald R. 1961. *A History of the United State Weather Bureau.* Urbana, Ill.: University of Illinois Press.

William B. Willcox, ed. 1976. "Letter from Franklin to William Brownrigg, November 7, 1773." In *The Papers of Benjamin Franklin, January 1 through December 31, 1773.* New Haven, Conn.: Yale University Press.

Web Sources

bigfoot.wes.army.mil/c803.html

cdiac.esd.ornl.gov/trends/temp/jonescru/jones.html

ftp://ftp.nws.noaa.gov/om/floyd.pdf

marfchp1.met.psu.edu/Flood/agnes.html

marfchp1.met.psu.edu/Flood/jan96.html

marine.rutgers.edu/geomorph/slr.htm

members.aol.com/rogertiii/quotes.htm

pubs.usgs.gov/circular/c1075/change.html#fig7

typhoon.atmos.colostate.edu/forecasts/index.html

www.aoml.noaa.gov/general/lib/reason.html

www.cpc.ncep.noaa.gov/data/teledoc/nao.html

www.crh.noaa.gov/dtx/i-summer.htm

www.dcnr.state.pa.us/stateparks/parks/hickory.htm

www.dep.state.pa.us/dep/Rachel_Carson/donora.htm

www.elnino.noaa.gov/

www.epa.gov/oar/urbanair/ozone/index.html

www.esig.ucar.edu/socasp/weather1/golden.html

www.essc.psu.edu/mara/results/overview_report/index.html

www.geo.duke.edu/Research/psds/cost.html

www.geo.duke.edu/Research/psds/psds_skidaway.htm

www.geo.unizh.ch/wgms/

www.gfdl.gov/~rt/glob_warm_hurr.html

www.gfdl.noaa.gov/~tk/climate_dynamics/
climate_impact_webpage.html

www.grida.no/climate/ipcc/regional/230.htm

www.heinzcenter.org/publications/erosion/

www.hq.nasa.gov/office/pao/History/weathsat.html.

www.ldeo.columbia.edu/NAO/

www.legislative.noaa.gov/mahlmantst0517.htm

www.library.upenn.edu/special/gallery/mauchly/
jwmintro.html

www.nacc.usgcrp.gov/

www.nap.usace.army.mil/cenap-dp/projects/de_projects.htm

www.nap.usace.army.mil/cenap-dp/projects/nj_projects.htm

www.ncdc.noaa.gov/ol/satellite/satelliteseye/cyclones/
pfctstorm91/pfctstorm.html

www.nealcomm.com/rlw/northeas.htm

www.nhc.noaa.gov/1999floyd.html

www.nhc.noaa.gov/aboutmodels.html

www.nhc.noaa.gov/aboutwindprofile.html

www.nhc.noaa.gov/HAW/basics/historic_storms.htm#agnes

www.nhc.noaa.gov/HAW/basics/historic_storms.
htm#connie

www.nhc.noaa.gov/HAW/basics/historic_storms.htm#floyd

www.nhc.noaa.gov/HAW/basics/historic_storms.htm#great

www.nhc.noaa.gov/HAW/basics/historic_storms.htm#hazel

www.nhc.noaa.gov/HAW/day3/inland_flooding.htm

www.noaanews.noaa.gov/stories/s444.htm

www.noaanews.noaa.gov/stories/s720.htm

www.nssl.noaa.gov/~trapp/symposium-BAMS.html

www.nws.noaa.gov/pa/special/history/

www.ogp.noaa.gov/enso/

www.publicaffairs.noaa.gov/pr97/jun97/noaa97-r228.html

www.spc.noaa.gov/history/early.html

www.state.nj.us/drbc/salt.htm

www.state.nj.us/drbc/thedrb.htm

www.udel.edu/dgs/breakwatertides.html

www.yogiberraclassic.org/quotes.htm

ADDITIONAL WEB RESOURCES

These web sites were active as of January 2002.

CHAPTER ONE
History of Weather Science and Observing in the Philadelphia Area

History of Weather Science and National Weather Service

www.nws.noaa.gov/er/gyx/timeline.html
www.nws.noaa.gov/pa/special/history/

Observation Systems and Weather Instruments

www.nws.noaa.gov/asos/
www.ems.psu.edu/wx/instrument.html

Modern National Weather Service

www.nws.noaa.gov
www.nws.noaa.gov/er/phi/

Recent Satellite Imagery

www.ghcc.msfc.nasa.gov/GOES/
www.goes.noaa.gov/
www.nws.mbay.net/satl.html
www.rap.ucar.edu/weather/satellite/
cimss.ssec.wisc.edu/goes/realtime/realtime.html
www.nnvl.noaa.gov/

Recent Radar Imagery

www.nws.noaa.gov/er/phi/radar.htm
www.srh.noaa.gov/elp/dop/dop.html
weather.noaa.gov/radar/national.html
weather.noaa.gov/radar/mosaic/DS.p19r0/
     ar.us.conus.shtml

Historical Satellite and Radar Images

www.ncdc.noaa.gov/pub/data/images
www.osei.noaa.gov/
rsd.gsfc.nasa.gov/goes/
orbit-net.nesdis.noaa.gov/arad/fpdt/pix/pix.html
www5.ncdc.noaa.gov/cgi-bin/hsei/
     hsei.pl?directive=welcome
www4.ncdc.noaa.gov/cgi-win/
     wwcgi.dll?WWNEXRAD~Images2
www.rap.ucar.edu/staff/pneilley/NIDS_archives.html

Satellite Tutorials and Information

cimss.ssec.wisc.edu/
noaasis.noaa.gov/NOAASIS/ml/genlsatl.html

Remote Sensing

seawifs.gsfc.nasa.gov/SEAWIFS.html
visibleearth.nasa.gov/

CHAPTER TWO
Basics of Weather and Weather Forecasting

Sunrise/sunset/season information

aa.usno.navy.mil/data/
www.srrb.noaa.gov/highlights/sunrise/gen.html

Maps of Pressure and Other Variables

www.ems.psu.edu/wx/usstats/pressstats.html
weather.unisys.com/surface/contour.html
www.aos.wisc.edu/weather/

Coriolis Effect

www.ems.psu.edu/~fraser/Bad/BadCoriolis.html
www,staff.fi.edu/~jnese/coriolis.html

Real-time and Historical Upper-air Maps

www.rap.ucar.edu/weather/upper/
weather.unisys.com/upper_air/index.html
weather.noaa.gov/fax/nwsfaxa.shtml
www.cpc.ncep.noaa.gov/products/intraseasonal/
    z200anim.html
virga.sfsu.edu/crws/jetstream.html

Cloud Catalogs

vortex.plymouth.edu/cloud.html
asd-www.larc.nasa.gov/SCOOL/cldchart.html
www.scienceclass.com/dayscape/pages/main.htm
www.photolib.noaa.gov/historic/nws/index.html

"Normals" and Departure from "Normal"

www.ncdc.noaa.gov/ol/climate/research/cag3/
    cag3.html
www.cdc.noaa.gov/USclimate/states.slow.html
www.worldclimate.com
www.washingtonpost.com/wp-srv/weather/
    historical/historical.htm

Philadelphia-area Climate

www.nws.noaa.gov/er/phi/clidat.htm
www.fi.edu/weather/data2/
www.nws.noaa.gov/er/box/clstns.htm
pasc5.met.psu.edu/PA_Climatologist/
climate.rutgers.edu/stateclim/
www.udel.edu/leathers/stclim.html
www.ncdc.noaa.gov/

General Weather Glossaries and Terminology

www.wrh.noaa.gov/Monterey/guide.html
k12.ocs.ou.edu/teachers/glossary/
www.crh.noaa.gov/hsd/hydefa-c.html
www.wrh.noaa.gov/Monterey/glossary.html

Weather Education Resources

www.ucar.edu/40th/webweather/
profhorn.meteor.wisc.edu/wxwise/
www.stormtrack.org/links/education.htm
met-www.cit.cornell.edu/education.html
www.nws.noaa.gov/om/edures.htm
www.nws.noaa.gov/om/educ/activitactivit.htm

Computer Model Forecast Maps

www.ems.psu.edu/wx/newx.html
members.home.net/cheric/models.htm
www.ems.psu.edu/wx/etats.html
www.aos.wisc.edu/weather/content/computer.html
weatheroffice.ec.gc.ca/charts/index_e.html
weather.cod.edu/forecast/
www.wx.rutgers.edu/models_menu.htm
twister.sbs.ohio-state.edu/models.html

Decoding Computer Model Text Forecasts

twister.sbs.ohio-state.edu/helpdocs/fous_help.html
snowfall.envsci.rutgers.edu/clubs/metclub/
    fouscode.html
www.nws.noaa.gov/tdl/synop/fwcexpln.htm
www.nws.noaa.gov/tdl/synop/avnexpln.htm
www.nws.noaa.gov/tdl/synop/purplemex.htm
www.nws.noaa.gov/tdl/synop/redmav.htm

One- to Two-Week Forecasts

www.hpc.ncep.noaa.gov/medr/medr.html
www.cpc.ncep.noaa.gov/products/predictions/
    814day/

Ensemble Forecasts and Information

www.met.utah.edu/jhorel/html/models/
    model_ens.html
weatheroffice.ec.gc.ca/ensemble/index_e.html
eyewall.met.psu.edu/super/index.html
www.cdc.noaa.gov/map/images/ens/ens.html
sgi62.wwb.noaa.gov:8080/ens/info/ens_detbak.html

CHAPTER THREE
Winter

Winter Weather Information

www.nws.noaa.gov/om/winter/index.shtml

www.nws.noaa.gov/om/wntrstm.htm

www.ncdc.noaa.gov/publications/blizzard96.html

About Snow and Wind Chill

www.its.caltech.edu/~atomic/snowcrystals/

nsidc.org/NSIDC/EDUCATION/SNOW/snow_FAQ.html

www.noaanews.noaa.gov/stories/s720.htm

Winter Flooding

www.nws.noaa.gov/oh/hrl/surveys/flood96/
    FL96cver.htm

Lake-effect Snow

www.comet.ucar.edu/class/smfaculty/byrd/

CHAPTER FOUR
Spring

Nor'easters

home.att.net/~noreaster909/pages/s1962.htm

www.intellicast.com/DrDewpoint/Library/1122/
    content.shtml

www.nhc.noaa.gov/1991unnamed.html#FIG1

Severe Weather Forecasting and Research

www.spc.noaa.gov/

www.nssl.noaa.gov/

Squall Lines and Microbursts

www.ems.psu.edu/~young/rozindex.htm

www.crh.noaa.gov/lmk/soo/docu/bowecho.htm

www.nssl.noaa.gov/~doswell/microbursts/
    Handbook.html

Severe Weather Databases

www.nssl.noaa.gov/hazard/

www.spc.noaa.gov/archive/

www.spc.noaa.gov/climo/

www4.ncdc.noaa.gov/cgi-win/
    wwcgi.dll?wwEvent~Storms

Tornadoes and Tornado Safety

www.tornadoproject.com/

www.chaseday.com/chaseday5.htm

www.ncdc.noaa.gov/ol/climate/severeweather/
    tornadoes.html

www.nws.noaa.gov/om/brochures/tornado.htm

www.outlook.noaa.gov/tornadoes/

www.fema.gov/library/tornadof.htm

Lightning

thunder.msfc.nasa.gov/

www.nssl.noaa.gov/researchitems/lightning.shtml

www.nofc.forestry.ca/~kanderso/ltgfaq.html

www.glatmos.com/lightinfo/recommendations.html

www.glatmos.com/nldn/nldn.html

Severe Weather Watches, Warnings, Terminology

kamala.cod.edu/spc/

www.met.tamu.edu/weather/warning.html?

www.srh.noaa.gov/oun/wxdata/severeindex.html

iwin.nws.noaa.gov/iwin/nationalwarnings.html

www.nws.noaa.gov/er/pit/branick2.html

Hail

www.chaseday.com/hail.htm

Flooding

marfcws1.met.psu.edu/

CHAPTER FIVE
Summer

Philadelphia Heat Wave Warning System

www.udel.edu/SynClim/phl.html

The Shore

www.marine.rutgers.edu/cool/

www.met.tamu.edu/class/Metr151/tut/seabr/
   seabreezemain.html

www.marine.rutgers.edu/cool/seabreeze/
   seabreeze.htm

co-ops.nos.noaa.gov/tp4days.html

www.geo.duke.edu/Research/psds/cost.html

www.heinzcenter.org/publications/erosion/index.htm

www.nap.usace.army.mil

www.epa.gov/globalwarming/publications/impacts/
   sealevel/barrier_islands.html

www.epa.gov/globalwarming/publications/impacts/
   sealevel/delaware.html

Ocean Water Temperatures

www.fnoc.navy.mil/PUBLIC/OTIS/otis.html

www.marine.rutgers.edu/cool/sat.data2.html

psbsgi1.nesdis.noaa.gov:8080/PSB/EPS/SST/SST.html

fermi.jhuapl.edu/avhrr/index.html

Drought

www.cpc.ncep.noaa.gov/products/
   monitoring_and_data/drought.html

enso.unl.edu/ndmc/

www.state.nj.us/drbc/hydrorep.htm

www.dep.state.pa.us/dep/subject/hotopics/drought/
   drou_map.htm

marfcws1.met.psu.edu/Maps/

Stratospheric and Ground-level Ozone

www.epa.gov/ozone/

jwocky.gsfc.nasa.gov/

www.epa.gov/oar/oaqps/gooduphigh/

see.gsfc.nasa.gov/edu/SEES/strat/class/S_class.htm

www.epa.gov/airnow/

www.dvrpc.org/transportation/ozone.htm

www.epa.gov/oar/aqtrnd96/brochure/sixpoll.html

www.dep.state.pa.us/dep/deputate/airwaste/aq/
   aqm/airmon.html

Flash Floods

www.fema.gov/library/floodf.htm

www.cira.colostate.edu/fflab/

CHAPTER SIX
Autumn

Real-time Hurricane Information

www.nhc.noaa.gov

hurricanealley.net/

kauai.nrlmry.navy.mil/sat-bin/tc_home

www.aoml.noaa.gov/hrd/tcfaq/tcfaqHED.html

www.wunderground.com/tropical/

www.hurricanehunters.com/

www.aoml.noaa.gov/hrd/

Historical Hurricane Information

weather.terrapin.com/hurricane/index.jsp

weather.unisys.com/hurricane/index.html

www.osei.noaa.gov/Events/Tropical/Atlantic/

Hurricane Preparedness

www.fema.gov/fema/trop.htm

www.nws.noaa.gov/om/hurricane/index.shtml

Tropical Glossary

www.nhc.noaa.gov/aboutgloss.html

CHAPTER SEVEN
Philadelphia's Future Climate

El Niño and La Niña
www.elnino.noaa.gov/
www.ogp.noaa.gov/enso/
www.ncdc.noaa.gov/ol/climate/elnino/elnino.html
www.pmel.noaa.gov/toga-tao/el-nino/nino-home.html
www.cdc.noaa.gov/ENSO/
topex-www.jpl.nasa.gov/science/el-nino.html
www.pmel.noaa.gov/tao/

North Atlantic Oscillation and Other Teleconnections
www.cdc.noaa.gov/map/wx/indices.shtml
www.cpc.ncep.noaa.gov/data/teledoc/nao.html
www.ldeo.columbia.edu/NAO/
www.met.rdg.ac.uk/cag/NAO/index.html
tao.atmos.washington.edu/pdo/

Monthly and Seasonal Forecasts
www.cpc.ncep.noaa.gov/products/predictions/30day/
www.cpc.ncep.noaa.gov/products/predictions/90day/
weatheroffice.ec.gc.ca/saisons/index_e.html
www.cpc.ncep.noaa.gov/data/

Global Climate Change/Global Warming
www.usgcrp.gov/ipcc/
www.epa.gov/globalwarming/
www.ncdc.noaa.gov/ol/climate/globalwarming.html
www.gfdl.noaa.gov/~gth/web_page/
    climate_and_weather.html
www.gfdl.noaa.gov/~gth/web_page/article/
    aree_page1.html
gcmd.nasa.gov/
www.gcrio.org/NationalAssessment/
www.ghcc.msfc.nasa.gov/temperature/
www.greeningearthsociety.org/climate/

MISCELLANEOUS

Unit Conversion and Weather Calculators
www.nws.noaa.gov/er/lwx/wxcalc/wxchart.htm
lwf.ncdc.noaa.gov/oa/climate/conversion/
    mppconversion.html
www.pmel.org/unitconv.htm

Maps and Mapping
nationalatlas.gov/index.html
mapping.usgs.gov/www/gnis/gnisform.html
fermi.jhuapl.edu/states/states.html
www.lib.utexas.edu/Libs/PCL/Map_collection/
    Map_collection.html
tiger.census.gov/
www.indo.com/distance/
terraserver.homeadvisor.msn.com/default.asp

Philadelphia-area Current Weather
iwin.nws.noaa.gov/iwin/pa/pa.html
iwin.nws.noaa.gov/iwin/nj/nj.html
iwin.nws.noaa.gov/iwin/de/de.html
nbc10.com [click on weather]
aws.com/nbc/wcau/default.asp
www.uswx.com/us/wx/PA/
www.wunderground.com/US/PA/Philadelphia/
    KPHL.html
www.ems.psu.edu/~nese/phillywx.htm

Historical Weather Maps
weather.unisys.com/archive/index.html
www.ncdc.noaa.gov/onlineprod/drought/xmgr.html

LIST OF ILLUSTRATIONS

Plates follow p. 120.

LIST OF TABLES

INDEX

Page numbers in *italics* indicate figures; page numbers in **boldface** indicate definitions.

Oceans, and global warming, 170
Ocean temperature
    average at Atlantic City, N.J., 192
    trends in global, 166–67
    during winter, 55
Ogletown, Del., *57*
Old Farmer's Almanac, 156
Old Wives' Summer. *See* Indian Summer
Ontario, Lake, 64, *64*
Overrunning, **33–34**
Ozone, ground-level, 115–18
    forecasting of, 117–18
    global warming and, 171, 173–74
    steps for reducing, 117–18
Ozone, stratospheric, 117–20
    depletion of Antarctic, 118–19, *119*
        (*see also* plate)
    trends in, 119–20
Ozone Action Day, 117–18
Ozone depletion area. *See* Ozone hole
Ozone hole, **119–20,** *119* (*see also* plate)
Ozone layer, **118–20**

Paine, Halbert, 11
Palmer Drought Severity Index
        (PDSI), 113–14, *113*
Parker Ford, Pa., hail in, *94*
Parsons, L. H., 9
Particulates, 115–16
PDO (Pacific Decadal Oscillation),
        161
Peirce, Charles
    *A Meteorological Account of the
        Weather in Philadelphia,* 9, 74,
        77–78
    weather observations of, 80, 110–11,
        130
Pemberton, Phineas, 5
    weather observations of, 76, *76,* 140
Pennsville, N.J., *146*
Pennsylvania Department of Environ-
        mental Protection, 19
Pennsylvania Dutch, and weather folk-
        lore, 28
Pennsylvania Gazette, 75
Pennsylvania Germans, and weather
        folklore, 28
Pennsylvania Hospital
    on Smithsonian weather network, 11
    weather observations from, 77–78
Pennsylvania legislature, 8
Pennsylvania State Weather Service, 12
Pennsylvania weather observing net-
        work, 8–9
Perfect Storm, 85–86, *86*
Perkiomen Creek, 146
Philadelphia Gazette, 77

Philadelphia International Airport,
        18–19, 66
Philadelphia Museum of Art, 106
Philadelphia Naval Hospital, weather
        observations from, 77
Philipsburg, N.J., record flood levels
        at, 136
The Philosophy of Storms (Espy), 9
Phoenixville, Pa., 109
Piedmont Plateau, *38* (*see also* plate),
        39, 55
Pilkey, Orrin, 120, 122–25
Pinatubo, Mount, 110
Pineapple Express, 158–59, *159*
Pine Barrens
    hurricanes and, 139
    lack of tornadoes in, 97
Pittsburgh, Pa., urban heat island of, *40*
Poconos, 39, 173
    average foliage peak time, *152*
        (*see also* plate)
    lightning in, 91–92
Polar high, 33
Polar stratospheric clouds, **119**
Polar vortex, **119**
Pollution, 115–18
    global warming and, 166
    episodes of, 115–16, *116*
    legislation and regulation of, 115–16
Port Richmond, Pa., tornado in, 99
Post Office Building, 19
Pottstown, Pa., *20, 127*
    record flood levels at, 136
Precipitation
    average patterns of, 41–42, *42*
    longest periods without, 130
    topography and, 41
    types of, 54–55
Pressure, 30–32
    highs and lows of, 30–32, *32*
    instruments to measure, 31. *See also*
        Barometer
    lowest on record, 74
    North Atlantic Oscillation and, 162
    relationship to weather, 34, *35*
Princeton University, 44
PSFS Building, 93
Psuty, Norbert, 122, 125
Psychrometer, 11
Punxsutawney Phil, 79

Quakers, and weather folklore, 28

Radar (Radio Detection and Ranging),
        23–25
    Doppler, 24–25. *See also* Doppler
        radar

Radiation
    infrared, 21, **164**
    microwave, 23
    ultraviolet, 117
Radiosonde, **19–20**
Rain, daily and monthly statistics,
        178–89
Rainbow, 28, *29*
Rain gauge, 4, 8, 11
Rain shadow, **37**
Rain-snow line, 53
Rain-snow ratio, 56, 64
Reading, Pa., 17, *20*
    record flood levels at, 136
Reedy Island, Del., record flood level
        at, 136
Reedy Point, Del., and 1992
        nor'easter, 86
Rehoboth Beach, Del.
    damage from March '62 nor'easter,
        86
    and 1878 hurricane, 141
    and 1998 nor'easter, 87
    damage from March '62 nor'easter,
        86
Relative humidity, **36–37**
    and comfort, 107
    dependence on temperature, *37*
Reverse laning, 139
Revolutionary War, notable winters
        during, 76–77
Ridge, **32,** *32*
    upper-air, 35, *35*
    See also High-pressure system
Ridley Creek, 146
Riegelsville, Pa., record flood level at,
        136
Rittenhouse, David, weather observa-
        tions of, 5, 76, 81
River flooding, 100–103
Roadside Weather Information System,
        19
Row homes, and heat, 106–8
Rowland, Sherwood, 118
Runyon, N.J., 109
Rush, Benjamin, 5

Safety
    flooding, 95
    lightning, 91
    tornado, 95–96
Saffir-Simpson scale, **132,** 141–46
Salt line, 112–14
Sandy Hook, N.J., 87, 124
Satellites, 20–23
    geostationary, 20–21
    types of imagery, 21–23